Fusion and Integration of Clouds, Edges, and Devices

This book provides an in-depth examination of recent research advances in cloud-edge-end computing, covering theory, technologies, architectures, methods, applications, and future research directions. It aims to present state-of-the-art models and optimization methods for fusing and integrating clouds, edges, and devices.

Cloud-edge-end computing provides users with low-latency, high-reliability, and cost-effective services through the fusion and integration of clouds, edges, and devices. As a result, it is now widely used in various application scenarios. The book introduces the background and fundamental concepts of clouds, edges, and devices, and details the evolution, concepts, enabling technologies, architectures, and implementations of cloud-edge-end computing. It also examines different types of cloud-edge-end orchestrated systems and applications and discusses advanced performance modeling approaches, as well as the latest research on offloading and scheduling policies. It also covers resource management methods for optimizing application performance on cloud-edge-end orchestrated systems.

The intended readers of this book are researchers, undergraduate and graduate students, and engineers interested in cloud computing, edge computing, and the Internet of Things. The knowledge of this book will enrich our readers to be at the forefront of cloud-edge-end computing.

Junlong Zhou is an Associate Professor at the School of Computer Science and Engineering, Nanjing University of Science and Technology, Nanjing, China. His research interests include edge computing, cloud computing, and embedded systems.

Kun Cao is an Associate Professor at the Jinan University, Guangzhou, China. His research interests include the Internet of Things, edge/fog/cloud computing, and cyber-physical systems.

Jin Sun is a Professor at the School of Computer Science and Engineering, Nanjing University of Science and Technology, Nanjing, China. His research interests include computer architecture and high-performance computing.

Keqin Li is a SUNY Distinguished Professor at the State University of New York, a National Distinguished Professor at Hunan University, China, and a Member of Academia Europaea (Academician of the Academy of Europe). He is among the world's top five most influential scientists in parallel and distributed computing in terms of single-year and career-long impacts based on a composite indicator of the Scopus citation database. He is an AAAS Fellow, an IEEE Fellow, an AAIA Fellow, an ACIS Fellow, and an AIIA Fellow.

Fusion and Integration of Clouds, Edges, and Devices

Junlong Zhou, Kun Cao, Jin Sun, and Keqin Li

CRC Press
Taylor & Francis Group
Boca Raton London New York

CRC Press is an imprint of the
Taylor & Francis Group, an **informa** business

This book is funded by the National Natural Science Foundation of China under Grants 62172224 and 62102164, the Natural Science Foundation of Jiangsu Province under Grant BK20220138, the China Postdoctoral Science Foundation under Grants 2021T140272 and 2021M691240, Guangdong Basic and Applied Basic Research Foundation under Grant 2024A1515010232, the Fundamental Research Funds for the Central Universities under Grant 30922010318.

Cover image designed by the authors.

First edition published 2025
by CRC Press
2385 NW Executive Center Drive, Suite 320, Boca Raton FL 33431

and by CRC Press
4 Park Square, Milton Park, Abingdon, Oxon, OX14 4RN

CRC Press is an imprint of Taylor & Francis Group, LLC

ISBN: 978-1-032-88457-8 (hbk)
ISBN: 978-1-032-88911-5 (pbk)
ISBN: 978-1-003-54028-1 (ebk)

DOI: 10.1201/9781003540281

Typeset in Latin Modern Roman
by KnowledgeWorks Global Ltd.

Contents

Preface

Motivation of the Book

With the rapid advancement of Internet of Things (IoT) technology, the amount of data generated by end devices has grown explosively. The transmission of such massive data would inevitably induce heavy network traffic and consume excessive network resources, resulting in high network latency and even service interruption. Emerging applications such as autonomous driving and augmented reality usually demand low network latency, but the quick growth in the number of end devices has made it challenging for cloud servers to execute a large volume of tasks in time. Although cloud servers have strong computing power, their quantity is limited, and they are commonly located far from end devices, resulting in significant transmission delays. To address this issue, edge computing has emerged, where computation can happen at the network edge. Edge servers are not as powerful as cloud servers, but their number is significantly more, and they are closer to end devices, reducing task response latency. However, complex tasks such as deep neural network training are difficult to execute on edge servers. Therefore, cloud-edge-end orchestration, integrating clouds, edges, and devices, has gained popularity in both academia and industry. This orchestration can effectively improve and balance system performance from many aspects, such as energy consumption, latency, security, and reliability.

Summary of Contents

Chapter 1 introduces the supporting technologies of cloud-edge-end orchestration, including cloud computing, edge computing, and IoT. For each supporting technology, we first present its basic concepts for better understanding. Regarding cloud computing, we also discuss its architectures, service quality metrics, and types. Edge intelligence has emerged as a prominent research topic that integrates edge computing and artificial intelligence (AI). We also study four key research dimensions of edge intelligence: edge offloading, edge caching, edge inference, and edge training. Concerning IoT, we mainly focus on the computing hardware and performance metrics of end devices.

Chapter 2 introduces the evolution of cloud-edge-end computing. It then presents the fundamental concepts, enabling technologies, and hierarchical/horizontal models of cloud-edge-end computing. The chapter finally concludes by discussing the research efforts of cloud-edge-end computing, including service and server placement methods, data analysis and training approaches, and resource management strategies.

Chapter 3 studies the performance models and optimization methods for cloud-edge-end orchestrated systems. The chapter begins with a discussion of the challenges faced in optimizing various aspects of cloud-edge-end orchestrated systems such as latency, energy consumption, security and privacy, and reliability. It then examines the various performance models associated with these systems and how they impact the overall performance. This includes models related to latency, energy consumption, security and privacy, and reliability. Finally, the chapter describes performance optimization methodologies that can improve the latency, energy consumption, security, and reliability of cloud-edge-end orchestrated systems.

Chapter 4 discusses the design and implementation of cloud-edge-end orchestrated applications in various domains such as IoT, cyber-physical systems (CPS), and smart cities. The IoT scenarios that are covered in this chapter include Intelligent IoT, time-sensitive IoT, and Internet of Vehicles. In addition to this, the CPS scenarios that are discussed include mobile CPS, medical CPS, automotive CPS, social CPS, and agricultural CPS. Furthermore, this chapter also presents various aspects of smart cities such as smart buildings, AI applications in smart cities, intelligent transportation systems, smart grids, and data analytics for smart cities.

Chapter 5 provides a summary of the book and suggests potential avenues for future research.

This book aims to provide a comprehensive introduction to cloud-edge-end computing, covering its background, fundamental concepts, supporting technologies, architectures and implementations, and the latest research. It should be a useful reference for researchers and engineers interested in cloud computing, edge computing, and IoT. We hope readers can find inspiration and value for their own studies.

Acknowledgments

We wish to express our thanks to students Yufan Shen, Xiangpeng Hou, Lu Yin, Linhua Ma, Tianjian Gong, Yizhou Shi, Xiaoyong Kou, and Tian Wang for their help in collecting related papers, to students Mingzhou Zhao, Sijie Lin, Youling Zeng, Changle Tao, Lan Lan, Jie Niu, Xiaozhu Song, Linhui Wang, and Shilong Zhu for their assistance with drawing pictures, and to students Yingying Zheng, Qixi Yin, Peixin Xu, and Shengchen Cai for their help with editing and proofreading.

This book is funded by the National Natural Science Foundation of China under Grants 62172224 and 62102164, the Natural Science Foundation of Jiangsu Province under Grant BK20220138, the China Postdoctoral Science Foundation under Grants 2021T140272 and 2021M691240, the Guangdong Basic and Applied Basic Research Foundation under Grant 2024A1515010232, and the Fundamental Research Funds for the Central Universities under Grant 30922010318.

Nanjing, Jiangsu, China   Junlong Zhou
Guangzhou, Guangdong, China   Kun Cao
Nanjing, Jiangsu, China   Jin Sun
New Paltz, NY, USA   Keqin Li

Foreword

Cloud computing has become a dominant computing platform in recent years due to its cost-efficient pay-as-you-go pricing, flexible on-demand resource delivery, and convenient service deployment and maintenance. It offers tremendous advantages to various application domains, such as artificial intelligence, big data analysis, industrial automation, and medical fields. The rapid development of the Internet of Things, cyber-physical systems, embedded computing, and edge computing has led to the creation of a new hybrid computing technology called cloud-edge-end computing. This technology has gained popularity in both academia and industry. Cloud-edge-end computing combines clouds, edges, and devices to fully utilize their different capabilities and improve the quality of services for various application scenarios. It is considered to be the next-generation dominant computing paradigm. However, many challenges must be addressed, such as scalable architecture design, lightweight network virtualization, energy-aware sustainable resource allocation, fault-tolerant computation offloading, and security and privacy protection. These issues have yet to be fully considered in current research work. This book aims to provide solutions to system performance optimization problems and in-depth case studies that demonstrate how to model and optimize the system performance of clouds, edges, and devices for cloud-edge-end computing.

This book first introduces the background and fundamental concepts of clouds, edges, and devices. It then presents the evolution, basic concepts, enabling technologies, architectures, and implementations of cloud-edge-end computing. Cloud-edge-end computing can offer users low-latency, high-reliability, and cost-efficient services by fusion and integration of clouds, edges, and devices. As a result, it has become widely used in various application scenarios. This book further explores different types of cloud-edge-end orchestrated systems and applications, and discusses various advanced performance modeling approaches. The book also contains the latest research on offloading and scheduling policies and resource management methodologies for optimizing the performance of applications on cloud-edge-end orchestrated systems. Finally, the book is rich in content and detailed in graphics. It is indeed a significant contribution to the field of cloud-edge-end computing.

This book is a joint effort and creation of four energetic and enthusiastic scholars in the international distributed computing community. The authors have published extensively on cloud computing, edge computing, and embedded systems in the last few years. They are undoubtedly well-regarded leading scientists in cloud-edge-end computing. I congratulate the authors on an excellent job and look forward to seeing the book in print.

Albert Y. Zomaya
Peter Nicol Russell Chair Professor of Computer Science
School of Computer Science
University of Sydney
Sydney, NSW, Australia

1

Introduction to Clouds, Edges, and Devices

1.1 Cloud Computing

1.1.1 Understanding Cloud Computing

Since the inception of cloud computing in 2007, it has gained significant traction through vigorous promotion by industry giants such as IBM, Google, Amazon, Microsoft, and Alibaba Cloud. Its services have widespread application across diverse fields, garnering extensive recognition and yielding substantial commercial value. By 2020, the global and Chinese cloud computing markets had reached remarkable scales of 312 billion US dollars and 178.18 billion RMB, respectively. The rapid progress in computer software, hardware, and network communication technologies has propelled the advancement of cloud computing. However, this progress has also exposed inherent challenges and limitations within the cloud computing paradigm. Consequently, researchers and experts continually explore novel computing models grounded in cloud computing to address these challenges and further augment its development and application.

- **Cloud Computing Definition**

This section presents an overview of how to understand cloud computing. To understand cloud computing, it's imperative to first grasp its fundamental concept and established definition. Various scholarly works offer distinct interpretations of cloud computing, with several prevailing definitions that characterize it [32, 121, 371].

- **Definition 1:** Cloud computing [32] is a distributed architecture that consists of virtualized computing resources. It dynamically delivers services following pre-established service-level agreements between providers and end-users.

- **Definition 2:** Cloud computing [121] embodies a service that currently provides the capability to promptly allocate resources on-demand within a short period, effectively mitigating issues related to both over-provisioning and under-utilization of said resources.

- **Definition 3:** Cloud computing [371] is a technology that allows users to access a pool of virtualized resources that can be dynamically reconfigured based on the current workload. This results in optimal utilization of resources. Before using the service, users and the service provider must agree on a Service Level Agreement (SLA). Based on this agreement, users will be charged on a time-based payment model for accessing the resources.

In the context of three distinct scholarly works, Definition 1 accentuates the real-time provisioning aspect, focusing on the ability to meet user demands instantly. Definition 2 underscores the significance of the SLA customarily established between users and service providers. Both definitions contribute unique insights into cloud computing. Definition 3 amalgamates the salient aspects of its predecessors, integrating their descriptions to more

DOI: 10.1201/9781003540281-1

TABLE 1.1
Advantages and corresponding problems of cloud computing.

Cloud Computing	Advantages	Problems
Security	Boost efficiency by distributing computation tasks across multiple machines to reduce their execution time.	Users are concerned about the security of sensitive data transmitted to the cloud.
Reliability	Cloud computing reduces hardware costs and financial risks by leasing resources, enabling users to prioritize core business functions.	Users, freed from hardware and software maintenance, still rely on the quality of cloud computing services for their own experience.
Maintainability	Specialized software management and maintenance services help users reduce daily operational costs.	Not all software applications suit cloud development. Transitioning legacy software to the cloud brings issues.
Interactivity	Users can dynamically provision cloud services to meet fluctuating business needs, scaling up for peak demand and down for non-peak periods.	Cloud service providers face scalability constraints, requiring collaboration, which is insufficient in the real world.

comprehensively illuminate cloud computing's attributes and intrinsic nature. By analyzing these perspectives of definition, one can deduce that cloud computing inherently possesses several key characteristics.

- **Service Resource Pooling:** Through virtualization technology, resources such as computing, memory, and network are allocated dynamically according to user needs.

- **Scalability:** Users can quickly and flexibly request and purchase service resources according to their needs anytime, anywhere, and enhance their processing capabilities.

- **Broadband Network Call:** Users use various client software to call cloud computing resources through the network.

- **Measurability:** The use of service resources can be monitored and reported to users and service providers, and fees can be charged based on specific usage types (such as bandwidth, number of active users, storage, etc.).

- **Reliability:** Cloud computing can automatically detect failed nodes, enable them to work normally through data redundancy, provide high-quality services, and meet the SLA requirements.

Cloud computing has many advantages compared to traditional computing paradigms, but it also brings new problems that need to be solved urgently. These factors constrain the popularization and application of cloud computing, as listed in Table 1.1.

- **Classification of Cloud Computing**

The types of cloud computing can be divided from different perspectives. This section introduces various types of cloud computing from a vertical perspective of services provided and a horizontal perspective of deployment. Combining with typical cloud computing service platforms, Table 1.2 summarizes and compares their differences and characteristics.

From the services aspect, cloud computing has three different levels of infrastructure abstraction and user responsibility: Infrastructure as a Service (IaaS), Platform as a Service (PaaS), and Software as a Service (SaaS).

TABLE 1.2

Comparison of representative cloud computing services and platforms.

Attributes	Amazon EC2	Google AppEngine	Microsoft Azure	Sales Force.com (CRM)
Types	IaaS	PaaS	PaaS	SaaS
Service Content	Storage, computing, and management.	Program run API and develop and deploy system platform.		Web application and services.
User Call Methods	Reliable underlying API and command-line tools.	Web API and command-line tools.		Mainly used in a simple browser way.
Platform	Linux, Windows	Linux, Windows.Net		Linux, Windows
Features	Provide virtualized storage, processing, computing services, and other infrastructure resources and frameworks.	Provide a development and operation platform for applications in the cloud environment.		Provide anytime, anywhere application usage.
Deployment Patterns and Languages	Customize Linux-based Amazon Machine Images (AMI) and Java language.	Python, Java	.Net supported languages	No deployment, invoked by browser.

- **IaaS** represents the foundational service tier closest to physical infrastructure, leveraging virtualization technology to deliver computing, storage, networking, and other resource services to users, enabling them to install operating systems and execute software. Notable examples include Amazon Elastic Compute Cloud (EC2) [17] and the Apache open-source project Hadoop [140].

- **PaaS** stands atop IaaS, offering users a software runtime environment and configuration that they deploy using tools and development languages provided by cloud vendors. With PaaS, users avoid managing underlying networks, storage, operating systems, and other technical complexities as these services are abstracted and presented as a transparent software development and runtime environment. This layer serves as a platform for creating and hosting web applications, exemplified by services like Google App Engine [19] and Microsoft Azure [24].

- **SaaS** comprises software applications built on infrastructure and platform layers. Users effortlessly access these services via a simple client interface, often through web browsers. They can tailor the application software services they need by communicating requirements to providers over the Internet and paying fees proportional to the quantity and duration of services utilized.

Regarding deployment forms, cloud computing can be classified into three types: private cloud, public cloud, and hybrid cloud. The private cloud is solely managed by individual organizations for internal use. Public cloud, managed by cloud service providers (CSPs), offers open-access services to the public. A hybrid cloud blends two deployment types together.

- **Multi-tier Services of Cloud Computing**

Cloud computing offers multi-tier services that transform physical resources into virtual machines through virtualization. However, this poses several challenges for various

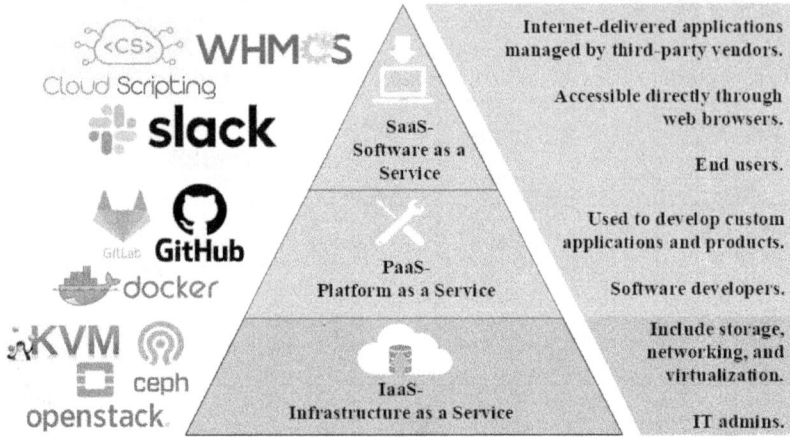

FIGURE 1.1
Cloud service models (IaaS, PaaS, and SaaS).

stakeholders, such as developers, providers, admins, and users. Middleware orchestrates task scheduling, resource/security management, and billing while heavily relying on third-party software and data processing to ensure safety and effectiveness. Additionally, energy consumption in data centers requires innovative resource scheduling techniques. Performance monitoring is also necessary to meet SLAs with providers, which helps in flexible payments. Finally, user-friendly interfaces are essential for easy service invocation. To help users select appropriate cloud services, establishing a global marketplace with clear metrics is crucial [32].

1.1.2 Cloud Computing Architectures

Cloud computing encompasses both client and server components, as well as three primary service delivery models. Cloud clients include software or hardware abstractions used to connect to cloud services. Cloud servers are the remote servers hosted by CSPs that provide various services and resources to clients over the internet. Presently, cloud computing primarily consists of three service delivery types, namely IaaS, PaaS, and SaaS, shown in Figure 1.1.

IaaS is a service in the cloud computing model that provides virtual resources such as basic computing, networking, and storage. In the IaaS framework, the main components include computing hardware (such as servers and virtual machines), network devices (such as routers and switches), and storage devices (such as disk arrays and object storage), typically deployed in data centers [340]. With IaaS, users can access and utilize these resources anytime and anywhere according to their needs [268]. Additionally, IaaS offers automatic resource adjustment capabilities to allocate resources based on application loads and demands, ensuring the needs of applications of different scales are met. The key advantages of IaaS include flexibility, scalability, high availability, and security. IaaS can achieve highly reliable and redundant system architectures by deploying resources across multiple data centers, ensuring uninterrupted business operations. Many cloud computing providers currently offer IaaS services, such as Amazon AWS, Google Cloud, and Microsoft Azure [175]. By leveraging IaaS's advantages and understanding its challenges, users can better use cloud resources, improving business efficiency and security. Furthermore, further research and development can enhance the performance, security, and reliability of IaaS to meet the ever-evolving user demands.

PaaS is a service built on top of IaaS. PaaS abstracts the underlying technical details such as network, storage, and operating systems, allowing users to focus on software development and deployment without the complexity of managing infrastructure. It empowers developers to establish the required software runtime environment and arrangement by utilizing software tools and programming languages offered by CSPs [268]. For example, users can leverage the provided programming languages, libraries, services, and tools to quickly and easily build and deploy web applications. Prominent examples of PaaS include Google App Engine and Microsoft Azure [175], which focus on enabling developers to create and utilize cloud software. Business intelligence, databases, development and testing tools, integration services, and application deployment are examples of PaaS applications. Business intelligence utilizes data analysis and reporting tools to aid enterprises in decision-making. Database services provide a flexible way to manage and access data. Development and testing tools assist developers in rapidly building and testing applications. Integration services facilitate application integration with external systems. Deployment services provide a convenient, efficient cloud application deployment method.

SaaS is a software delivery model where distributors or providers host and provide applications to customers over the Internet, allowing users to utilize and oversee cloud-based software easily. In SaaS, applications are hosted by service providers on their cloud platforms. SaaS providers are responsible for maintaining and ensuring the system's up-to-date operation. Users do not need to handle or monitor the underlying cloud infrastructure, including networks, servers, operating systems, storage, and specific applications [268]. Customers can access these applications through web or application interfaces on different client devices anytime and anywhere. This arrangement allows multiple users to use the applications flexibly, giving customers a certain level of autonomy. For instance, they can share applications with their clients and pay for them as required. The SaaS model also simplifies application upgrades and maintenance. Examples of SaaS include email and office productivity applications like Gmail and Hotmail [175], customer relationship management (CRM) applications such as Salesforce and IBM, and billing and finance applications. Users can access email applications, word editors and processors, spreadsheet applications, presentation applications, and more through web application interfaces for email and office productivity.

Numerous studies have been conducted on IaaS, PaaS, and SaaS. We summarize and list them in Table 1.3, with detailed explanations.

- **Studies for IaaS**

Most IaaS-related studies focus on network resource management, system performance, and security in IaaS. Regarding network resource management in IaaS, Colajanni et al. [73] examined the behavior of IaaS providers and introduced a network-based nonlinear mathematical model considering resource utilization, monetary costs, and energy consumption as optimization objectives. Liu et al. [233] proposed an online multi-workflow scheduling framework and a deadline-aware heuristic algorithm to reduce virtual machine leasing costs and minimize the probability of deadline violations while improving resource utilization efficiency.

For system performance in IaaS, Chang et al. [47] introduced a hierarchical random modeling method for IaaS Cloud Data Center (CDC) performance analysis under heterogeneous workloads, considering the virtual CPU quantity of different customer job requests. Additionally, a model is established to evaluate the impact of job arrival rate, buffer size, maximum vCPU quantity of PMs, and Virtual Machine size distribution on CDC performance. The authors [46] introduced a method based on discrete event system modeling and proposed an approximate analytical model based on $M/G/m/m+K$ queues for evaluating cloud computing performance and issues. This model can address the problem of lack of

TABLE 1.3
A summary of IaaS, PaaS, and SaaS-related work.

Services Model	Ref.	Focus	Key Ideas
IaaS	[73]	Manage resources to maximize profits	Network-based nonlinear model
	[126]	Assess the security of cloud computing services	Cloud-Trust evaluation model and measurement metrics
	[47]	Accurate performance analysis	A comprehensive model of PMs under heterogeneous workloads
	[46]	Find methods to assess the performance of systems	Use M/G/m/m+k queuing model to calculate the probability distribution of job quantities
	[233]	Minimize the cost of virtual machine rental	Propose an Online multi-workflow scheduling called NOSF
PaaS	[284]	Compatibility issues between different CSPs	Federated multi-cloud PaaS infrastructure to enable resource sharing
	[270]	High network latency between cloud and end users	Propose a novel NFV-based hybrid cloud and fog IoT PaaS architecture
	[375]	Protect security and privacy through data access control	PaaS framework incorporating semantic reasoning
	[334]	Quality of Service improvement	Minimize mutual interference between applications
	[314]	Address interoperability issues in cloud computing	Propose a unified description model and a generic PaaS application configuration and management API
	[76]	Flexible resource allocation, reducing hosting costs	Propose design decisions aiming to enhance PaaS systems
	[195]	Access control and security issues	Propose a novel model-driven approach and architecture
SaaS	[262]	Manage resources through forecasting of SaaS service demand	Introduce a new method for predicting SaaS service request numbers
	[16]	Energy consumption in data centers	Formalize the microservice-based SaaS deployment problem as a combinatorial optimization issue
	[104]	Personalized frameworks for the growing number of SaaS services	Propose an integrated, personalized framework with SaaS
	[103]	Address information flow vulnerabilities	Propose Security Diagnosis as a Service (SDaaS) to assess the security posture of SaaS applications

performance assessment and problem diagnosis in cloud computing environments, providing multiple system performance metrics such as task response time and resource utilization. Furthermore, it exhibits excellent scalability and adaptability, making it suitable for cloud computing systems of varying scales and complexities.

Aiming at the IaaS platform's security, a cloud security assessment model named Cloud-Trust has been introduced. This cloud architecture reference model encompasses a wide range of security controls and best practices [126]. Cloud-Trust enables quantitative evaluation of the confidentiality and integrity of IaaS cloud computing services while also considering the role of CSPs. This model offers two primary high-level security metrics: (1) the probability of APT accessing high-value data and (2) the probability of APT detection by cloud tenants or Cloud Computing Security (CCS) security monitoring systems. Moreover, Cloud-Trust can measure the effectiveness of particular CCS security measures, encompassing optional security functionalities provided by prominent commercial CSPs. This research provides a robust tool for the cloud security domain, aiding cloud tenants in making more informed security decisions.

- **Studies for PaaS**

Most PaaS-related studies focus on cross-platform, cloud service quality, resource management, and security in PaaS. Regarding the cross-platform in PaaS, the authors [284] proposed a federated multi-cloud PaaS infrastructure to ensure interoperability and portability across different cloud environments. It utilizes a configurable architecture and distinctions in underlying cloud environments, providing a unified open programming model. The core technology of multi-cloud infrastructure is Service Component Architecture (SCA). SCA can encapsulate and synthesize services and components from different cloud environments, ensuring application interoperability and collaboration. In response to the challenges posed by the specific and proprietary APIs of existing cloud platforms, which make the interaction between different clouds difficult and hinder cloud collaboration and integration, Sellami et al. [314] proposed a unified description model that enables applications to be independently represented of the target PaaS. Additionally, it introduces a general PaaS application configuration and management API (COAPS API). This solution offers a universal description and management method for PaaS applications, enabling application providers to switch between PaaS more easily and promote cloud collaboration and integration.

For cloud service quality in PaaS, Mouradian et al. [270] introduced a novel Network Function Virtualization (NFV)-based hybrid cloud and fog Internet of Things (IoT) PaaS architecture aimed at addressing the issue of high latency associated with Internet connections. This architecture supports application development, deployment, management, and orchestration. It utilizes Virtual Network Functions (VNFs) to implement application components modeled as graphs, catering to IoT use cases. To address the limitations of QoS management in cloud services due to the lack of understanding of the hosted application architecture and dynamics, as well as the challenges in mapping applications to cloud resources caused by virtual machine interference, cloud cluster heterogeneity, diverse resource requirements, and network asymmetry, the authors proposed a deployment strategy. It introduces two key parameters, namely the isolation index and closeness, to assess the level of interdependence among application components [334]. The goal is to minimize mutual interference among applications and reduce costly runtime adaptation procedures to meet service-level agreements. This approach improves the performance and predictability of cloud-native applications.

Regarding resource management in PaaS, Costache et al. [76] explored the challenge of managing resources in PaaS systems to meet diverse workloads while minimizing costs. It categorizes design decisions in resource management and analyzes existing PaaS systems across various application domains. Future research opportunities are identified, including mechanisms for meeting user service-level objectives, designing PaaS solutions for general application types using containerization, and efficient data processing for emerging applications.

As for the security issues in PaaS, Verginadis et al. [375] proposed the PaaSword access control framework, depended on the Extensible Access Control Markup Language standard, which incorporates semantic reasoning capabilities to support context-aware access control policies and inject security policies into cloud applications. PaaSword can generate new knowledge based on the context, effectively resolving syntax mismatches, providing a universal policy representation, and enhancing feasibility checks of policies. A model-driven architecture is proposed to address the security and access control challenges in a multi-cloud environment to ensure users' security requirements are met [195]. By extending Cloud Application Management for Efficient Loading and introducing new security sub-DSL, it supports the adaptive configuration of multi-cloud applications and transforms the model-driven repository into a multi-tenant repository, thereby restricting access to organization-specific information. It also provides management APIs to enhance security configuration and access control for multi-cloud applications.

- **Studies for SaaS**

Most SaaS-related studies focus on resource management, energy consumption, personalized services, and security in IaaS. For resource management in SaaS, Matoussi et al. [262] introduced a new method to forecast the number of requests to SaaS services. Based on the method, they successfully balanced execution time and prediction accuracy. They achieved this by dynamically adjusting the computation time to adapt to workload variations, providing multi-step ahead predictions, and considering the correlation between SaaS service and virtual machine workloads.

Aiming to reduce energy consumption for SaaS, Alzahrani et al. [16] proposed a microservice-based SaaS deployment method that considers the energy consumption problem in data centers. To address this problem, the authors proposed an energy consumption model and formalized the microservice-based SaaS deployment problem as a combinatorial optimization problem. The authors used a genetic algorithm and repair mechanism to solve this combinatorial optimization problem. This study also demonstrates the effectiveness and scalability of the proposed genetic algorithm.

To provide personalized services in SaaS, the problem addressed in [104] is the lack of existing personalized frameworks in the context of the increasing number of SaaS services and the demand for personalization from users. A SaaS-based integrated, personalized framework is proposed to provide efficient, flexible, and integrated customized services. This framework integrates multiple technologies and models to achieve personalized preference collection and corresponding delivery of SaaS services. By using techniques such as client-side personalization, semantic integration, user modeling, data mining, and recommendation engines, customized service experiences are provided to users. Additionally, the framework can dynamically combine multiple cloud services to provide personalized services.

Regarding the SaaS platform's security, Elsayed et al. [103] proposed a new security diagnostic as a service framework (SDaaS) to address the security risks and information flow vulnerabilities in SaaS applications. SDaaS employs the method of quantitative static information flow analysis, combined with techniques such as system dependency graph construction and information flow control, to evaluate the security posture and identify latent security flaws within SaaS applications. It can promptly identify and address potential information flow vulnerabilities, providing an effective solution for protecting data integrity and confidentiality in SaaS applications in the cloud.

TABLE 1.4

A summary of related research on SLA.

Ref.	Optimization Objective(s)				Key Idea
	SLA Violation	Cost	Storage	Resource Availability	
[475]	✓				Propose a prediction method combining machine learning and resampling techniques
[266]				✓	Introduce a new method based on MDD
[261]	✓				Propose the method of using GNNs to improve the accuracy of SLA violation prediction in cloud services
[462]			✓		Propose a multi-layer SLA-driven deduplication framework for cloud storage systems – MUSE framework
[507]		✓			Propose a profit-optimized resource scheduling method based on SLA
[325]		✓		✓	Introduce a SLA-aware adaptive cloud resource management technology STAR

1.1.3 Cloud Service Quality Metrics

In this section, we introduce two cloud service quality metrics. They are service-level agreement and reliability.

- **Service Level Agreement**

  SLA is a set of guiding principles and protocols for service quality and availability between suppliers and customers. It can clarify the responsibilities and obligations of both parties, provide reference and measurement standards for service providers, and protect customers' rights.

  This section will give a detailed introduction using a widely applied SLA model as an example [397]. The model defines the service fees that are charged for service requests

$$C(r, R) = \begin{cases} ar, & \text{if } 0 \leqslant R \leqslant \frac{ar}{s}, \\ ar - \gamma \left(R - \frac{ar}{x_0} \right), & \text{if } \frac{ar}{s} < R < \left(\frac{a}{r} + \frac{c}{s} \right) r, \\ 0, & \text{if } R \geqslant \left(\frac{a}{\gamma} + \frac{c}{s} \right) r, \end{cases} \qquad (1.1)$$

where R is the response time, r is the workload of requests, a is the price per unit of workload, γ is a coefficient representing the compensation intensity for low service quality, c is a constant related to the SLA, and s is the expected service processing speed for customers. Based on the workload r of the service request and the response time R, there are several scenarios for the fees:

- When the response time is within the promised shortest time, the customer will be charged the base fee.

- When the response time exceeds the promised shortest time but is still within a certain threshold, the fee will decrease linearly because the quality of service has not fully met the requirements.

- When the response time exceeds the above threshold, the service provider may not charge a fee, indicating compensation for services that do not meet the SLA.

Numerous studies have been conducted on SLA. We summarize and place them in Table 1.4, with explanations provided in the following text.

SLA Violation Prediction. This type of work is all about conducting work focused on predicting SLA violations. Zeng et al. [475] proposed a method that combines machine learning and resampling techniques to predict SLA violations for batch processing big data analysis applications workloads on the Alibaba Cloud platform. They used four machine learning algorithms (logistic regression, artificial neural network, random forest, and extreme gradient boosting) combined with 12 resampling strategies (oversampling, undersampling, and combinations) to address data imbalance issues. Experimental results showed that the model combining extreme gradient boosting and near-neighbor sampling best detected SLA violations. In addition, the authors also developed a mathematical model to quantify service providers' profits and discussed the impact of prediction technology on profits. Maroudis et al. [261] specifically focused on enhancing prediction accuracy through graph neural networks (GNNs). They first categorized existing SLA breach prediction techniques into three types: time series, feature vectors, and graph representations. Next, they introduced an innovative composite model that combines the strengths of GNNs and random forests to leverage contextual information and client characteristics.

Islam et al. [178] investigated Apache Spark job scheduling algorithms based on SLA in a hybrid cloud environment. They proposed two algorithms for optimizing resource costs and reducing job deadline violations. Both algorithms consider the prices of different virtual machine instances to improve cost efficiency in Spark clusters deployed in hybrid clouds.

Profit Optimization: Zhao et al. [507] proposed a profit-driven resource scheduling method based on SLA for the cloud environment's big data analytics service (AaaS) platform. This method efficiently automates query admission control and resource scheduling, maximizing the AaaS provider's profit while ensuring service quality and reducing query response time. Through data sharding techniques, users can balance accuracy, response time, and cost when handling large datasets. The algorithm considers task dependencies, data locality, and resource utilization to achieve flexible resource leasing and releasing, thereby reducing costs and improving efficiency.

The authors in [325] proposed an SLA-aware adaptive cloud resource management technology, STAR, aimed at reducing the rate of SLA breaches and optimizing service quality parameters to enable more effective delivery of cloud services. STAR considers execution time, cost, reliability, and availability as QoS parameters to minimize SLA violations and enhance user satisfaction.

Storage Optimization: MUSE, a multi-tier SLA-driven deduplication framework for cloud storage systems that balances I/O performance and storage cost, is proposed [462]. MUSE introduces a deduplication SLA that quantifies service quality and employs a multi-tier deduplication strategy that dynamically adjusts deduplication behavior based on workload characteristics. MUSE offers a more granular quality of service agreement between the service provider and the client through a combination of different performance/storage cost levels. Through experimental comparison, MUSE significantly optimizes the balance between I/O performance and space cost compared to other deduplication solutions, thereby enhancing the service quality of deduplicated cloud storage systems.

Resource Availability: Mo et al. [266] proposed a novel method for analyzing resource availability in cloud computing systems based on Multi-Valued Decision Diagrams (MDD) to reduce SLA violations. This method is suitable for cloud systems with heterogeneous, multi-state computing nodes and involves constructing an efficient MDD model to analyze the availability of system resources. The authors presented a comprehensive introduction to utilizing MDD to depict system states, introduced a novel MDD construction algorithm, and proposed a resource availability evaluation algorithm. Case studies and benchmark tests demonstrate the efficiency of the MDD method, comparing it to traditional methods such as continuous-time Markov chains and generic generating functions.

- **Reliability**

   With the increase in the number of users and the growing complexity of data center architectures, operational failures of cloud computing systems occur frequently, which causes significant losses. Therefore, guaranteeing the reliability of a cloud computing system is crucial for the system to provide reliable cloud computing services. Reliability can be defined as a system's ability to provide uninterrupted and faultless services within an agreed-upon time scale. To enhance cloud computing system reliability, Adams et al. of Microsoft have proposed three design principles [3].

   - **Elasticity Design Principles:** Cloud computing services can tolerate the failure of system components and promptly take corrective measures under conditions that do not require human intervention so that the user does not notice an interruption in service. When the service is not working, the system cannot collapse and still provide some of its functions.

   - **Data Integrity Design Principles:** For user-hosted data integrity to be maintained during a system failure, the service must be able to manipulate, store, and discard data consistent with normal operation.

   - **Recoverable Design Principles:** The system should automatically restore service as soon as possible when an abnormal situation occurs. When service is interrupted, the system maintainer should restore it as soon as possible and to the maximum extent possible.

   Cloud computing systems consist of various physical resources, software, and heterogeneous virtual machines, all affecting the system's reliability. In this paper, we assume the execution of multitasking workflows in a cloud computing system environment and categorize system reliability into two parts: execution and transmission.

   Execution Reliability: Tasks may be executed with soft errors caused by transient failures that do not damage the hardware. A Poisson distribution is usually used to model the reliability of task execution on multitasking resources. Then the classical form of the reliability function is shown below,

$$R_{\text{exe}} = e^{-\lambda \times \nu \times T_{\text{exe}}}, \tag{1.2}$$

where λ is the constant failure rate of the virtual machine (VM), ν is the vulnerability factor of each task, and T_{exe} is the execution time of a single task on the VM.

   Transmission Reliability: During transmission, the task may be subject to bit errors due to environmental and link disturbances. The transmission reliability model for a single task can be expressed as

$$R_{\text{trans}} = e^{-\gamma \times T_{\text{trans}}}, \tag{1.3}$$

where γ is defined as the constant bit error rate of the channel transmission and T_{trans} is the transmission time of the task. Suppose that failures are independent. The system reliability is given by the product of the task's execution and transmission reliability, i.e.,

$$R_{\text{sys}} = R_{\text{exe}} \times R_{\text{trans}}. \tag{1.4}$$

   System failures are categorized into three types: resource failures, service failures, and other failures, which are the main factors affecting reliability. Resource failures are physical resources that go into an incorrect state, such as hardware, software, power, and network failures. Service failures refer to situations in which service providers cannot provide or users cannot obtain services that satisfy the quality specified in the SLA. Other failures generally refer to unpredictable failures caused by natural or man-made reasons.

TABLE 1.5

A summary of related research on reliability.

Ref.	Categories	Key Ideas	Contributions
[154]	Fault removal	A fault injection language	Build complex failure scenarios in a simple way
[462]	Fault tolerance	A framework for multi-tiered and SLA-driven deduplication	Optimize I/O performance and balance space cost
[510]		A method to optimize the placement of redundant Virtual Machines (VM)	Minimize the consumption of network resources
[528]		A algorithm for dynamic fault-tolerant dispatching	Improve utilization of cloud resources
[486]		A fault-tolerant system for transparent VM clustering	Get VM-level incremental checkpoints
[353]	Fault forecast and avoidance	A strategy for reliability and cost-aware job scheduling	Forecasting software failures in cloud VMs
[21]		A predictive model-based probabilistic scheduler	Calculate the probability of a task executing without failure

Currently, most fault tolerance work focuses on resource failures. Researchers have designed various techniques and approaches to manage resource failures that tend to occur in cloud environments and ensure service availability. The main fault management techniques are categorized into three types: fault removal, fault tolerance, and fault forecast and avoidance. The relevant literature will be presented next in the above three categories. See Table 1.5 for more details.

Fault Removal: Fault removal techniques are used to discover and remove potential faults in cloud computing systems using traditional software testing and verification methods. Fault injection is a more typical approach to improve the test coverage of a software system by introducing faults into the code. A fault injection language, FAIL, and its clustered version tools, FCI, are proposed by Hoara et al. [154] for software fault injection in distributed applications. Since fault removal techniques are challenging to implement effectively in complex cloud computing systems, researchers have paid less attention to this technique than to fault tolerance and prediction techniques.

Fault Tolerance: Fault tolerance is a well-known technique for addressing the reliability issues of cloud services. It involves utilizing one or more backups in case of failure to ensure the continuity of cloud services. One of the popular fault tolerance strategies is using a primary backup model. Zhu et al. [528] introduced a dynamic fault-tolerant scheduling algorithm for real-time workflows in virtualized clouds through an extended primary backup model to achieve hardware failure tolerance and increase system schedulability and resource utilization. Replication is a fault tolerance mechanism for the redundant deployment of computing resources. Zhou et al. [510] presented a new network topology-aware redundant VM placement optimization method to improve the reliability of cloud services under k-fault tolerance constraints. A multi-layer and SLA-driven deduplication framework was presented by Yin et al. [462] to realize a dynamic tuning mechanism in cloud storage systems. Another common fault tolerance mechanism is checkpointing, which periodically saves the execution state of a running task and enables the task to recover from the most recently saved state after a failure occurs. Zhang et al. [486] coordinated acquisition of incremental checkpoints for VM images to transparently back up the entire operating system and restore the whole virtual cluster to the last correct state in case of failure.

Fault Forecast and Avoidance: Due to the complexity and high overhead of implementing fault tolerance techniques, many researchers have begun to adopt fault forecast and avoidance mechanisms to guarantee the reliability of cloud services. Tang et al. [353] used

TABLE 1.6
Comparison of cloud types.

Cloud Types	Security Level	Scope	Organization
Public cloud	Low	General public and industries	Cloud service providers
Private cloud	High	Individual institutions	Individual institutions
Hybrid cloud	Medium	Public and institutions	Public and institutions

a parallel two-dimensional long short-term memory neural network with compute unified device architecture support to predict software failures in cloud virtual machines and proposed a reliability and cost-aware job scheduling strategy for cloud computing systems. Aral et al. [21] proposed a predictive model and probabilistic scheduler in a decentralized multi-cloud environment based on blockchain to predict the probability of faultless execution and correctly computed task outputs by relying on historical logs.

1.1.4 Types of Clouds

Cloud computing is proliferating with the rapid development of big data, which plays a vital role in meeting the growing demand for storage and infrastructure. Cloud computing deployment models can be divided into three types: public cloud, private cloud, and hybrid cloud.

- **Public Cloud:** Public cloud is a cloud infrastructure for many users, individuals, or businesses. The basic feature is multi-tenancy, designed to serve multiple users rather than a single customer. Characteristics of the public cloud include a flexible payment model, elastic scalability, high security, and high reliability, so it is widely used in enterprise IT infrastructure, software development, data storage, and processing.

- **Private Cloud:** Private cloud is a cloud computing environment specifically tailored to an individual organization or entity. It is usually constructed, owned, and controlled by the organization itself and can be deployed in an internal data center or hosted in a third-party data center. Private Cloud offers greater security and privacy because it is built on a separate hardware infrastructure and is protected by strict access control and security policies.

- **Hybrid Cloud:** Hybrid cloud is a mixture of public and private clouds. This model enables organizations to flexibly deploy and run workloads on a public or private cloud based on specific business requirements, leveraging the advantages of public and private clouds for optimal resource utilization and business agility. Hybrid clouds are highly suitable for carrying out big data actions on non-sensitive data in public clouds and securing sensitive data in private clouds, thus helping in multiple deployments.

Table 1.6 briefly compares public, private, and hybrid clouds about security level, scope, and organization. Figure 1.2 shows the relationship between the three cloud deployment types.

- **Public Cloud**

In the age of big data, many people are turning to the public cloud to lay out their business software to get lower storage costs, high I/O performance, and more flexible resource allocation. Some common CSPs are Google Cloud, Amazon, Microsoft Azure, etc.

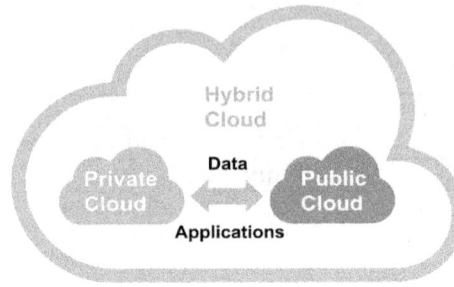

FIGURE 1.2
Types of cloud computing.

TABLE 1.7
A summary of related research on public cloud. I: I/O performance, S: Security, D: Data integrity, W: Workload, T: Time, C: Cost, R: Reliability. ✓ indicates that the factor is considered in the literature.

Ref.	Key Ideas	Optimization Goals						
		I	**S**	**D**	**W**	**T**	**C**	**R**
[222]	A global sharing paradigm that maximizes I/O resources	✓						
[384]	A anonymous and secure aggregation Scheme that maximizes security		✓					
[383]	A security model that uploads and checks data integrity			✓				
[382]	A precise reward scheme achieving maximal security		✓					
[381]	A blockchain-based fair payment smart contract that maximizes data security			✓				
[1]	Estimate QoS degradation by detecting the virtual machines interference				✓			
[407]	A resource allocation algorithm to minimize the total cloud operating cost						✓	
[287]	A new framework for enabling rapid, fully online learning on resource allocation policy				✓			
[264]	A modeling and optimization methods for ensuring service reliability							✓
[65]	A optimal configuration search algorithm that minimizes the execution time and cost					✓	✓	

A business or organization can rent cloud computing resources and services through CSPs. They only pay for the services and resources they subscribe to and consume. The following describes related works on resource allocation, QoS, and security in public clouds. An overview of these works in terms of key ideas and optimization objectives is presented in Table 1.7.

Public clouds facilitate the sharing of storage resources, and how to allocate resources is critical. A global shared resource paradigm for coded storage systems is proposed by Li et al. [222]. It allows efficient sharing of I/O resources across multiple logical volumes while guaranteeing load balancing and fault tolerance. Woo et al. [407] investigated the benefits of distributing distributed application components over multiple public clouds and proposed a resource allocation algorithm while meeting the required SLA. Penney et al. [287] presented a new model for a fast, completely online policy for resource allocation in public cloud environments. Combining the performance prediction model and the multi-objective optimization algorithm, Cheng et al. [65] proposed a novel optimal configuration search algorithm called AB-MOEA/D.

In addition to optimizing resource allocation, CSPs should also ensure the QoS of user applications, such as data reliability and workloads. In [383], the authors developed a novel agent-based model for checking data uploads and remote data integrity in public key cryptography. Wang et al. [381] presented the concept of non-interactive publicly demonstrable data occupation and designed a smart contract based on blockchain for equitable payment while guaranteeing the reliability and integrity of data in cloud storage. Pons et al. [1] designed Cloud White to detect and estimate the quality of service degradation for delay-critical tasks in public clouds. It can estimate the performance decrease due to interference between virtual machines in the case of multiple colocated delay-critical VM management systems. In [264], Meng et al. developed a reliability model and optimization approach from a system service perspective. This method systematically avoids different kinds of failures in cloud service execution through artificial intelligence prediction and can effectively restore the reliability of cloud computing systems.

There has been extensive research on secure clouds and related technologies to ensure computing security. Many encryption schemes, public auditing schemes, and authentication approaches have been developed recently. Wang et al. [384] presented the notion of an anonymous security aggregation solution in public cloud computing based on fog. It saves bandwidth by connecting fog nodes to the public cloud server while ensuring the security of PCS and the privacy of end devices. In [382], the authors developed the new notion of a traceable privacy-preserving motivation method for crowd computing in public clouds and an exact incentive solution based on bilinear pairings for the first time. The solution provides both traceability and anonymity for crowd computing in public clouds.

- **Private Cloud**

  Unlike public clouds, private clouds do not provide computing, storage, and network resources for external use and are intended only for internal use by the business or organization. Due to increasing enterprise requirements for data security and compliance, private clouds are experiencing a boom in growth. Private clouds are widely used in industries with high data privacy requirements, such as large enterprises and financial institutions. Table 1.8 summarizes the key ideas, optimization goals, and contributions of related studies on private clouds.

  Since private cloud providers' primary objective is to maximize their profits, cost optimization has been the subject of much research. Ghanbari et al. [122] solved a problem of allocating resources in a private cloud using a dynamic model based on a feedback loop to minimize the provider's cost. In [249], the authors designed and evaluated a cost-effective cluster auto-scaling and dispatching approach for big data in private clouds, which increases resource efficiency and preserves the required QoS. Chen et al. [54] first classified the baseboard resources in cloud data centers into additive and non-additive and presented a framework called Heuristic Assisted Dueling Double Deep Q-Networks, which can maximize time-averaged revenue.

  Computing resources are not effectively utilized in many cloud data centers, and performance loss due to resource competition is a significant challenge. To minimize the negative impact of competition for the same compute resources in virtual environments, Yokoyama et al. [464] presented a model based on affinity to dispatch virtual machines hosting and running scientific applications in private cloud environments.

  A private cloud is a rising virtual environment that provides more secure data communication. Rajasoundaran et al. [301] presented a secure cloud job service based on machine learning for providing multi-layered safety for this zero-trust cloud job service system.

TABLE 1.8
A summary of related research on private cloud.

Ref.	Key Ideas	Optimization Goals	Contributions
[464]	Affinity-based virtual machine scheduling model	Maximize the throughput of jobs	Host and run scientific applications
[122]	Feedback-based loop for optimization problem	Minimize the costs	Demonstration of the superiority of the "adaptive" models over the "static" models
[54]	Heuristic Assisted Dueling Double Deep Q Network.	Maximize time-averaged revenue	Solve a virtual network embedding problem
[249]	Automatic cluster scaling and scheduling solution for big data	Maximize the resource utilization and QoS	Implement the energy-efficient resources provision
[85]	A framework for multi-data center cloud architecture based on IoT	Maximize security	Decrease the effects of VM relocation
[301]	Secure cloud work services based on machine learning	Maximize security	Assist the scheduler to do job scheduling
[447]	A framework for an automated power dispatch system	Maximize security	Protect against local extreme natural disasters

In [85], Dhaya et al. developed an Internet of Everything-based framework for private multi-data center cloud architecture to leverage the inclusive security architecture of multi-data centers. They also introduced a distributed VM clustering model to limit the VM migration process and reduce its impact on the connection set of private cloud data centers. A framework for a power dispatch automation system based on a virtual private cloud is proposed by Yang et al. [447], which can provide safer and more reliable operations.

- **Hybrid Cloud**

A hybrid cloud can provide organizations greater flexibility, scalability, and security while utilizing the advantages of public and private clouds. In this environment, organizations have their private cloud that offers and manages many resources internally. When business applications or platforms need to expand and require extra resources that a private cloud cannot provide, they can use resources provided by an external public cloud. Table 1.9 summarizes the methods, application scenarios, and optimization goals of related works in hybrid clouds.

In cloud environments, various service demands make cost-effective task scheduling and resource allocation challenging. In [49], Charrada et al. proposed a forward-backward reconfiguration algorithm for emerging applications based on services deployed in a hybrid cloud, which minimizes the hosting and communication costs. To elastically and accurately allocate virtual machines, Shin et al. [322] presented a real-time supervisory approach for multiple self-adaptive resource allocations to achieve the goal of cost minimization under deadline constraints. In [178], efficient scheduling algorithms are proposed by Islam et al., who leverage diverse VM instance pricing of hybrid cloud deployment clusters to optimize the cost of VM use for native and cloud resources and maximize work deadline satisfaction. Yuan et al. [467] studied the time variance in hybrid green IaaS clouds and developed a Multi-queue Scheduling method to maximize profits.

Dispatching several massively synchronous workflow applications on a heterogeneous computing system such as a hybrid cloud is a fundamental np-complete problem essential

TABLE 1.9

A summary of related papers on hybrid cloud. L: Latency, C: Cost, E: Elasticity, S: Security, T: Throughput. ✓ indicates that the factor is considered in the literature.

Ref.	Methods	Application Scenarios	Optimization Goals			
			L	C	S	T
[49]	Forward-Backward-Refinement algorithm	Service-based application explosion		✓		
[270]	PaaS architecture	Hybrid cloud/fog systems	✓			
[98]	Sequential cooperative game algorithm	Multi-objective scheduling	✓	✓		
[322]	Real-time supervisory program for multi adaptive resource allocation	Industrial internet of things		✓		
[467]	Multi-queue scheduling method	Hybrid green IaaS clouds		✓		✓
[310]	Demand response program	Thermostatically controlled loads			✓	
[178]	Cluster-level scheduling algorithm	Spark jobs based on service-level agreements		✓		

for satisfying all QoS demands. In [98], Duan et al. proposed a multi-objective algorithm for communication and memory-awareness to optimize executing time and financial cost while satisfying two constraints: bandwidth and storage demands of the network. A new architecture for PaaS based on Network Function Virtualization is proposed by Mouradian et al. [270] for hybrid cloud/fog systems. This scheme reduces the high latency caused by the cloud's connection to the end user's wide area network. In addition, to achieve privacy preservation, A framework for a hybrid cloud is designed by Sang et al. [310] for demanding response schemes based on thermostatic control of the dominant workloads.

1.1.5 Conclusions and Discussions

In this chapter, we systematically studied the relevant concepts of cloud computing, including what cloud computing is, its architecture, measurement of cloud service quality, and types of cloud computing. Cloud computing is a computing model based on the Internet that provides on-demand resources and services to users, offering flexible, scalable, and cost-effective computing capabilities. There are various types of clouds in cloud computing: IaaS, PaaS, SaaS-based on service models, private cloud, public cloud, and hybrid cloud based on deployment models.

Resource scheduling strategies in cloud platforms are employed to meet different application scenarios and requirements. A static allocation strategy is suitable for stable and predictable workloads, where resources are pre-allocated to specific tasks. A dynamic scaling strategy adjusts resource allocation in real-time based on the current resource demands, catering to workloads requiring elastic scaling. The load-balancing strategy evenly distributes tasks across multiple servers, enhancing system performance and fault tolerance. A predictive scheduling strategy utilizes historical data and predictive models to forecast future resource demands and allocate resources preemptively.

Cloud computing finds wide-ranging applications. For enterprises and organizations, it offers flexible resources and services that can be rapidly scaled up or down to meet diverse business needs. IaaS enables the creation of virtual machines, storage, and networking on cloud platforms, facilitating the establishment of flexible IT infrastructures. PaaS provides a platform environment for application development and deployment, empowering developers to build and scale applications quickly. SaaS offers cloud-based applications that users can access and utilize directly via the web.

Cloud computing has also brought convenience to individual users. Cloud storage services allow individuals to share and synchronize files across different devices, granting them access to their data anytime, anywhere. Cloud media services enable online streaming and sharing of media content such as music and movies. Furthermore, cloud computing provides powerful computing and storage resources for fields like artificial intelligence and big data analytics, driving innovation and advancements in scientific research.

Nevertheless, cloud computing faces challenges. Security is a primary concern as cloud platforms must safeguard user data and privacy. Cost and performance must also be balanced to ensure efficient resource utilization and meet user requirements.

In summary, cloud computing transforms our work and lifestyle into a revolutionary computing model. By offering diverse cloud services and resource scheduling strategies, it provides users with flexibility, scalability, and cost-effectiveness. With ongoing technological advancements and innovation, cloud computing will continue to play a crucial role across various domains, bringing more convenience and opportunities to users.

1.2 Edge Computing

1.2.1 Understanding Edge Computing

With the start of the Internet of Things (IoT) era and the fast development of intelligent technology, the number of networked devices and the amount of data created and used by network devices are increasing quickly. Cisco forecast in 2020 that there would be 29.3 billion networked devices by 2023, compared to 18.4 billion in 2018. The network's data traffic is growing quicker than the number of network end devices because of the rapid increase in video applications and video quality enhancement. In Cisco's previous forecast [72], the data amount generated by all people, machines, and things would reach 850 Zettabytes (ZB) by 2021, and only 10% would be helpful for storage. This trend will bring a huge challenge for traditional cloud computing. In cloud computing, the cloud data center must receive all data. Meanwhile, the cloud server must execute all applications. It is challenging for bandwidth to transfer so much data to the cloud data center. And since only 10% of data is valid, transferring all data would waste a lot of bandwidth. This is unfriendly for the limited bandwidth. Although the computing capacity of the cloud servers is powerful, it is still limited. Processing so many applications may damage real-time performance, which is unacceptable for time-sensitive applications. Meanwhile, people gradually intensified their focus on the security and privacy of data. The migration of all data to the cloud center may lead to data leakage and other security problems.

Data execution and storage must be decentralized to the edge of the network, close to end devices. For this reason, industry and academia have proposed a novel computing paradigm called edge computing. Shi et al. [321] proposed an edge computing definition. In their point, edge computing is an enabling technology that can provide computing capability to download data from cloud data centers and upload data from end devices at any computing and storage nodes between end devices and cloud data centers. Figure 1.3 presents the edge computing architecture. Edge computing has the following advantages compared with cloud computing:

- **Low Bandwidth:** The edge servers are closer to the end devices. Edge servers can process many data. Only the data cannot be executed at the edge servers,

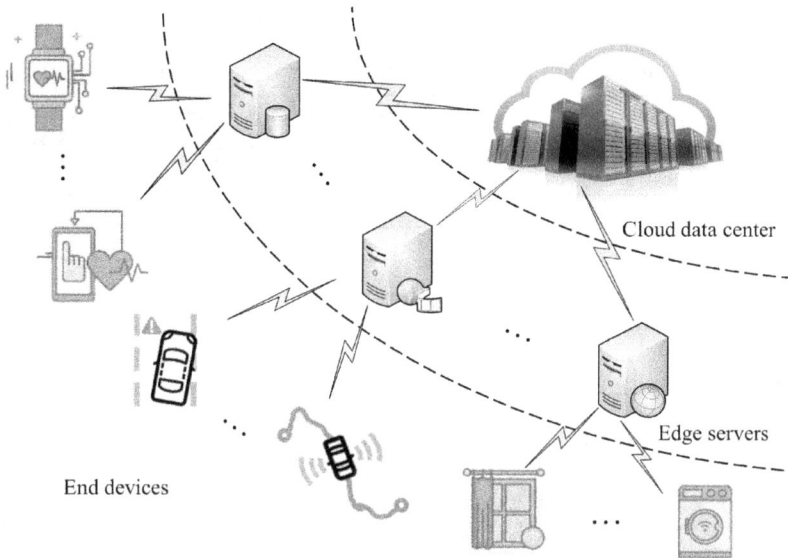

FIGURE 1.3
Edge computing architecture.

and the results that need to be stored for a long time would be transferred to cloud data centers. This greatly relieves the pressure on the bandwidth for the center network. Thanks to edge computing, the total volume of data transmitted and received by data centers worldwide amounted to only 20.6 ZB in 2021 [72], much less than the generated data amount.

- **Short Latency:** Despite their lower computational power compared to cloud servers, the number of edge servers is significantly greater than that of cloud servers. Each edge server needs to process significantly less data compared to cloud servers. The data transfer distance of edge computing is shorter. The above factors help edge computing have lower latency than cloud computing when executing the same application. This phenomenon can be verified by the experiment concluded by Yi et al. [460]. They performed a face recognition application in both edge computing and cloud computing. The face recognition time on edge and cloud servers is almost the same, roughly 2.5 ms for both. However, the total reaction time of edge computing is 169 ms, while the total reaction time of cloud computing is 900 ms.

- **High Security:** In edge computing, the data transmission distance is shorter. The security-critical data can be stored at the edge servers rather than transferred to the cloud data centers. This can decrease the probability of tampering attacks and information disclosure. Edge computing has a relatively simple network structure. Thus, the probability of routing attacks can also be reduced. All of these help edge computing have higher security and privacy.

Table 1.10 shows more aspects of comparison between cloud computing, edge computing, and only executing tasks on devices (Devices Only). While edge computing has some advantages over cloud computing, it cannot completely replace cloud computing. They can complement each other.

A major standard for determining the potential of a technology is whether the industry applies it. Edge computing has found applications in numerous scenarios, such as the

TABLE 1.10
Comparison of cloud computing, edge computing, and devices only.

Item	Cloud Computing	Edge Computing	Devices Only
Distance	Remote	Near	/
Latency	Medium	Shortest	Longest
Security	Lowest	Medium	Highest
Bandwidth Demand	High	Low	/
Number of Devices	Billions	Millions	One
Device Energy Consumption	Medium	Least	Highest
Type of Computer Networks	WAN	LAN	/

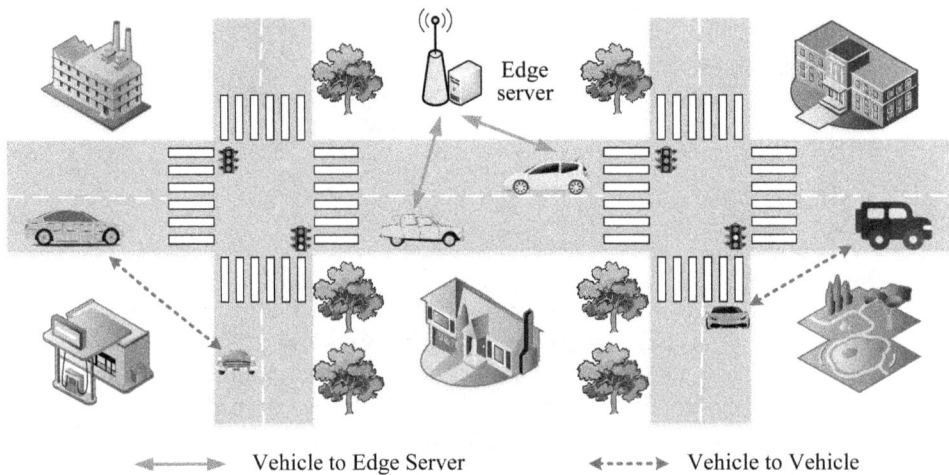

Vehicle to Edge Server   Vehicle to Vehicle

FIGURE 1.4
Illustration of IoV and ITS.

Intelligent Traffic System. Thus, edge computing is judged as an excellent computing paradigm. Several real-world scenarios using edge computing will be presented.

- **Internet of Vehicles (IoV) and Intelligent Traffic System (ITS)**

    IoV is a subset of IoT, in which various vehicle devices communicate with other entities using wireless communication technology. It can provide drivers a more comfortable and efficient experience than traditional vehicles. The vehicles in IoV are Connected and Autonomous Vehicles (CAVs), which support intelligent functions such as autonomous driving. While some functions, such as temperature control, can tolerate a relatively large latency, others, such as autonomous driving, are latency-critical. If the instructions, such as braking or steering, cannot be judged and completed in time, there will be a traffic accident. And the more vehicles on the road, the more serious the accident will be. ITS applies advanced sensing, communication, control, and other technologies to the transportation system. ITS can relieve transportation blockages and improve transportation safety. Electronic Toll Collection (ETC) is a typical example of ITS. Figure 4.3 presents an illustration of IoV and ITS. The information on ITS has temporal and spatial locality. For example, the road condition in the morning is useless for the vehicles passing this location at night. And the road condition in one city is useless for the vehicles in the other cities. Therefore, edge computing is more suitable than cloud computing for ITS. Hou et al. [157] integrated the mobile and fixed edge servers to deliver low-latency services. They also took into consideration that the communication links may not be reliable. The communication links may be broken, which could lead to accidents. A reliable reprocessing mechanism has been proposed to ensure the IoV services can be completed with high reliability. The heuristic-based fault-tolerant algorithm determines the task allocation and maximum reliability under the latency constraint. Hu et al. [163] first used the real-world traffic dataset to confirm that integrating resources from different vehicles is feasible. Considering the dependence of tasks in applications, they proposed a latency-aware real-time scheduling framework LRSF to minimize the application's total latency and maximize the system's resource utilization.

- **Smart Home**

Nowadays, the smart home system has gained increasing popularity. The scenarios of smart homes are mainly in residential or office buildings. Smart homes can utilize advanced sensing and other technologies to intelligently control IoT devices in the room, such as the lighting system, which enables people to have a comfortable, environmentally friendly, and convenient living and office environment. According to QuestMobile [297], in February 2023, the smart home application recorded 265 million monthly active users, accounting for about 80% of the total smart devices users. Although the smart home's latency requirement is not as high as that of ITS, its applications are still latency-critical. Meanwhile, the information in the smart home system is more private, especially the images or videos in residences. Few people don't resent others obtaining the images in their homes. Meanwhile, the smart home system is characterized by spatial locality. Transmitting all kinds of information from each room to the cloud data centers wastes storage and bandwidth resources and is unfriendly to the data owners. Edge computing can keep private data at the edge servers and only transmit the essential data to the cloud data centers. Here, the edge servers are near users and even at home. This helps to improve the privacy of the smart home system. It is unfriendly for program designers and device users without a unique operating system. Cao et al. [35] discovered this problem and designed an edge operating system for the smart home, which is called $EdgeOS_H$. The authors designed the communication, data management, and self-management models in this paper and provided the programming interface. $EdgeOS_H$ can handle various applications in the smart home in just one operating system to simplify management. Verma et al. [377] introduced a distant system implemented within the smart home environment to monitor patient health status. IoT devices can monitor patients' health conditions in real-time in this monitoring system. Edge/fog servers perform a simple real-time analysis of the condition's severity. Specific data are selected to transmit to cloud data centers for further analysis. The proposed system has high accuracy and lower latency.

- **Smart City**

ITS and the smart home are both subsets of the smart city. ITS only applies in the traffic industry, while the smart home only applies in residential or office buildings. The smart city refers to utilizing advanced communication and information technologies to enhance residents' quality of life. As the parent set of ITS and the smart home, the information of the smart city has temporal and spatial locality, and the applications in the smart city have latency requirements. Edge computing is more suitable than cloud computing for smart cities. Many researchers have investigated the usage of edge computing paradigms in smart city scenarios. Xu et al. [435] considered the privacy leakage problem when end devices offload tasks to the edge servers in smart cities. They suggested an intelligent offloading method (IOM) to attain the lowest energy consumption, shortest reaction time, and workload balance while guaranteeing privacy protection. Khan et al. [191] made an exhaustive survey on smart cities empowered by edge computing. They introduced some critical requirements for edge computing-based smart city architectures and proposed several research challenges.

- **Virtual Reality (VR)**

   VR technology indicates that simulation technology can create a virtual world that can give users an immersive feeling. It has been widely used in education, medicine, and other fields. Images' clarity and refresh rate must be relatively high to ensure users can be immersed in the virtual world. In other words, the VR applications have hard real-time constraints. As we all know, the latency of edge computing is lower than that of cloud computing. It is natural to combine VR technology and edge computing. Li et al. [218] developed a systematic mobile VR platform for multi-users called Multi-User Virtual Reality (MUVR). MUAR makes full use of the unique features of VR frame rendering. The edge server stores the results of VR frame rendering, and when someone else needs these results, it can transmit them to users directly without rendering the frame again. Also, MUAR will not transmit the complete data for all VR frames. It transmits the complete data for "reference frames". For other frames, it only transmits the data differently from the "reference frames". The computation and communication resources can be utilized most efficiently in these ways. Zhang et al. [483] proposed an Unmanned Aerial Vehicle (UAV)-assisted edge computing framework that can provide high-quality performance for mobile 360-degree video VR applications. In their proposed edge computing framework, UAVs can move and provide communication and computation capabilities to VR applications. The authors proposed an approximation strategy to maximize VR users' quality of experience (QoE) by determining the place of UAVs and assigning computation and communication resources.

- **Industrial Internet of Things (IIoT)**

   The expansion of IoT into the industrial area has given rise to IIoT. IIoT only focuses on the industrial area. By connecting machines with sensors, computers, people, and others, IIoT helps industrial production become increasingly efficient and intelligent. The amount of data produced by machines in IIoT is vast, and most is private data. Also, the applications in IIoT are latency-critical, and the IIoT has the characteristic of spatial locality. Industrial Internet Consortium affirmed the role of edge computing in the IIoT scenario and published *Introduction to Edge Computing in IIoT* Write Book [128]. In the Write Book, they illustrated the benefits of utilizing edge computing instead of cloud computing in IIoT and gave examples. Apart from the typical advantages of edge computing, such as high privacy, the demand for communication resources of edge computing is much lower. It is friendly for the industries in remote areas. Network infrastructure tends to be deployed in densely populated urban areas, and less is deployed in remote areas. However, industries such as mining, oil and gas, and chemicals can only be located in remote areas. For these industries, deploying edge computing infrastructure nearby is much easier and more efficient than connecting them to the cloud data centers. To address the development challenges and low reusability rate when mainstream mobile edge computing (MEC) framework supports IIoT, Hou et al. [158] suggested a novel MEC platform for IIoT, called IIoT-MEC. This framework can enable the development of IIoT to be mainly focused on the software and have a high rate of development reusability.

   In the past few years, edge computing has been widely developed. At the same time, Artificial Intelligence (AI) technology is developing rapidly. Combining these two promised technologies can help to thoroughly release the prospect of big data at edge [522]. This combination generates a new interdiscipline, i.e., Edge Intelligence (EI), which has drawn the attention of many researchers. The applications can hardly be completed on just one end device or edge server. The applications can offload from end devices to edge servers. Thus, edge offloading has emerged as a critical research area within the field of EI. The

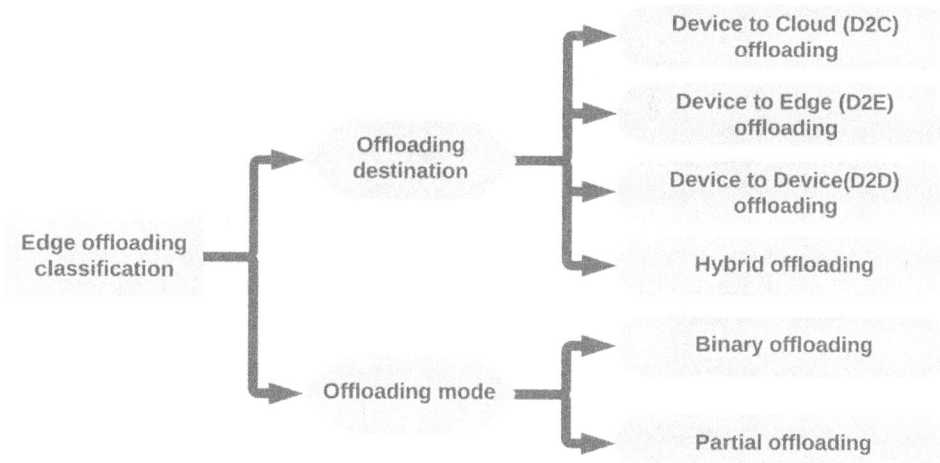

FIGURE 1.5
Classification of edge offloading.

fundamental options in AI applications are data collection, inference, and training. They are essential for EI and correspond to edge caching, edge inference, and edge training, respectively [427]. We will introduce these four research topics in detail as follows.

1.2.2 Edge Offloading

Edge offloading refers to transferring computing tasks from user devices or terminals to more powerful servers. Because most end devices have limited resources, resource-intensive tasks must be offloaded, either fully or partially, to servers with powerful computing resources (such as edge servers or remote cloud servers) and other devices with surplus computing resources. As shown in Figure 1.5, the edge offloading can be broadly classified from two perspectives: one is from the different destinations for computing offloading [427]: Device to Cloud (D2C), Device to Edge (D2E), Device to Device (D2D) and Hybrid offloading. The other is based on offloading mode, i.e., Binary and Partial offloading.

- **Offloading Destination**

    Edge offloading can be categorized into four types: D2C, D2E, D2D, and hybrid offloading.

    - **D2C Offloading:** D2C Offloading refers to transferring computation-intensive tasks to remote cloud servers, alleviating computation and storage constraints, and extending device battery life. For example, Wu et al. [409] introduced an energy-efficient offloading decision algorithm based on Lyapunov optimization, which dictates the optimal way for executing applications locally or forwarding them directly for remote execution in a cloud infrastructure.

    - **D2E Offloading:** D2E offloading transfers computation tasks to edge servers, which can reduce network bandwidth consumption and improve service response time compared to D2C. For instance, Zhan et al. [476] have demonstrated that offloading resource-intensive tasks to the MEC servers within roadside units significantly reduces latency and energy usage for vehicle applications.

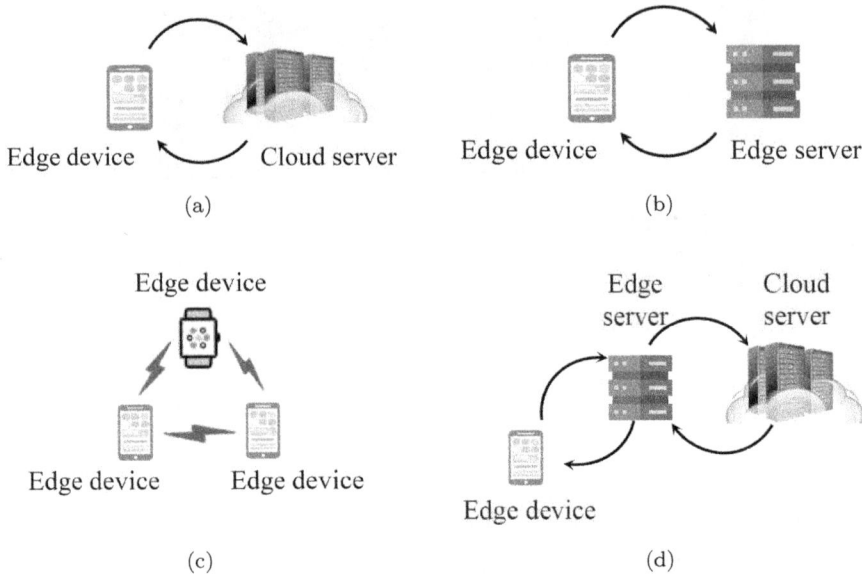

FIGURE 1.6
The different destinations for computing offloading. (a) D2C offloading. (b) D2E offloading. (c) D2D offloading. (d) Hybrid offloading.

- **D2D Offloading:** D2D offloading can fully utilize the idle end-side device resources in the edge network, transferring computation-intensive tasks from one device to another to reduce computation latency and energy costs. As an illustration, Luo et al. [250] designed a VEC collaborative task data dispatching program in which computational tasks of a vehicle could either be executed locally or migrated to other vehicles with unused computational resources.

- **Hybrid Offloading:** Hybrid offloading refers to transferring computing tasks to multiple devices and servers for execution. This method maximizes the utilization of resources on each device to enhance task execution efficiency and user experience. For example, Liu et al. [234] put forward a framework for task offloading contingent upon various mobile applications' requirements. This framework allows for the adaptive allocation of offloaded tasks by mobile devices to MEC and Cloud resources to elevate the user experience.

- **Offloading Mode**

  Based on the divisibility of the task, the offloading mode is categorized into two types: binary and partial offloading.

  - **Binary Offloading:** Binary offloading is also called "0-1 offloading". It means that the entire computational task is either executed locally or offloaded to another location. The terms "0" and "1" indicate whether the task is offloaded. Typically, "0" signifies that the computational task is entirely processed locally, while "1" indicates that it is offloaded to a different location [91, 476]. Binary offloading is simpler and suitable for atomic tasks that cannot be split.

TABLE 1.11

The summary of related work on edge offloading.

Ref.	Offloading Mode	Optimization Objective(s)			
		Latency	Energy Consumption	Computation Efficiency	Payment
[91]	Binary		✓	✓	
[106]	Partial	✓			
[135]	Binary	✓	✓		
[163]	Partial	✓		✓	
[229]	Binary				✓
[251]	Partial	✓			✓
[351]	Partial	✓			
[476]	Binary	✓	✓		
[501]	Partial	✓			✓

- **Partial Offloading:** Partial offloading involves splitting the computational task into several subtasks and dispatching a subset to computing devices. During the division of the task, it is crucial to consider the interdependence of the data associated with each subtask. If they are independent, they can be split arbitrarily. If there are dependencies in the task data, the execution order of the sub-tasks after splitting must meet the requirements. For example, Tang et al. [351] broke down computationally intensive tasks into a series of sequential subtasks, where each subsequent subtask is generated upon completion of the preceding subtask by the same vehicle.

Offloading modeling is crucial for studying the overall performance of edge computing. Its main objective is to find optimal offloading decisions and resource allocation strategies to optimize system performance. Table 1.11 describes most of the offloading schemes studied in the literature cited in this section and compares offloading mode and optimization objective(s).

As indicated in Table 1.11, the researchers have designed various offloading algorithms for different offloading modes and offloading objective(s). In a scenario where numerous user devices and several mobile edge computing servers face resource constraints, Ding et al. [91] introduced a decentralized computing offloading strategy algorithm to minimize the weighted sum of execution time and energy consumption. In a Vehicular Edge Computing (VEC) scenario with roadside units, Fan et al. [106] explored a method that synergizes task offloading with resource allocation. This method aims to reduce the collective delay incurred in task processing across all vehicles while simultaneously adhering to each vehicle's specific delay tolerance and waiting time limits. In the same application scenario as [106], Tang et al. [351] developed a dynamic model for task offloading among a fleet of mobile vehicles, which is aimed at decreasing the cumulative delay and waiting time for the vehicles involved. An innovative algorithm known as Dynamic Framing Offloading is introduced to determine the best offloading decisions for consecutive subtasks, founded on the Double Deep Q-Network principles. In a typical VEC setting, Zhan et al. [476] designed a novel DRL-based offloading algorithm to realize the same optimization goals as in [106]. In [135], an online task allocation algorithm using dynamic Lyapunov optimization is presented in addition to energy harvesting technology. The primary objective of this algorithm is to reduce the weighted aggregate of mobile device's computation latency and energy expenditure.

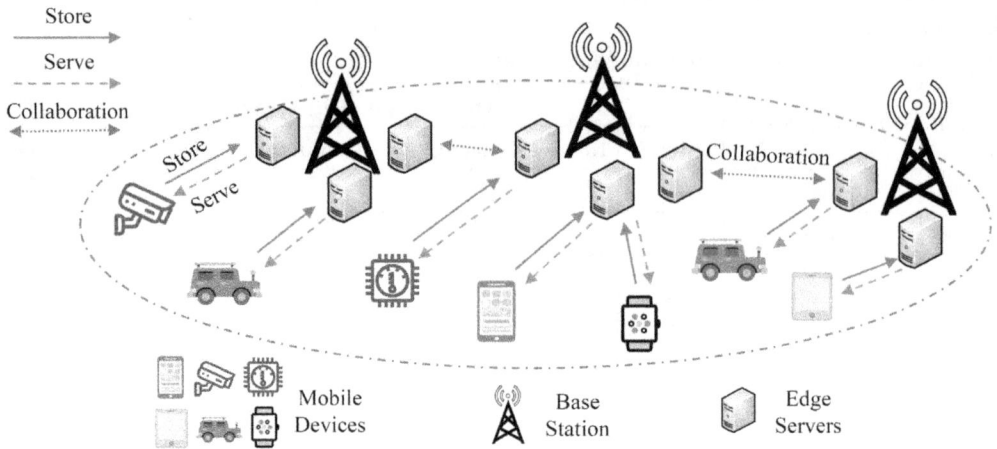

FIGURE 1.7
Illustration of edge caching.

Considering the dependencies between offloading application tasks in VEC, Hu et al. [163] designed a greedy-based job offloading scheduling algorithm to minimize the total job delay and maximize the queue resource utilization. Liu et al. [229] investigated the efficient task offloading for end users by leveraging scattered computation and communication resources through a multi-path and multi-hop approach in edge computing. They introduced a binary optimization framework that extended the concept of multi-hop wireless edge computing task offloading. Through game theory analysis, Luo et al. [251] introduced an algorithm for computation offloading grounded in particle swarm optimization. From the perspective of multi-objective optimization, they conduct a thorough analysis of the delay and cost associated with VEC computational offloading to reduce both delay and expense. Zhang et al. [501] explored the allocation of task offloading for vehicles in demand and the pricing strategies for edge servers and the cloud. Viewing the situation through the lens of a market economy, they conceptualize the interactions between vehicles, edge servers, and the cloud as a dynamic Stackelberg game that includes aspects of competition and cooperation.

In summary, edge offloading leverages the computational power of various layers in the network, promoting faster, more efficient, and decentralized data processing. The development of edge-offloading technology brings more convenience and opportunities to future networking and computing models, fostering the advancement of digital intelligence.

1.2.3 Edge Caching

Edge caching stores data, applications, or content in edge computing devices near users. It aims to provide fast access and delivery, reducing latency and bandwidth usage. Edge caching reduces reliance on remote data centers or cloud services, enabling more efficient data transfer and processing [427]. Figure 1.7 demonstrates the main processes of edge caching. Data is gathered at the edge, and edge devices such as edge servers store environmental information. For example, the traffic light captured by cameras could be cached on edge servers for the high-precision maps in intelligent connected vehicles [203]. Table 1.12 shows the common representations of edge caching.

While edge caching is always effective in enhancing user experience, it is not feasible to cache every service from the data center to the edge cloud because edge servers usually

TABLE 1.12

The common representation methods of edge caching.

Ref.	Scenario	Optimization Targets	Strategy
[124]	MEC Environments	Reduce user access latencies	An innovative and scalable hierarchical caching policy
[203]	Intelligent connected vehicles	Minimize delay and protect privacy	A cooperative edge caching strategy utilizing federated deep reinforcement learning
[213]	Wireless Infostation Networks	Maximize the edge cache hit rate	A hybrid content placement caching strategy
[255]	MEC Environments	Minimize service response time and outsourcing traffic	An iterative algorithm named ICE
[360]	Mobile Edge Networks	Improve the cache efficiency	An efficient cooperative caching framework
[416]	MEC Environments	Maximize the caching revenue	Online Mobile Edge Data Caching
[474]	Vehicular Edge Computing	Reduce the support data fetching delay	A cooperative caching approach that employs division of service areas, categorization of servers, and segmentation of storage capacity
[515]	MEC Environments	Maximize the fairness and minimum caching costs	A FEDC-opt optimization method based on integer programming

have limited and expensive caching capacity. Therefore, a suitable caching strategy becomes crucial for edge caching. In an intelligent connected vehicle scenario, satisfying the demand for minimal latency and privacy in high-precision map caching, Li et al. [203] introduced a cooperative edge caching strategy utilizing federated deep reinforcement learning. The goal is to realize a dynamically adaptive edge caching system while safeguarding user privacy. The service provider determines the quantity and positioning of service instances designated for caching. Xia et al. [414] explored the core issue of deploying service instances from distant data centers to edge clouds within a multi-tier edge cloud architecture. They developed an efficient heuristic approach aimed at addressing the varied service demands of users while the service provider determines the number of service instances allowable for caching and their respective caching locations. The investigation of a novel Mobile Edge Data Caching (MEDC) issue through the lens of application providers is conducted in [416]. They proposed online MEDC, an approach to develop MEDC policies for application providers that do not require information about future data requirements. However, many studies have overlooked fairness when developing edge data caching strategies for application providers, although it is worth considering. Zhou et al. [515] first attempted to solve the fair edge data caching problem, formulated it as a constrained optimization problem, and proposed an optimization method based on integer programming.

In the absence of collaboration among edge nodes, suboptimal use of the available edge resource capacities is typical. Many scholars have proposed different collaborative caching strategies to address this issue. Aiming to reduce the service response time and the volume of outsourced traffic, Ma et al. [255] investigated collaborative service caching and workload distribution and designed a two-layer iterative algorithm named ICE. Tian et al. [360] proposed an efficient collaborative caching framework to maximize the edge cache hit ratio and minimize the associated communication and computational overheads. In particular, they designed a DRL-CA algorithm for cache admission, which identifies a subset of boundary attributes from a vast array of requests to enhance the efficiency of the caching process. In VEC, the fulfillment of offloading tasks necessitates not just the task data transmitted by the soliciting vehicle but also supplementary data that aid in the seamless completion of the task. In [474], a collaborative caching approach is investigated to decrease the delay

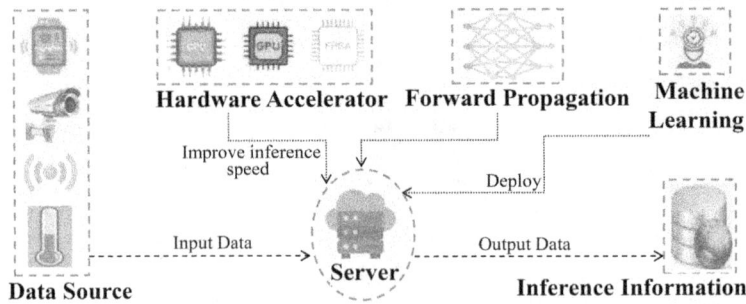

FIGURE 1.8
Process of edge inference.

fetching support data by utilizing service area partitioning, server grouping, and storage space partitioning. This ensures that the required support data is stored either in the requesting server or its proximate servers.

As research progresses, scholars have found that storing frequently accessed content in edge nodes can significantly minimize user access latency. Saibal Ghosh et al. [124] introduced an innovative and scalable hierarchical caching policy for MEC that considers data size and access frequency. The goal of enhancing throughput is achieved by intelligently calculating the popularity of the content and caching it at the best possible location. Li et al. [213] cached specific popular content files in the infostations and make them available to the requesting user in case of a cache hit via dedicated short-range communication. A hybrid content placement caching strategy is proposed to enhance the edge cache hit rate within the aggregate storage constraints of the infostations.

In conclusion, edge caching enables users to access content quickly and efficiently, improving throughput and reducing latency. This technology not only refines user experience but also alleviates congestion in the central network infrastructure. With an increasing need for these types of services, the significance of edge caching within the evolving landscape of network technology is set to become even more evident.

1.2.4 Edge Inference

Edge inference involves deploying a pre-trained machine learning model on edge devices and generating model output by processing input data through calculations. Figure 1.8 shows the specific processing of edge inference. Edge devices collect raw data from sensors, cameras, or other sources. They then load pre-trained machine learning models into them. The edge devices execute forward propagation computations once the data and models are ready. During this process, the data starts from the input layer and undergoes calculations through a series of neural network layers, ultimately obtaining the results from the output layer, thereby acquiring the inference information. To enhance inference speed and computational efficiency, edge devices typically utilize hardware accelerators to perform the computational tasks of Deep Neural Network (DNN) models.

Researchers are actively seeking innovative solutions that address security, model optimization, and computational synergy to overcome resource limitations in edge-based deep learning. Table 1.13 offers a cintegratingnalysis of edge inference, highlighting advancements in enhancing inference accuracy, resource optimization, and model performance within edge environments.

From a security standpoint, edge computing is dedicated to integrating robust measures designed to safeguard the reliability and efficacy of data analysis and computation. Guo et al. [133] introduced a confidential CNN inference scheme with edge assistance,

TABLE 1.13

Analysis of related work on edge inference.

Ref.	Inference Strategy	Contribution	Performance Evaluation Metrics			
			Accur-acy	Laten-cy	Perfor-mance	Resource utilization
[133]	Edge-assistance	Propose a privacy-preserving CNN inference scheme that achieves efficient inference	✓		✓	
[139]	Collaborative inference	Introduce various model-parallel techniques to enable collaborative inference among IoT devices	✓	✓		
[148]	Hardware accelerator collaborates with the host CPU	Introduce the SECDA-TFLite toolkit designed for the efficient creation of hardware accelerators for FPGA-based DNNs in edge inference			✓	
[450]	Collaborative model segmentation and compression	Enhance fast and accurate inference in the end-edge-cloud system by utilizing combined model segmentation and compression			✓	✓
[451]	Collaborative inference	Propose a pipelined cooperation scheme for edge devices that maximizes throughput	✓		✓	
[491]	Edge nodes are tasked with parallel processing of inference jobs	Propose a dynamic architecture for convolutional neural network inference tailored for edge computing environments	✓		✓	
[174]	Dynamic model adjustments and resource allocation	Facilitate the acceleration of time-critical DNN inference by optimizing both model adjustments and resource allocation jointly			✓	
[209]	Edge/cloud collaborative inference	Enhance the trade-off between precision and computational/communication expenses by considering the complexity of the inference				✓
[212]	Distributed inference	Propose a distributed scheduling strategy to coordinate inference in deep learning for resource-constrained IoT edge clusters		✓		✓
[277]	Coordinating data and DNN adaptation decisions among multiple users	Attain flexible assurances on end-to-end inference timing defined as a SLO through collaborative DNN adaptation	✓			
[389]	Adaptable service relocation with selection of multiple exit points	Propose a user-focused strategy for dynamic migration of DNN inference services to maximize user benefits	✓			✓
[448]	End-to-edge collaborative inference	Propose an efficient parallel model deployment protocol and an Advanced Multi-UE Integrated Optimization Strategy grounded in model deployment		✓		

safeguarding data and model privacy through data encryption and transferring computations to edge servers. This approach provides secure and efficient inference while minimizing computational burden for users. Another proposed by Hadidi et al. [139] addresses the compute intensity and privacy challenges in DNN applications by leveraging the collective computing capabilities of IoT devices for real-time inferencing. The technique centers on a collaborative network of IoT devices, utilizing model-parallelism strategies such as output splitting for fully connected layers and channel or filter splitting for convolutional layers. These methods effectively distribute the computational load, enhancing efficiency and safeguarding privacy through localized processing.

From a model optimization perspective, edge intelligence is a pivotal research area striving to enhance the efficiency and performance of edge computing. This is realized through advanced techniques such as FPGA-based accelerators, optimizing resource allocation in IoT edge devices, and distributing CNN inference tasks. Haris et al. [148] created a streamlined development toolkit for edge inference using FPGA-based DNN accelerators, minimizing design duration and enabling enhanced DNN inference capabilities on FPGA chips. Their SECDA-TFLite toolkit demonstrates significant performance enhancements across CNN and BERT-based models. Yang et al. [450] presented CNNPC, which resolves the clash between cloud computing and edge devices. Using integrated model segmentation and compression, CNNPC facilitates swift and precise CNN inference, leading to notable speed enhancements while maintaining accuracy. Additionally, Yang et al. [451] introduced the PICO scheme to tackle computing limitations in IoT edge devices for CNN inference. This approach enhances inference execution time by segmenting edge devices and neural layers into multiple phases, improving throughput and reducing inference latency. Furthermore, Zhang et al. [491] presented ADCNN, a system designed to address distributed execution of CNN inference tasks at the network edge. ADCNN applies fully decomposed spatial partitioning and compression methods to boost performance and cut resource expenses, yielding significant speedups without compromising inference precision.

From a computational collaboration standpoint, edge computing research emphasizes innovative strategies that leverage collaborative optimization and distributed scheduling to improve model efficiency and performance. Huang et al. [174] introduced a joint acceleration optimization scheme that employs Multi-exit DNNs to decrease inference time. By strategically exiting tasks prematurely at suitable depths, they aim to minimize task completion time. Through the concurrent enhancement of model modification and resource allocation, their approach achieves a substantial speedup of up to 6.01 times compared to existing methods. To address the challenges of precision and resource limitations in deep learning models for edge computing, Li et al. [209] introduced AppealNet. This collaborative architecture is both effective and precise in edge/cloud operations. AppealNet employs a dual-neural network structure to maintain a harmonious relationship between accuracy and computation/communication expenses. It utilizes edge devices for processing simple inputs with compact models while transferring complex inputs to cloud-based models. This collaborative strategy leads to more than a 40% reduction in energy consumption without compromising accuracy. For implementing DNN inference in IoT edge clusters, Li et al. [212] introduced DISSEC, a spatially partitioned distributed scheduling approach. DISSEC optimizes inference execution latency by amalgamating and segmenting convolutional layers for autonomous execution, efficiently leveraging edge device capabilities. The authors presented a technique for illustrating interdependencies among partitioning tasks and employ a heuristic-driven exploration algorithm to formulate a distributed concurrent scheme. To tackle the obstacle of model transfer in DNN inference tasks for edge computing, Yang et al. [448] presented an efficient parallel model uploading method. Their multi-UE joint optimization algorithm comprises a DNN partitioning sub-algorithm based on pruned binary trees and a resource allocation sub-algorithm with asynchronicity. This approach enhances DNN inference efficiency, reducing execution latency by as much as 64.5% when compared to current methods.

From the perspective of inference services, research in edge network intelligence and dynamic migration seeks to enhance deep learning performance in edge computing. This is achieved by effectively managing the trade-off between accuracy and latency through adaptive data and DNN strategies. Nigade et al. [277] introduced Jellyfish, an innovative edge deep learning inference service platform in addressing the need for prompt inference delivery in evolving edge networks. Jellyfish is designed to harmonize precision and response time through data and DNN adjustments. By coordinating decisions across multiple users using cooperative adaptation strategies, Jellyfish ensures latency requirements are met with a success rate of approximately 99% while preserving exceptional accuracy. Within the realm of mobile edge computing, Wang et al. [389] introduced a user-centric management approach for dynamic DNN inference service migration. Their methodology optimizes overall user satisfaction by utilizing a versatile multi-exit mechanism. Incorporating dynamic programming, model predictive control, and an intelligent migration decision-making algorithm minimizes computational burdens, leading to notable enhancements in user satisfaction.

1.2.5   Edge Training

Edge training involves transferring the training process of machine learning or deep learning models from conventional centralized cloud servers to edge devices. During edge training, the model is trained locally on the devices, avoiding transmitting all data to the cloud for training. This minimizes data transfer, latency, and privacy issues. As illustrated in Figure 1.9, edge devices input the samples into locally deployed models for training. Subsequently, the gradient information is sent to the cloud server, where it is employed to aggregate the model parameters. Then, the cloud server forwards the updated parameters

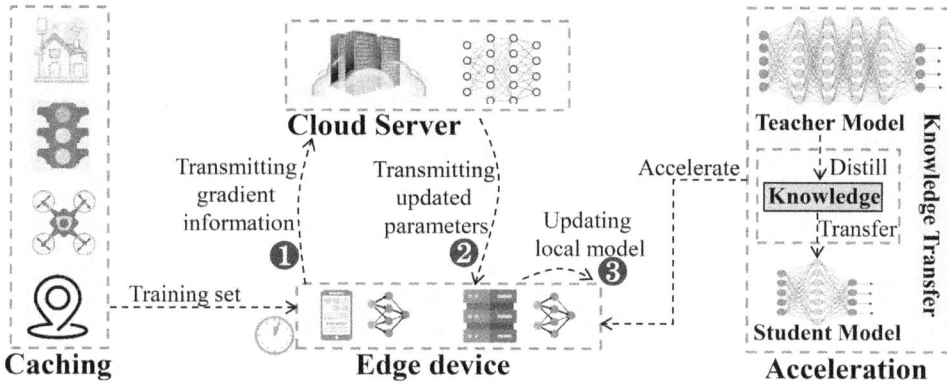

FIGURE 1.9
Process of edge training.

TABLE 1.14
Analysis of related work on edge training.

Ref.	Training Strategy			Contribution	Methodology
	Distributed Learning	Transfer Learning	Model Enhancement		
[296]	✓	✓	✓	Propose a two-stage framework for collaborative training benween cloud and edge devices	Apply homomorphic encryption and model partitioned training
[316]	✓		✓	Propose a distributed deep learning training method using shielded reinforcement learning	Introduce a shield mechanism to detect and prevent job collisions
[403]	✓			Propose a resource-efficient federated learning mechanism	Introduce hierarchical aggregation with cluster aggregation
[437]				Propose a privacy-aware task offloading method (POM)	Use an improved Strength Pareto Evolutionary Algorithm to optimize strategies of task offloading
[111]				Propose a cybersecurity training platform that is open-source and tailored for simulating scenarios in edge-IoT environments	Adopt a design and implementation approach
[357]	✓	✓		Propose a method for distributing the training of deep neural networks on IoT edge devices	Generate small networks, apply knowledge distillation technique, and perform distributedtraining on edge devices
[459]	✓			Propose the Eco-FL framework, which enables edge collaborative pipeline training	Use edge collaborative pipeline training and grouping-based hierarchical aggregation
[418]	✓			Propose a method for selecting vehicles and optimizing resources in federated learning	Adopt a formulation of min-max optimization problem, considering vehicle position and velocity

to the edge devices. The edge devices use the returned parameters to update their local models.

Table 1.14 presents an analytical overview of edge training in computation, highlighting scholarly endeavors to bolster data security, optimize resources, ensure the safe training of models, and refine task scheduling in edge computing settings.

From the perspective of data security and resource optimization, smart edge computing has become a critical area of ongoing research, focusing on developing strategies that ensure efficient edge training despite limited resources. Qu et al. [296] proposed a two-phase model training framework to address privacy and resource challenges in intelligent edge computing. The initial phase involves collaborative federated pre-training, encouraging edge server participation, and ensuring data authenticity. The second phase focuses on privacy-preserving model segmentation training, utilizing homomorphic encryption for model improvement and protection. This framework provides a comprehensive solution to the intricacies of privacy and resource management in intelligent edge computing. Wang et al. [403] addressed limited bandwidth and privacy concerns in edge computing through a resource-efficient federated learning method. The method employs hierarchical aggregation to divide edge nodes into clusters, enabling local updates and global aggregation. An optimization algorithm is adapted to dynamic network conditions, reducing communication resources while maintaining high accuracy. Xu et al. [437] tackled the privacy-conscious task offloading challenge in generative adversarial network training processes for edge computing. They introduced a privacy-focused task offloading strategy called POM, utilizing an improved strength Pareto evolutionary algorithm to boost collaborative performance and uphold privacy standards.

From the model training security angle, these research efforts are committed to establishing robust cybersecurity training platforms and refining effective DNN training methodologies to ensure network security and data privacy. Ficco et al. [111] unveiled Leaf, a cybersecurity training platform explicitly crafted for edge IoT settings. The platform equips users with tools and services to simulate complex cybersecurity scenarios and edge applications, serving educational, training, and security evaluation functions. With its flexible modeling capabilities, users can grasp network security threats and attack vectors and validate various security measures, including prevention, detection, mitigation, and recovery solutions. This platform is invaluable for tailored training and exercises in Edge-IoT environments. Tanghatari et al. [357] focused on training DNNs on IoT edge devices using distributed learning while safeguarding data privacy and reducing the burden on cloud servers. They introduced a transfer learning-based method that employs knowledge distillation to convey the knowledge from a main network to a smaller generated network. The approach results in a marginal average accuracy decrease of about 3.5% while ensuring data privacy, in contrast to training the network on the cloud with access to all datasets from edge devices.

From the task scheduling perspective, the research focuses on accelerating local model training and refining resource allocation within edge federated learning and vehicle edge computing, thereby enhancing algorithm training accuracy and ensuring efficient use of computational resources. Sen et al. [316] presented a technique for training distributed deep learning models in edge computing networks through protected reinforcement learning. By integrating multi-agent reinforcement learning and distributed coordination mechanisms, edge nodes autonomously schedule tasks and avoid conflicts. The approach notably reduces training time and resource usage and minimizes task conflicts. Ye et al. [459] addressed the challenges posed by constrained, heterogeneous, and dynamically changing computing resources in IoT devices. They proposed Eco-FL, an adaptive federated learning method that accelerates local model training through edge collaborative pipeline training. Eco-FL utilizes a hierarchical architecture based on groupings for model aggregation and employs adaptive scheduling to handle runtime resource fluctuations. Compared to current methods, Eco-FL enhances training accuracy, reduces local training time, and improves local training throughput. In vehicular edge computing, a study by Xiao et al. [418] focused on the challenge of vehicle selection and resource optimization within the framework of federated learning. The paper introduced a min-max optimization problem considering the vehicle's spatial position and speed. They utilized a greedy algorithm to dynamically select vehicles based on image quality and optimize resource allocation and local model accuracy, achieving a balance between minimal cost and equity.

1.2.6 Conclusions and Discussions

The edge computing paradigm has recently captured substantial interest in academic and industrial domains. This chapter aims to offer insights into the fundamental concepts of edge computing and its benefits in low bandwidth, short latency, and high security. We highlight five real-world scenarios: IoV and ITS, smart home, smart city, VR, and IIOT. For each scenario, we explore the latest challenges and solutions associated with edge computing.

Meanwhile, combining edge computing with AI can fully utilize the benefits of edge big data. We offer a comprehensive introduction to EI across four key dimensions: edge offloading, edge caching, edge inference, and edge training. Edge offloading means transferring intensive computational workloads from resource-constrained devices—typically those with limited processing power, storage capacity, and battery life—to more powerful servers at the network's edge. This method not only eases the burden on devices with limited resources but also reduces latency and boosts efficiency by minimizing data transmission distances compared to cloud offloading. Edge caching pertains to storing data closer to the end-user within the network's edge. Storing frequently accessed data in local caches reduces the overall traffic load on the network and shortens data access time, thereby enhancing user experience and network performance. Edge inference involves executing machine learning algorithms locally on edge devices. It enables the direct deployment of trained models onto edge devices, facilitating real-time decision-making based on processed data. This reduces communication with the cloud, offering improved latency, bandwidth usage, and privacy preservation. Lastly, edge training involves training machine learning models directly on edge devices, enabling data generation and model training to happen locally. This allows for continuous model updates without relying on constant communication with the cloud. This is particularly beneficial in scenarios with strict privacy requirements or limited network connectivity.

In conclusion, edge computing offers a hopeful solution to overcome the constraints of centralized computing. It offers enhanced efficiency, reduced latency, improved privacy, and robust performance through edge offloading, caching, inference, and training. As an innovative computing paradigm, edge computing is positioned to accelerate progress across diverse fields, bringing about new opportunities and convenience for users.

1.3 End Device in IoT

1.3.1 Internet of Things

The Internet of Things (IoT) is a vast network comprising physical objects equipped with sensors, software, and connectivity, facilitating real-time data acquisition and sharing. These objects range from everyday devices like smartphones, smart home appliances, and wearables to industrial systems such as manufacturing machinery and infrastructure components. The interconnectedness of these devices allows for seamless communication, data sharing, and automation, resulting in improved efficiency, productivity, and convenience. Moreover, the development of IoT has resulted in two notable progressions: the emergence of the Artificial Internet of Things (AIoT) and its implementation within the Industrial Internet of Things (IIoT) domain.

- **AIoT**

The AIoT represents the convergence of AI and IoT, bringing together the power of intelligent algorithms and the extensive network of interconnected devices. More advanced sensors and improved computing capabilities for deep learning enable AIoT devices to collect and transmit data and perform complex analyses in real-time. This empowers AIoT devices to make informed strategies, automate processes, and provide valuable insights to users. Regarding system architecture, AIoT typically consists of three layers: intelligent end layer, edge intelligence layer, and cloud computing layer (shown in Figure 1.10). The intelligent end layer involves sensors and actuators that perform multimodal perception and distributed execution. These devices gather diverse data from the physical environment, execute data preprocessing and lightweight computing, and transmit service requests to the

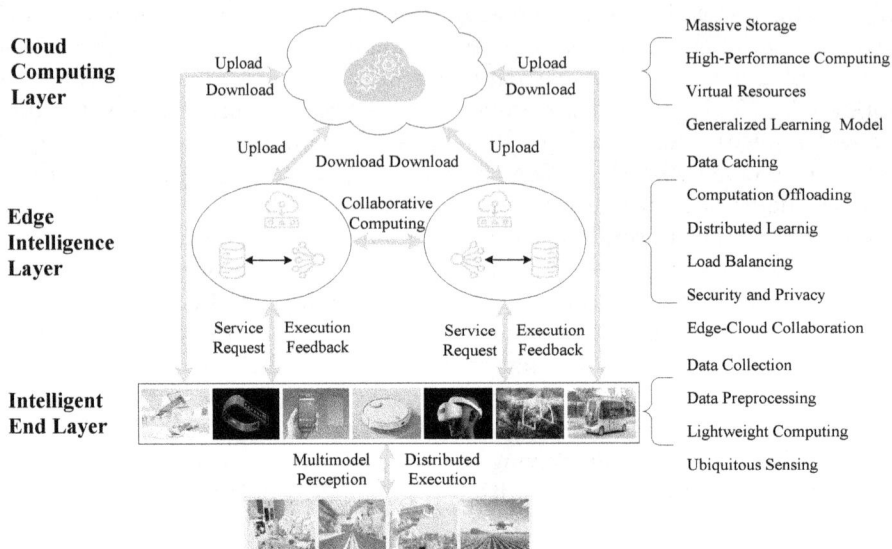

FIGURE 1.10
Architecture of AIoT.

edge intelligent layer. The edge intelligence layer deploys computational and intelligent processing capabilities on edge servers near the end. Edge servers cache data and requests from the Intelligent End Layer and provide real-time execution feedback through collaborative computing. The cloud computing layer supports data and requests uploaded from the edge and the end. The cloud server leverages the processed data to provide resource-intensive services, such as massive storage, high-performance computing, and a generalized learning model. At present, key areas of AIoT include data analytics and deep learning algorithms for intelligent data processing, edge computing for real-time and low-latency decision-making, and security and privacy frameworks for protecting sensitive data. Additionally, AIoT research explores integrating AI technologies with edge devices, cloud platforms, and hybrid architectures to optimize system performance and resource utilization. Table 1.15 summarizes the research of AIoT based on optimization objectives, devices involved, technologies used, and improvement.

- **IIoT**

    The IIoT refers to integrating IoT technologies in industrial settings, enabling the digital transformation of traditional manufacturing processes, supply chains, and infrastructure. IIoT combines sensors, actuators, and intelligent systems to create a network of interconnected devices and machines that can capture, analyze, and exchange data in real-time. This connectivity and data exchange enables enhanced operational efficiency, predictive maintenance, and optimized resource utilization. IIoT system architecture involves three layers: device layer, edge layer, and cloud service layer, as shown in Figure 1.11. The device layer comprises diverse sensors, smart vehicles, machines, and other devices. These devices gather significant sensing data transmitted to the Edge Layer while awaiting control directives. The edge layer's primary function is accepting and analyzing data from the device layer, supplying delay-aware services such as data analysis, process optimization, and real-time control. The cloud service layer retrieves enormous volumes of data from the edge layer via the public

TABLE 1.15
Summary of research related to AIoT.

Objective	Device	Technology	Improvement	Ref.
Cost Efficiency	Edge-Cloud	Continuous learning	Improve the reduction of accuracy due to model compression	[182]
	Edge	Intelligent edge based on RL	Distribute AI tasks between multiple edge devices	[525]
	Edge	Edge caching	Minimize the need for remote communication	[490]
	End	Multi-agent Deep RL	Optimize multi-UAV's trajectory to minimize user's consumption	[388]
Security	End	NSCP prediction based on CNN	Achieve security prediction in real-time	[429]
	End	Back propagation, Bayesian's classification	Determine the legitimacy of device during the access phase	[304]
Data Quality	End-Edge	Task assignment based on swarm intelligence	Match appropriate sensing tasks with competent workers	[246]
	End-Edge	Mobility prediction based on fuzzy control	Facilitate multi-task allocation considering real-world imprecision	[482]
Scalability	End	Multi-agent Deep RL	Improve communication among agents in the context of ATSC	[70]

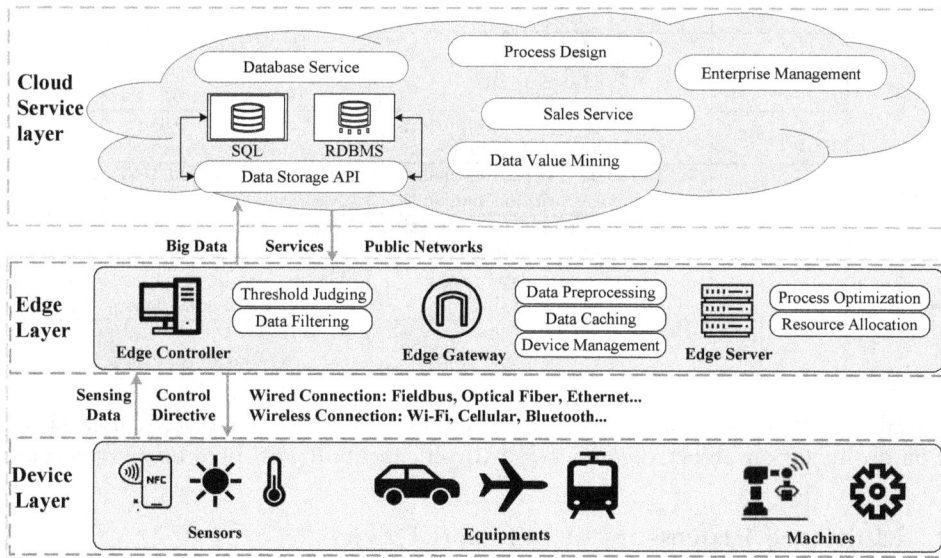

FIGURE 1.11
Architecture of IIoT.

network, mines potential value from the data, and realizes upper applications such as product design, sales service, and enterprise management. Key research areas in IIoT include connectivity and interoperability standards for seamless integration of diverse industrial systems, edge computing for localized data processing and latency reduction, and digital twin technologies for virtual modeling and simulation of industrial processes. Additionally, IIoT research focuses on security and privacy frameworks to protect critical infrastructure, sensitive data, and energy-efficient and sustainable solutions for industrial environments.

TABLE 1.16
Summary of researches related to IIoT.

Objective	Device	Technology	Improvement	Ref.
Cost Efficiency	End-Edge	Soft actor-critic based on distributional RL	Optimize edge workload distribution to enhance resource utilization	[477]
	Edge	Software-defined networking (SDN)	Create programmable network to ensure the efficient communication	[83]
	Edge	Mobile edge computing, Deep RL	Minimize WAN transmission expenses for cooperative MEC providers	[58]
	Edge	Distributed Q-learning, Heterogeneous network	Guarantee the quality and fairness of network power allocation	[115]
Security	Edge	State transition model	Achieve IIoT-based continuous object tracking for production safety	[252]
	Edge	State transition model	Achieve efficient hazardous gas tracking	[181]
	End	Physical-layer key generation	Ensure encrypted, low-latency communication	[190]
	End-edge	Artificial bee colony	Protect IIoT devices from Sybil attack upon detection	[487]
	Edge	Federated learning based on deep RL	Manage privacy data of IIoT devices efficiently	[31]
	Edge-Cloud	Residual neural network, FL-based DL	Develop a trustworthy FDL service and preserve privacy data	[160]
	Edge	Multi-access edge computing, Federated learning	Preserve the privacy of multi-industrial system	[290]
Scalability	Edge	Blockchain based on hash consensus and mechanism	Achieve lightweight IIoT traffic classification service at the edge	[386]
	Edge	Blockchain, Alternating direction method of multipliers	Obtain the optimal data sharing solution	[184]
Robustness	End	Anomaly detection with data denoising	Distinguish between noise and anomalous data	[143]
	End	Partial domain adaptation with momentum algorithm	Increase the certainty of fault diagnosis	[170]

Table 1.16 also offers a comprehensive consolidation and analysis of the research of IIoT based on optimization objectives, devices involved, technologies used, and improvement.

1.3.2   Multicore Processors of IoT End Devices

This section introduces three processor models used in IoT end devices: the single-core processor, the homogeneous multi-core processor, and the heterogeneous multi-core processor.

- **Single-Core Processor**

A single-core processor refers to a processor that contains only one processing unit, called cores, for executing instructions and processing computational tasks, such as the Intel Pentium 4 670 and AMD Sempron 2800+ processors. In the single-core processor, all the computing tasks are handled by a single core (shown in Figure 1.12), which sequentially executes instructions and completes one instruction per clock cycle. A clock cycle refers to the basic unit of time used by the system's clock, during which only one basic operation, including fetching instructions from memory, decoding the instructions, executing the

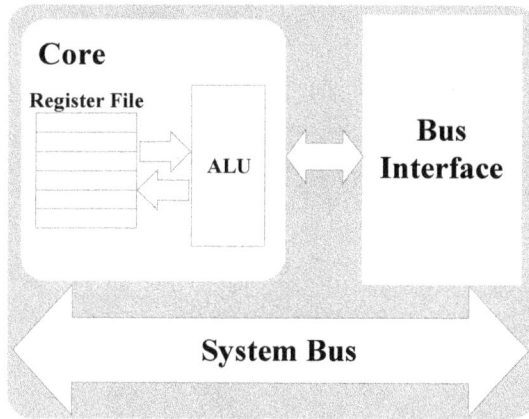

FIGURE 1.12
Diagram of the single-core processor.

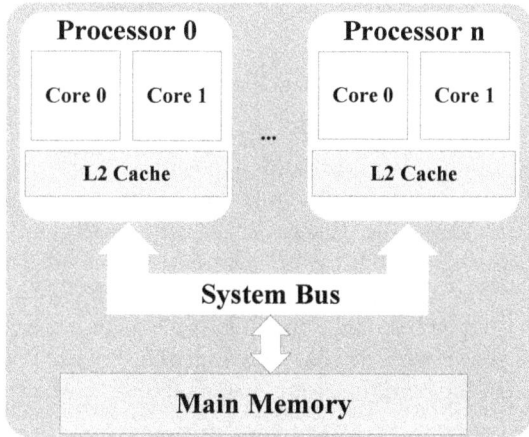

FIGURE 1.13
Diagram of the multi-core processor.

instructions, and storing the results, can be completed. As applications required greater processing power, single-core processors, once common in desktop computers, hindered performance due to their slower speed. However, single-core processors find utility in specific niche scenarios, particularly in hobbyist computers like the Raspberry Pi 1 Model B and single-board microcontrollers.

• **Homogeneous Multi-Core Processor**

A homogeneous multi-core processor refers to a processor that contains multiple identical but independent processing units on a single integrated circuit (shown in Figure 1.13), and these processing units can operate precisely the same program commands and have the same capabilities available, enabling improved processing performance. The commands include CPU operations like addition, data movement, and branching. However, a single processor can execute these instructions simultaneously on separate cores, resulting in enhanced speed for programs that utilize multithreading or other parallel computing techniques, thus improving overall performance. These multi-core processors are commonly employed to build

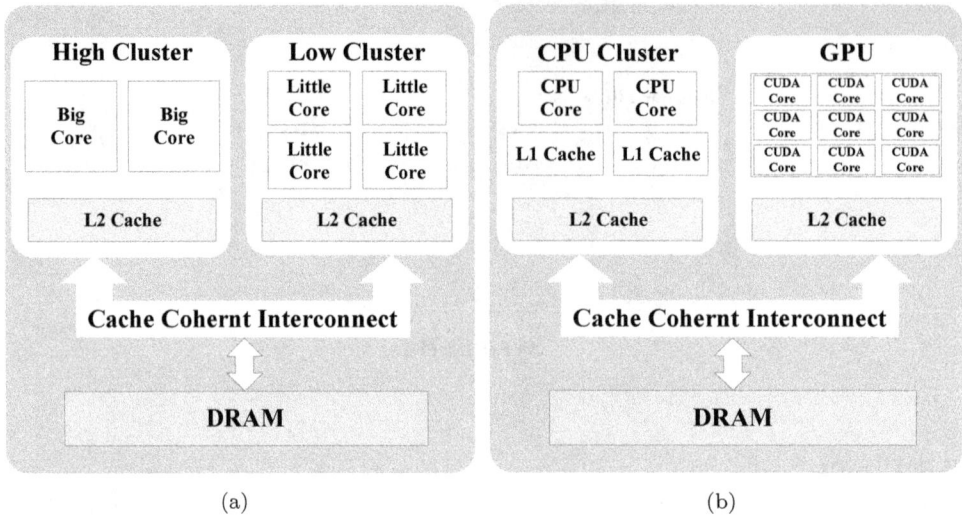

FIGURE 1.14
(a) Diagram of a big.LITTLE style heterogeneous multi-core processor. (b) Diagram of a NVIDIA Tegra style heterogeneous multi-core processor.

a system-on-chip (SoC) that incorporates multiple processors, forming a multi-processor system-on-chip (MPSoC). Examples of current homogeneous multi-core processors include TI OMAP4460, Intel Core Duo, and Freescale i.MX6 Dual/Quad series. In a modern PC processor, the homogeneous cores exhibit no variation in power consumption when executing tasks. Regardless of the core on which a task is scheduled, it can be expected to be completed simultaneously. It is beneficial for IoT applications that involve processing large amounts of data or simultaneous execution of multiple real-time tasks. Examples include video surveillance systems, industrial automation controllers, and intelligent transportation systems.

- **Heterogeneous Multi-Core Processor**

  The heterogeneous multi-core processor is a processor containing multiple different processing units. These processing units can have the same architectures but different computing performance or even have entirely different architectures, including general-purpose processors, network processors, media processors, GPUs, DSPs, FPGAs, and more. Each processing unit can be tailored to various requirements to enhance the applications' real-time performance or computational capabilities. Common examples of heterogeneous multi-core processors include TI DaVinci DM6000 series (ARM9+DSP), Xilinx Zynq7000 series (dual-core Cortex-A9+FPGA), ARM big.LITTLE, and NVIDIA Tegra series.

  In a big.LITTLE design, energy-efficient and slower processor cores (referred to as "LITTLE"), and more powerful and power-hungry cores (referred to as "big") are coupled. The objective is to develop a multi-core processor that can adapt effectively to dynamic computing requirements while consuming less power than relying solely on clock scaling. As we can see in Figure 1.14(a), due to the different performance between the "LITTLE" clusters and the "big" clusters, there is a requirement to implement run-state migration. One of the most powerful approaches is global task scheduling, which ensures simultaneous utilization of all physical cores, allowing for allocating high-priority or computationally intensive threads to the "big" cores. Meanwhile, threads with lower priority or lower computational intensity, such as background tasks, can be executed by the "LITTLE" cores.

NVIDIA Tegra is a common integrated CPU-GPU processor designed for mobile devices like smartphones, personal digital assistants (PDAs), and mobile Internet devices (MIDs). GPU offers extensive parallel computing capabilities, while the CPU provides general-purpose computing capabilities. The processor comprises a GPU processor and multiple homogeneous CPU cores, all with DVFS (dynamic voltage and frequency scaling) capabilities. While NVIDIA's GPUs consist of multiple cores, users cannot directly assign tasks to specific GPU cores. Consequently, the GPU is treated as a unified processor, and its idle state is determined only when none of its cores are actively engaged in task processing.

1.3.3 Performance Metrics of IoT End Devices

This section mainly introduces the performance metrics of IoT end devices, including reliability, power, and temperature. Additionally, energy harvesting and dissipation are also introduced.

1.3.3.1 Reliability

Life reliability and soft error reliability assess the reliability of IoT devices from different perspectives. Life reliability mainly concentrates on the likelihood of an edge device successfully performing its intended functionality during its lifecycle. It is typically used for evaluating long-term operational equipment, such as electronic and mechanical products or industrial systems. Soft error reliability pertains to temporary or transient faults brought about by external elements like radiation or electromagnetic interference that a system or device experiences during operation [517]. Typically, these faults do not cause permanent damage but are crucial to using edge devices and may have an impact. Below, we present definitions and models of reliability.

- **Lifetime Reliability**

Lifetime reliability relates to a device, system, or component's capacity to maintain regular operation and carry out its intended functions during its lifespan. It embodies the stability and reliability of equipment during actual use. Equipment with high lifetime reliability is unlikely to break down over a prolonged period, reducing the need for repairs and replacements and thus cutting operating costs. The assessment of life reliability encompasses numerous factors, such as equipment failure rate and mean-time-to-failure (MTTF), which are employed to evaluate life reliability [257]. The failure rate of edge equipment indicates the likelihood of equipment failure within each time unit (usually measured in hours, years, or other time units) during the equipment's operational lifespan. The failure rate is a crucial metric in reliability engineering. It measures the frequency with which a device or system experiences failures over time. The failure rate of edge devices can be expressed as

$$\lambda(t) = \frac{f(t)}{T}, \tag{1.5}$$

$\lambda(t)$ denotes the failure rate at time t. $f(t)$ denotes probability density function of the time-to-failure distribution at time t. And T denotes the total time of observation. The value of the failure rate can help us understand the frequency of equipment failure at different points in time, help develop maintenance plans, predict equipment life, and take appropriate preventive measures to improve equipment reliability.

The MTTF of edge equipment is a critical reliability indicator. It denotes the average time during which the equipment can function without failure. MTTF is typically expressed in hours, years, or other time units and is a critical metric for determining equipment's lifetime reliability. The MTTF can be calculated as

$$\text{MTTF} = \frac{\sum_{i=1}^{M}(T_i \times P_i)}{\sum_{i=1}^{M} P_i}, \tag{1.6}$$

where M denotes the number of different failure modes. T_i denotes the average time for the ith failure mode. P_i denotes the probability of occurrence of the ith failure mode.

Edge devices generally have four failure modes: electro-migration (EM), time-dependent dielectric breakdown (TDDB), stress migration (SM), and thermal cycling (TC). Below, we will detail these four failure mechanisms that mainly reduce the reliability of edge devices.

EM is the movement of metal atoms within a metal wire due to the flow of electric current [331]. Structural changes in the wire occur when metal atoms accumulate in one area after moving from another. The migration process results in local thinning of metal lines, which can create electromigration voids (EM voids). When the current density rises, EM voids can lead to line breakage and eventual device failure. The resulting MTTF_{EM} model under the EM model is as follow

$$\text{MTTF}_{\text{EM}} = \frac{B_{\text{EM}}}{\rho^n} \times e^{\frac{E_{a_{\text{EM}}}}{\kappa T}}, \tag{1.7}$$

where B_{EM} denotes a constant defined by the metal interconnect's physical properties. $E_{a_{\text{EM}}}$ denotes the activation energy of electromigration. ρ denotes the current density, and n denotes an empirically determined constant. κ denotes Boltzmann constant and T denotes the temperature. The following reliability solutions can be implemented to mitigate the effects of electromigration: choosing a material that is more resistant to electromigration (such as copper instead of aluminum), optimizing circuit layout and reducing current density, reducing resistance and heat by increasing the thickness of the metal layer, designing the current distribution to be uniform to reduce local current density, and controlling operating temperature and avoid high-temperature environments.

TDDB is a common reliability issue found in integrated circuits and semiconductor devices [332]. Its occurrence significantly impacts the reliability of insulating materials or layers. TDDB occurs when the insulation properties of dielectric materials gradually deteriorate over time under the influence of electric field stress. Ultimately, this degradation leads to the breakdown or damage of the dielectric. Therefore, it is crucial to identify and address this issue to ensure optimal operational performance and longevity of electronic components. When the electric field stress persists, carriers (usually electrons or holes) in the dielectric gradually accumulate under the influence of the electric field. In addition, electric fields can induce localized thermal heating in the dielectric, leading to chemical reactions and material degradation. This poses a serious problem as it can result in device failure and performance deterioration. The resulting $\text{MTTF}_{\text{TDDB}}$ model under the TDDB model can be expressed as the following equation

$$\text{MTTF}_{\text{TDDB}} = B_{\text{TDDB}}\left(\frac{1}{V}\right)^{(e-gT)} e^{\frac{A+D/T+CT}{\kappa T}}, \tag{1.8}$$

where B_{TDDB} denotes the fitting constant, V denotes the voltage supplied, e, g, A, D, and C denote empirical fitting parameters. Several reliability solutions can be implemented to reduce the impact of TDDB: selecting insulation materials that are resistant to TDDB, designing the device to reduce the strength of the electric field, controlling operating temperature and avoiding high-temperature environments, and evaluating and monitoring TDDB through testing and simulation.

SM is a reliability issue in semiconductor devices, primarily associated with stress and material migration in metal wires or lines [331]. Technical experts employ different mechanisms to mitigate the stress migration problem and enhance device reliability. Stress migration is a process that leads to fatigue and damage in metal wires. This stress can arise due to temperature variations, mechanical stress, or electric field effects. These stresses may induce metal atoms to migrate within the wire, leading to local thinning and potential circuit interruptions, performance degradation, and device failure. The resulting MTTF$_\text{SM}$ model under the SM model can be expressed as

$$\text{MTTF}_\text{SM} = B_\text{SM}|T - T_0|^{-n} e^{\frac{E_{a_\text{EM}}}{\kappa T}}, \tag{1.9}$$

where B_SM denotes a fitting constant. T denotes the operating temperature of the metal layer, and T_0 denotes the temperature at which the metal layer is deposited during fabrication. n denotes an empirically derived constant and E_{a_EM} denotes the activation energy. It is possible to implement several reliability solutions to reduce the impact of SM: choosing a metal material that is more resistant to stress migration (such as copper), reducing the risk of stress migration by optimizing circuit layout and reducing current density, using proper packaging and thermal design to maintain proper operating temperatures.

TC is a prevalent issue for reliability testing in electronic systems and semiconductor devices [71]. It requires multiple heating and cooling cycles within the device or component's operating temperature range, which can lead to reliability problems. This thermal cycling mimics the temperature fluctuations the device encounters under regular operation and shutdown or power-up situations, resulting in thermal expansion and contraction of both devices and components, potentially leading to thermo-mechanical strain. The resulting MTTF$_\text{TC}$ model under the TC model can be expressed as

$$\text{MTTF}_\text{TC} = \left(\frac{1}{T_\text{ave} - T_\text{amb}}\right)^p, \tag{1.10}$$

where T_ave denotes the average operating temperature of the system. T_amb denotes the ambient temperature. And p denotes the Coffin-Manson exponent, an empirically derived material-specific constant. Several reliability solutions can be implemented to reduce the impact of TC: choosing materials and components that are more resistant to thermal cycling (such as using high-temperature solder), reducing thermal stress through design optimization, and controlling the operating temperature to avoid frequent temperature cycling.

- **Soft Error Reliability**

  Soft error reliability is crucial in evaluating a device's ability to operate correctly when subjected to unpredictable external or internal disturbances. The possibility of soft errors causing data corruption, device performance degradation, or even catastrophic failure makes it a major concern for semiconductor devices and electronic equipment design. The soft error reliability of a single core within a given time interval is the probability of no soft errors occurring during that interval. The formula of single-core soft error reliability is as follows [258]:

$$r = e^{-\lambda(f) \times \mu \times \Delta t}, \tag{1.11}$$

where $\lambda(f)$ denotes the average failure rate that dependent by the frequency f. Δt denotes the interval and μ denotes the utilization. $\lambda(f)$ can be derived as

$$\lambda(f) = \lambda(f_\text{max}) \times 10^{\frac{d(1-f)}{1-f_\text{min}}}, \tag{1.12}$$

where $\lambda(f_\text{max})$ denotes the average failure rate when the core is running at its maximal frequency f_max. f_min is the core's minimal frequency. d $(d > 0)$ is a constant determined by hardware, and it indicates the sensitivity of failure rates to frequency changes. For the system-level reliability of the integrated GPU-CPU platform, we define its soft error reliability as

$$R = \prod_{i=1}^{G} R_{\mathrm{GPU}}(\rho_{i_g}) \times \prod_{i=1}^{C} R_{\mathrm{CPU}}(\rho_{i_c}), \tag{1.13}$$

where G and C denote the number of GPU and CPU cores of the edge device, respectively. $R_{\mathrm{GPU}}(\rho_{i_g})$ and $R_{\mathrm{CPU}}(\rho_{i_c})$ are the reliability of GPU ρ_{i_g} and CPU ρ_{i_c}, respectively.

The significance of dependable functioning for edge devices extends beyond device performance. For this reason, adequate attention should be devoted to the reliability of edge devices, and relevant modeling, testing, and monitoring measures ought to be employed to ensure stable functionality in diverse environmental scenarios while satisfying distinct use cases.

1.3.3.2 Power and Temperature

The power consumption of IoT devices typically stems from CPU and GPU cores. The power consumption and temperature of the core are usually highly coupled. Below, we describe the power model and thermal model in detail.

- **Power Model**

Regarding a multi-core processor system chip based on CPU and GPU, the primary power usage typically stems from the CPU and GPU [156]. This is primarily because these components carry out computing and graphic-based responsibilities, which demand significant power, thus consuming most of the overall system power. We can simplify by equating the total power consumption of the MPSoC to the combined power consumption of the CPU and GPU. Other parts, such as the memory controller or interconnect structure power consumption, can be ignored. Effective cooling measures must be implemented to ensure stable performance, as the CPU and GPU generate heat during operation. This is an integral factor in the design and management of MPSoC. Since both GPU and CPU are based on CMOS technology, the same power consumption model can be applied to their cores. This means similar methods and models can be used to analyze and manage GPUs' and CPUs' power consumption and temperature, following similar electronic and power consumption characteristics. This enables a consistent approach when designing and optimizing both processors to manage their power consumption and heat dissipation better. The power dissipation of a CMOS circuit is typically described as arising from two main components: dynamic switching power P_{dyn} and static leakage power P_{sta}. That is, the total power consumption of the CPU-GPU-based edge device can be expressed as $P = P_{\mathrm{dyn}} + P_{\mathrm{sta}}$. Below, we analyze the power consumption of these two parts.

Static power dissipation is a crucial factor for consideration in edge devices. It pertains to the amount of power wasted due to conductive paths in the transistors within a semiconductor device during periods of idleness or dormancy. This power wastage is not caused by any computational tasks or data transfer that the device may perform but rather by the transistor's subthreshold or reverse leakage current. Static power leakage is an inherent property of CMOS semiconductor devices. When a transistor's gate voltage approaches or falls below its threshold voltage, it allows current to flow, resulting in power dissipation. The static leakage power dissipated in each chip transistor depends on temperature, voltage, and characteristics. Static power P_{sta} and temperature Θ exhibit a linear relationship [96], and it can be modeled as

$$P_{\mathrm{sta}} = \kappa_1 \Theta + \kappa_2, \tag{1.14}$$

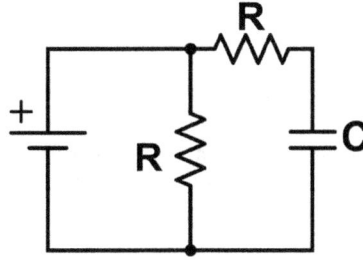

FIGURE 1.15
RC circuit diagram

where κ_1 and κ_2 are system constants depending on chip technology. Θ denotes the running temperature. Managing static leakage power is critical to increasing battery life, reducing device heating, and increasing energy efficiency, especially in resource-limited, power and thermal-constrained environments such as edge devices. Therefore, there is increasing attention to static leakage power in edge computing and IoT devices.

The dynamic switching power of edge devices refers to the power consumption caused by switching operations within the processor (such as switching transistors) when the device performs different computing tasks. This power consumption is related to the processor's operating frequency and task load and usually peaks when the device handles heavy workloads. An edge device consumes dynamic switching power as its processor performs tasks proportional to its operating frequency. Edge devices usually need to perform computing tasks in different application scenarios, so the operating frequency of their processors may dynamically change under other circumstances. Dynamic power consumption can be calculated as a function of the operating frequency [96], i.e.,

$$P_{\mathrm{dyn}} = \kappa_0 (f)^{\eta}, \tag{1.15}$$

where κ_0 and η are system constants determined by the semiconductor technology employed.

- **Thermal Model**

The above power model can be employed to develop the thermal model for the single processor. Since the operating system scheduler controls task execution using processor cores, each core can be viewed as an individual entity and modeled as an RC circuit, as shown in Figure 1.15. The equivalence between the RC template mature model and the RC circuit model is shown in Table 1.17. The core induces heat when it is active [96]. Using the

TABLE 1.17
The equivalence between RC temperature model and RC circuit model [326].

Thermal Quantity	Unit	Electrical Quantity	Unit
P, Heat flow, power	W	I, Current flow	A
T, Temperature difference	K	V, Voltage	V
R, Thermal resistance	K/W	R_{ele}, Electrical resistance	Ω
C, Thermal mass, capacity	J/K	C_{ele}, Electrical capacity	F
$t = R \cdot C$, Thermal RC constant	s	$t_{\mathrm{ele}} = R_{\mathrm{ele}} \cdot C_{\mathrm{ele}}$, Electrical RC constant	s

RC thermal model [327], the differential equation for temperature can be stated by Fourier's law [45], Θ concerning time

$$d\Theta(t)/dt = [P(t)/C] - [(\Theta(t) - \Theta_{env})/RC], \tag{1.16}$$

where Θ_{env} denotes the environment temperature. R and C denote thermal resistance and capacitance, respectively. By combining P_{dyn} and P_{sta} in Equation (1.16), we can convert Equation (1.16) as a typical linear differential equation

$$d\Theta(t)/dt = a - b\Theta(t), \tag{1.17}$$

where $a = \kappa_0(f)^\eta/C$, $b = (1 - \kappa_1 R)/RC$. Solving Equation (1.17) presents the temperature at time t as

$$\Theta(t) = a/b + \big(\Theta(t_0) - a/b\big)e^{-b(t-t_0)}, \tag{1.18}$$

where t_0 is the initial time. A deep sleep state is a low-power mode of the processor in which the processor is almost completely idle and performs almost no computing tasks. Since the processor activity in this state is very limited and the difference in power consumption is several orders of magnitude compared to when the processor is busy, it is reasonable to assume that the power consumption is negligible. This is a valid assumption because the processor's current draw is small during deep sleep and can be ignored. Based on this assumption, we can consider the processor to be cooling. The processor generates no additional heat in deep sleep mode because its power consumption is almost negligible. In the model, we can set the parameter $a = 0$ in Equation (1.18), representing the cooling model in the deep sleep state. It can be expressed as

$$\Theta(t) = \Theta(t_0)e^{-b(t-t_0)}. \tag{1.19}$$

This assumption is important for power consumption and temperature management of many edge devices. Deep sleep states are typically used to extend battery life, reduce device heat, and reduce energy consumption when the device is inactive. We can better optimize deep sleep mode by treating the processor as cool to improve edge device performance and efficiency.

The heterogeneous multi-core thermal model is designed to consider thermal effects in multi-core processor systems. Because of multi-core processors and other energy-hungry nodes, thermal issues challenge the environment and affect the interrelationship between nodes. We focus on the CPU and GPU cores because they are the primary power consumption nodes relative to other components in the device, so the thermal effect is relatively significant [179]. Therefore, we can reasonably ignore the thermal effects of the power dissipation of other components. In a system that integrates CPU and GPU cores, each core's temperature depends on its current temperature, its power consumption, and the temperature of its neighboring cores. This is because there is thermal conduction between these cores, and heat is transferred from one core to another, affecting their temperatures. Previous research has shown that we can estimate the temperature of each core using a model that achieves acceptable accuracy while considering the temperature and power consumption of adjacent cores [519]. This means that we can use this model to predict the temperature of each core at a future slot (e.g. $t + \Delta t$). This is important for thermal management as it allows us to understand the thermal interactions between cores better and take appropriate measures to maintain device performance and stability. This model can help us better handle thermal effects in heterogeneous multi-core processor systems, ensuring that the device remains within a safe temperature range while operating. Referring to the multiprocessor thermal model [156], with an beginning temperature $\Theta(t_0)$, we can express the temperature

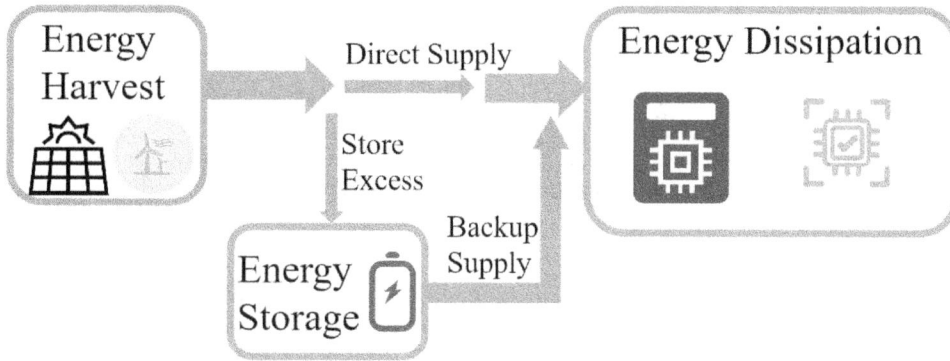

FIGURE 1.16
Architecture of energy-powered IoT system.

of a core after t time units as

$$\Theta_m(t_0 + t) = a + \big(\Theta(t_0) - a\big)e^{-bt} + \sum_{n=1}^{M_m} \gamma_{mn}T_n(t_0 + t), \qquad (1.20)$$

where $\Theta_m(t_0 + t)$ and $\Theta_n(t_0 + t)$ $(m \neq n)$ denote the temperature of the mth core and the nth core, respectively. γ_{mn} denotes the heat dissipation coefficient between two cores. M_m denotes the number of adjacent cores of the nth core.

1.3.3.3 Energy Harvesting and Dissipation

Renewable energy-powered IoT devices represent the forefront of sustainability and innovation, using renewable energy to power many applications, from smart cities to environmental monitoring and agriculture. These technologies emerged in the second half of the 20th century and were used primarily for electricity production. However, the idea of applying renewable energy to IoT devices has only matured in more recent decades. As solar cell and wind energy technology continues to improve and battery technology evolves, IoT devices can more reliably capture and store energy from the environment. In addition, the popularity of energy-saving IoT technology and wireless communication standards also supports the implementation of renewable energy-driven IoT devices. IoT devices powered by renewable energy have been successfully applied in many fields, including smart buildings, smart agriculture, environmental monitoring, smart transportation, healthcare, and many more areas. These devices allow us to monitor and manage various systems in real-time while reducing reliance on traditional electricity. As shown in Figure 1.16, renewable energy-driven IoT devices usually consist of three modules: energy harvesting (EH) module, energy storage (ES) module, and energy dissipation (ED) module [516]. These modules work together to enable sustainable, energy-self-sufficient IoT device operation. Below, we will introduce a detailed description of these three modules.

- **Energy Harvesting Module**

  Energy harvesting modules convert energy in the natural environment into electrical energy by collecting renewable energy sources. In this book, we concentrate on two fundamental types of renewable energy: solar energy and wind energy. Solar energy typically involves solar panels that absorb sunlight to produce electricity. Wind energy uses wind turbines to convert wind power into electrical energy. These technologies allow devices to operate autonomously outdoors without artificial power supply. EH module is designed to automatically capture and utilize energy from the environment, ensuring that equipment always has an available energy source. This helps extend equipment uptime, lowers maintenance costs, and reduces reliance on the grid. We use E_{harv} to represent the total energy collected by IoT devices, where the energy from solar energy is represented by E_{solar} and the energy from wind energy is represented by E_{wind}. It can be expressed as the following formula

$$E_{\text{harv}} = E_{\text{solar}} + E_{\text{wind}}. \tag{1.21}$$

The solar energy collection is related to time and solar energy collection power [439]. We define its collection time range as $[t_n, t_{n+1}]$, and its collection energy E_{solar} is defended as

$$\begin{aligned} E_{\text{solar}} &= \int_{t_n}^{t_{n+1}} P_{\text{solar}} dt \\ &= \int_{t_n}^{t_{n+1}} A_{\text{solar}} \times \Gamma_{\text{solar}} \times \psi_{\text{solar}} dt, \end{aligned} \tag{1.22}$$

where P_{solar} is solar energy collection power, A_{solar} is effective area of solar panel, Γ_{solar} is intensity of solar radiation, and ψ_{solar} is solar cell efficiency. Likewise, the wind energy harvesting energy E_{wind} depends on the wind turbine blade area A_{wind}, the wind speed V_{wind}, and the efficiency of the wind turbine ψ_{wind} [457]. It can be expressed as

$$\begin{aligned} E_{\text{wind}} &= \int_{t_n}^{t_{n+1}} P_{\text{wind}} dt \\ &= \int_{t_n}^{t_{n+1}} A_{\text{wind}} \times \rho \times A \times V_{\text{wind}}^3 \times \psi_{\text{wind}} dt, \end{aligned} \tag{1.23}$$

where ρ denotes the air density, A denotes the cross-sectional area of the wind turbine. These formulas can be used to estimate the power collected from solar and wind energy, helping to determine the performance and energy yield of a solar panel or wind turbine.

- **Energy Storage Module**

    Energy storage modules typically employ super-capacitors or batteries to store harvested energy [441]. We use E_{bat} to represent it. These energy storage devices act as buffers, storing energy to cope with uncertainties in weather conditions or changes in the day-night cycle. Super-capacitors can charge and discharge quickly for instantaneous energy needs, while batteries can store large amounts of energy for long-term needs. ES modules can also include smart energy management systems that monitor energy storage and release. This ensures that the device can efficiently call upon the stored energy when energy is needed.

- **Energy Dissipation Module**

    Energy dissipation modules include heterogeneous multi-processor system-on-chips on devices such as GPU and CPU. These processors perform various tasks on IoT devices, such as data processing, communication, sensor operation, etc. We use E_{dis} to represent it. The MPSoC's task is to ensure that the device effectively integrates available and stored energy. This includes task scheduling, energy management strategies, and the implementation of energy optimization algorithms to maximize equipment runtime and performance.

    Through the synergy of these modules, renewable energy-powered IoT devices can work efficiently under changing environmental conditions, minimizing reliance on traditional power sources and thereby reducing operating costs and environmental impact. Such systems support sustainable development and allow the deployment of IoT devices in remote or difficult-to-access areas.

1.3.4 Conclusions and Discussions

The IoT connects the physical and digital worlds through an interconnected network of devices, signifying a significant technological advancement. This chapter provides a comprehensive overview of end devices in IoT, offering valuable insights into their operation, performance, and optimization. The IoT's extensive network comprises diverse devices embedded with sensors, software, and connectivity options, enabling real-time data collection and exchange. This connection only enhances efficiency, productivity, and convenience across various domains and fosters innovative advancements and applications. It has given rise to two significant paradigms, AIoT and IIoT, which have played pivotal roles in enhancing the capabilities and applications of IoT.

The emergence of AIoT represents a significant leap forward by combining AI with IoT, leveraging intelligent algorithms within an extensive network of devices. This convergence enables sophisticated data analysis, real-time decision-making, and process automation, ultimately leading to more informed decisions and valuable insights. The structured architecture of AIoT, encompassing intelligent terminals, edge intelligence, and cloud computing layers, forms the foundation for these capabilities, ensuring optimal performance and resource utilization.

Similarly, the IIoT applies IoT technologies in industrial settings, revolutionizing traditional practices and establishing a synergistic network of devices and machines. This integration facilitates improved operational efficiency, predictive maintenance, and resource optimization, supported by a three-layered system architecture.

The integration of multicore processors in end devices is a testament to technological advancements in this domain. It ensures that devices are smarter, more efficient, and reliable. Evaluating and ensuring the optimal functioning of IoT systems relies heavily on IoT devices' performance metrics. This section highlights key performance metrics, including reliability, power, temperature considerations, energy harvesting, and dissipation. These metrics serve as crucial indicators, guiding developers and engineers in optimizing device performance, ensuring longevity, and maintaining energy efficiency. Emphasizing the delicate balance between performance and energy consumption underscores its significance in the sustainable deployment of IoT devices.

2

Introduction to Cloud-Edge-End Computing

2.1 Evolution of Cloud-Edge-End Computing

2.1.1 Mobile Cloud Computing

With the development of communication and computer-related technologies, mobile devices such as smartwatches and smartphones have become increasingly popular daily. As a common computing platform, mobile devices offer many convenient applications, allowing us to work and entertain almost anytime, anywhere. However, mobile devices face the issue of limited resources, such as memory capacity, network bandwidth, processor speed, and battery capacity [409]. Despite continuous technological advancements, mobile devices still struggle to meet the increasing demands for system resources, limiting the execution of complex applications.

Cloud computing allows users to access a wide range of resources stored in the cloud via the Internet, providing efficient and scalable computing capabilities. It addresses the resource limitations often encountered by mobile devices. Mobile cloud computing (MCC) is the combination of cloud computing and mobile computing technologies. MCC utilizes cloud resources to overcome the resource constraints of mobile devices. By merging the resource benefits of cloud computing with the convenience of mobile devices, resource-intensive tasks are transferred from resource-limited devices to cloud servers for execution. The results are then returned to the mobile devices, reducing their workload and extending battery life. The architecture of MCC is illustrated in Figure 2.1. Mobile devices connect to the mobile network through base stations, wireless access points, or satellites and then access cloud servers via the mobile network to offload complex tasks. Once executed by the cloud servers, the mobile devices receive the results.

The promotion of MCC has led to a wide expansion of its application scenarios. Many compute-intensive and power-consuming applications, such as interactive games, virtual reality, and natural language processing, are now frequently running on mobile devices. This meets our daily needs for work and entertainment on mobile devices. Furthermore, mobile cloud computing can be combined with Internet of Things (IoT) devices to achieve remote control and management of IoT devices, thereby promoting the development of IoT technology.

Task offloading is a crucial method for addressing the issue of limited resources on mobile devices in MCC. It involves transferring compute-intensive or storage-intensive tasks from mobile devices to cloud servers for execution. Utilizing the powerful computing capabilities and storage space of cloud computing helps overcome the resource limitations of mobile devices, thereby improving application performance [409]. However, offloading all tasks to the cloud is impractical because of the limited network resources. Therefore, making reasonable offloading decisions to achieve optimal benefits is very important. Kuang et al. [196] studied the issue of multi-user computation offloading in orthogonal frequency division multiple access systems and introduced a game-theoretic offloading strategy for multi-user MCC systems. This includes the beneficial offloading threshold algorithm and

DOI: 10.1201/9781003540281-2

Mobile Devices Access Devices Cloud Server

Tasks

Internet

Results

FIGURE 2.1
The architecture of MCC.

the beneficial offloading group algorithm. This mechanism can identify an optimal group of offloading tasks to maximize benefits while saving energy consumption.

MCC is an advancement of cloud computing technology specifically designed for mobile devices. Unlike traditional cloud computing, MCC focuses on enabling users to access cloud services using their mobile devices anytime and anywhere. However, because mobile devices operate in dynamic and constantly changing network environments, these changes can impact decisions related to offloading tasks. Lu et al. [247] presented a method for MCC task offloading. The method is based on iterative greedy algorithms and Tabu search, aiming to balance multiple objectives in a dynamic environment, such as system load and energy consumption. Considering network conditions, Wu et al. [409] designed a dynamic mobile cloud offloading decision algorithm. They utilized the Lyapunov optimization method to achieve an equilibrium between energy consumption and response time. Similarly, Li et al. [220] suggested a mobile cloud offloading strategy based on Lyapunov optimization to balance system offloading utility and queue congestion. They also introduced the Lagrangian optimization method and a multi-stage stochastic programming approach to establish the ideal offloading tasks for deterministic and random WiFi connections. Rahimi et al. [298] modeled mobile applications as location-time workflows, where the user's movement patterns are translated into mobile service usage patterns. They used a layered cloud to improve the performance and scalability of mobile applications. They proposed a mobility-aware service allocation algorithm to achieve near-optimal service allocation.

In task offloading, a significant amount of data is transferred and stored between mobile devices and the cloud. This data often contains important and sensitive user privacy information. Therefore, MCC prioritizes the security and backup of data centers, similar to cloud computing. Additionally, due to the vulnerability of mobile devices in insecure networks, MCC must ensure the security of data transmissions between mobile devices and remote clouds. Elgendy et al. [102] introduced a particle swarm optimization (PSO)-based dynamic offloading method, which includes a security layer using advanced encryption standard encryption technology to protect the transmitted data from attacks in the cloud. Li et al. [216] presented a lightweight data-sharing scheme based on attribute-based encryption (ABE) technology. This scheme ensures secure data sharing and effectively addresses the limitations of resources and low performance on mobile devices. Introducing attribute description fields also implements lazy revocation, reducing the cost of user revocation.

In MCC, cloud servers typically consist of multiple nodes responsible for receiving tasks offloading from mobile devices and executing them. In the event of a node failure, selecting alternative nodes to take over the tasks is essential to ensure reliability. Chen et al. [51] proposed a framework called "k-out-of-n", which involves distributing data across multiple nodes to achieve fault tolerance and optimize energy consumption for efficient data access.

TABLE 2.1

Summary of MCC-related work.

Ref.	Optimization Objectives and Constraints				Optimization Method
	Latency	Energy Consumption	Cost	Dependability	
[196]		✓	✓		Game theory based task offloading strategy.
[247]	✓	✓			Iterative greedy algorithm and Tabu search based task offloading method.
[409]	✓	✓		✓	Lyapunov optimization based dynamic offloading algorithm.
[220]			✓	✓	Lyapunov optimization based offloading strategy.
[298]	✓	✓	✓	✓	Simulated annealing based heuristic method.
[102]	✓	✓			PSO based dynamic offloading method.
[216]			✓		ABE based lightweight data sharing scheme.
[51]		✓		✓	Multi-node data distribution based framework.
[302]	✓	✓			Two servers for alternating operations based optimization method.

The framework addresses both energy efficiency and fault tolerance in data storage and processing. To reduce battery consumption and task execution time on mobile devices, Aldmour et al. [302] introduced an optimization method, which utilizes two servers for alternating operations and incorporates a decision engine system to optimize decision-making. The first server receives tasks and stores them in the cloud, while the second server receives tasks and saves file copies to prevent the need for re-uploading.

The combination of mobile computing and cloud computing has given rise to MCC. MCC effectively addresses the issue of resource constraints on mobile devices, enabling them to support more complex applications. Additionally, MCC reduces the power consumption of mobile devices, provides reliability, and offers a good user experience for mobile users. Research and practice in MCC are continuously evolving. In recent years, some scholars have been dedicated to optimizing MCC to provide a better user experience, as shown in related studies such as Table 2.1. In summary, the development trend of MCC demonstrates strong vitality and potential, promising to offer more convenient experiences in our daily lives.

2.1.2 Mobile Edge Computing

Mobile edge computing (MEC) complements mobile cloud computing by sinking computing resources from the central cloud to the network edge, which is closer to users. While MCC can offload compute-intensive tasks of mobile users to cloud servers with high computing power, the significant distance between mobile users and cloud servers results in additional communication consumption and transmission latency, reducing the users' quality of experience. The emergence of MEC addresses the low-latency needs of mobile users. The architecture of MEC is illustrated in Figure 2.2.

MEC migrates computing and storage capabilities from distant cloud center servers to nearby edge servers. This helps to increase task execution speed and reduce the energy consumption of mobile devices. The edge servers in MEC can establish reliable wireless communication with mobile devices and partner with the central cloud to offload tasks, providing mobile users with satisfactory Quality of Service (QoS) [110]. The advancement of 5G communication technology has led to the growth of MEC. MEC is characterized by

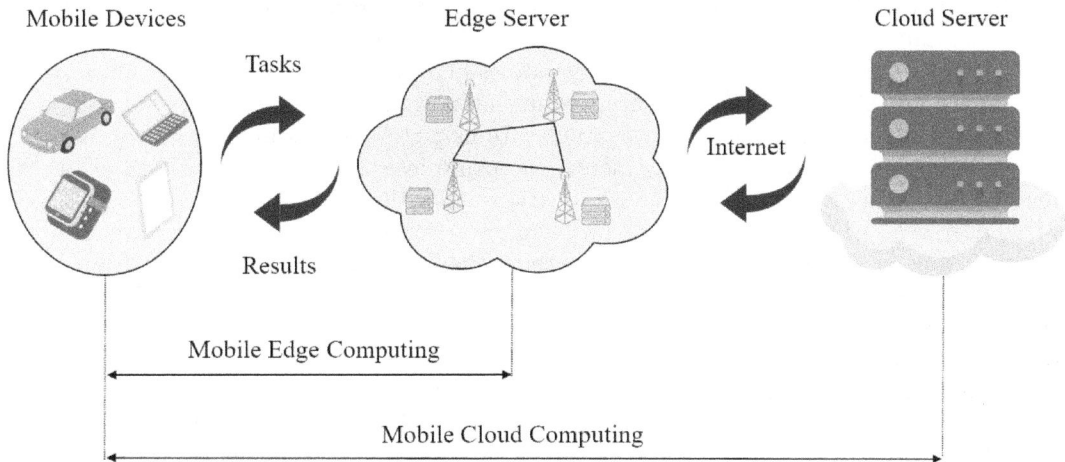

FIGURE 2.2
The architecture of MEC and MCC.

low latency and high transmission rates, providing strong application support. It enables low-latency data processing by performing computations and handling data closer to the user. Compared to MCC, MEC offers faster response speeds and reduced network bandwidth requirements. As a result, MEC is better suited to support applications with strict real-time demands, such as autonomous driving. Its implementation in the IoT and industrial internet sectors can facilitate intelligent transportation and automated industrial production. Additionally, MEC is beneficial for Virtual Reality and Augmented Reality applications, as its low-latency features can deliver a more seamless experience.

Like MCC, MEC also utilizes task offloading to relieve mobile devices' resource constraints. MEC task offloading transfers tasks from mobile devices to edge servers for execution. The purpose of task offloading is to leverage the computational resources of edge servers to relieve the computational burden on mobile devices. This, in turn, enhances performance, reduces energy consumption, and decreases mobile devices' response time. Energy harvesting technology is believed to be a feasible way to obtain energy from the environment to provide additional energy to mobile devices. Zhang et al. [479] studied the dynamic energy-delay trade-off problem with energy harvesting devices in MEC systems. They introduced an online dynamic Lyapunov optimization-based offloading algorithm to determine task allocation. This algorithm optimizes task allocation scheduling through dynamic task offloading and energy harvesting devices to achieve high energy efficiency and low-latency communication. In MEC, many tasks have dependencies, which can impact the performance and feasibility of task offloading. Zhao et al. [504] explored the problem of task offloading with dependencies and service caching in MEC. They proposed a convex programming-based algorithm and a favorite successor-based algorithm to optimize task offloading and reduce the completion time of applications. Similarly, Yang et al. [444] considered priority constraints among tasks and focused on solving the task offloading problem in MEC through the evolutionary multitasking optimization method. They introduced a multitask evolutionary algorithm, which includes segmented knowledge transfer and auxiliary task update, transforming the task offloading problem into a multi-task optimization problem. By creating auxiliary tasks, they enhanced the evolutionary search for task offloading, thereby improving the scalability and efficiency of the optimization algorithm in handling task offloading problems.

Due to the relatively limited computational resources in MEC compared to MCC, resource allocation is a key issue in MEC. Resource allocation involves effectively distributing computing, storage, and communication resources between edge servers and mobile devices to enhance overall system performance, reduce latency, and optimize energy consumption. The key to resource allocation is to balance various factors, including the diverse needs of mobile devices and real-time requirements, to ensure overall performance. Ding et al. [90] focused on optimizing service resource allocation under the economic budget constraints of MEC providers. They transformed the problem into a series of real-time linear programming sub-problems and developed an algorithm using the Lyapunov optimization method. This algorithm determines the best service resource allocation strategy for each user device to improve QoS while reducing economic costs. Chen et al. [63] focused on the energy constraints of mobile edge devices. They studied the joint optimization issue of task scheduling and energy management. The researchers introduced an integrated strategy for task scheduling and energy management to optimize the efficiency of MEC systems equipped with mixed energy sources. In a similar study, Huang et al. [172] addressed the impact of unstable energy supply on QoS while considering energy constraints. They proposed a neural network-based adaptive frequency adjustment solution that uses historical data to learn request arrival and energy harvesting patterns. This solution intelligently adjusts the server processor frequency using deep reinforcement learning techniques to balance service sustainability and QoS.

In practical applications, task offloading must be considered with resource allocation to achieve optimal allocation. Zhao et al. [506] investigated the optimization problem of energy-aware task offloading and resource allocation for latency-sensitive services in MEC systems. They divided the original problem into three sub-problems: task offloading ratio selection, transmission power optimization, and subcarrier and computational resource allocation. They achieved minimum energy consumption by sequentially addressing these sub-problems through an iterative algorithm. Zhou et al. [520] investigated methods for simultaneously carrying out device attachment, resource distribution, and computation offloading to reduce energy expenditure in ultra-dense IoT networks with many devices and tasks. They proposed an improved hierarchical adaptive search algorithm to solve the non-linear and mixed-integer formulation of the problem that aims to reduce energy consumption across the network.

Overall, MEC and MCC complement each other by providing robust computational support for mobile applications and adapting to various scenarios and requirements. MEC features low latency, making it suitable for scenarios with high real-time demands, while MCC can meet scenarios requiring large-scale computation and storage resources. By combining the strengths and weaknesses of both, more efficient and reliable services can be offered for a variety of mobile applications. Table 2.2 lists some relevant studies on MEC conducted by scholars in recent years.

2.1.3   Cloud-Edge Computing

As the IoT becomes more prevalent and network infrastructure continues to advance, the number of consumers and industrial IoT devices is increasing. The extensive interconnection of devices and the large amount of data consume significant bandwidth during transmission, causing traditional cloud computing to experience bandwidth limitations and delays. As a result, there is a growing focus on evaluating the advantages and disadvantages of both cloud computing and edge computing and considering how they can complement each other. This has led to the emergence of cloud-edge computing, which involves a close collaboration between cloud computing and edge computing, effectively allocating tasks between the two. By extending cloud computing to the edge, tasks such as computation and analysis previously carried out in the cloud are now performed at the edge server better to meet the diverse requirements of various application scenarios.

Evolution of Cloud-Edge-End Computing 53

TABLE 2.2
Summary of MEC-related work.

Category	Ref.	Optimization Objectives and Related Constraints				Optimization Method
		Latency	Energy Consumption	Task Dependencies	Economic Cost	
Task offloading	[479]	✓	✓			Online dynamic Lyapunov optimization based offloading algorithm.
	[504]	✓		✓		Convex programming and favorite successor based method.
	[444]	✓	✓	✓	✓	Evolutionary multi-task optimization based algorithm.
Resource allocation	[90]				✓	Lyapunov optimization method based algorithm.
	[63]	✓	✓			Joint task scheduling and energy management scheme.
	[172]	✓	✓			Neural network based adaptive frequency adjustment solution.
Task offloading and resource allocation	[506]	✓	✓			Problem decomposition and iterative algorithms based solution.
	[520]	✓	✓			Improved hierarchical adaptive search algorithm.

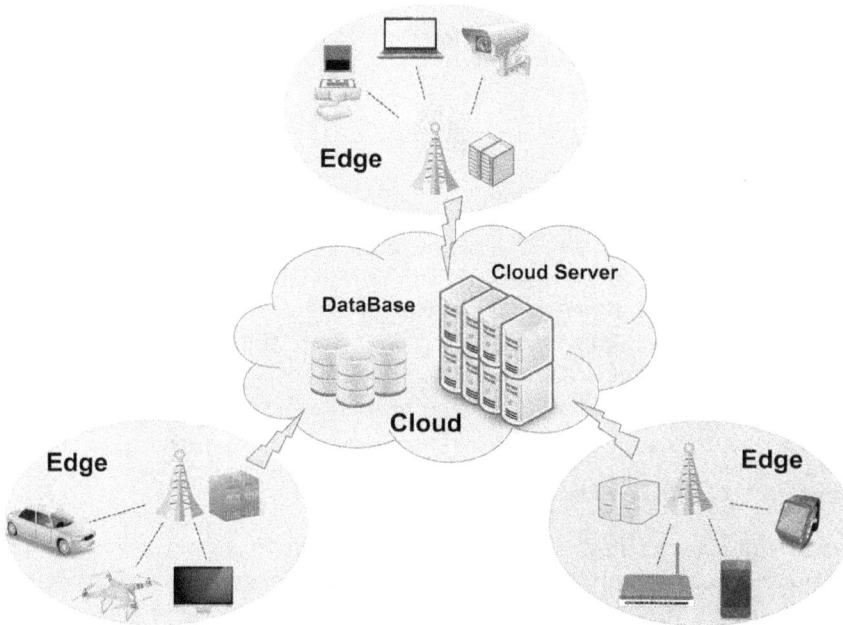

FIGURE 2.3
The architecture of cloud-edge computing.

Cloud-edge computing comprises two parts: edge computing and cloud computing. The schematic diagram of cloud-edge computing is illustrated in Figure 2.3. Edge computing is responsible for delivering real-time processing and execution of immediate and short-term data and local tasks to the cloud. On the other hand, cloud computing handles tasks beyond the capabilities of edge nodes, and processes non-real-time and long-term data through big data analytics. This approach ensures that edge computing can respond better to local demands and effectively manage the application's life cycle. Cloud-edge computing achieves cloud-edge complementation and resource fusion by establishing a unified and efficient collaboration framework for resources, data, applications, services, and other collaborative areas.

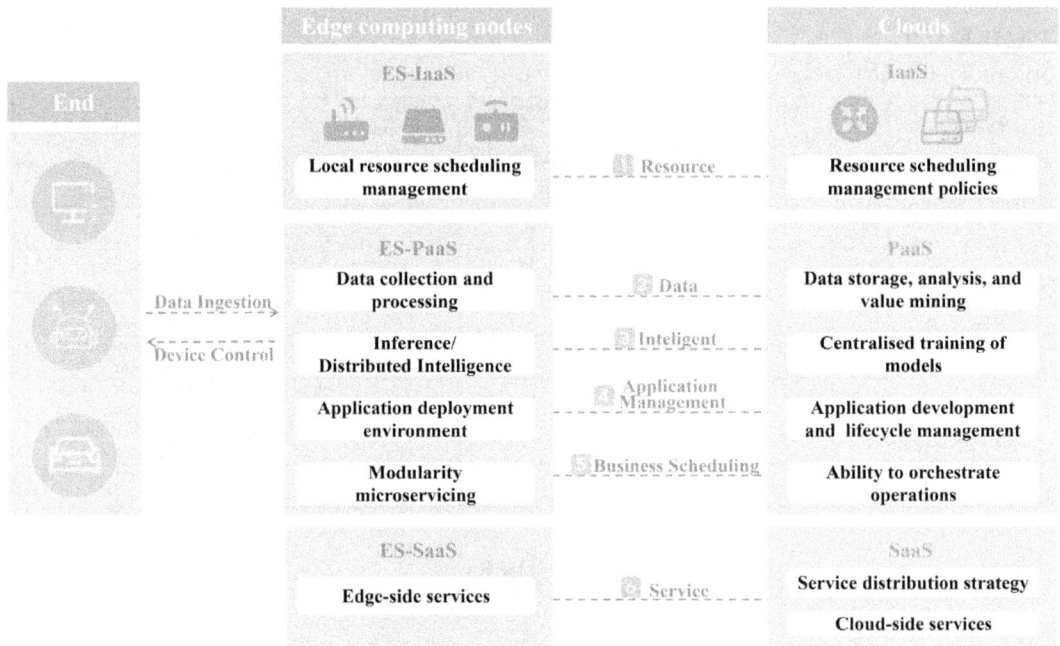

FIGURE 2.4
Three levels of cloud edge collaboration.

Cloud-edge computing encompasses the interaction between various cloud nodes and edge nodes, forming a multi-dimensional architecture. Cloud computing is typically classified into three levels: IaaS, PaaS, and SaaS, while edge computing is divided into EC-IaaS, EC-PaaS, and EC-SaaS. The cloud-edge computing involves collaboration at all levels, including infrastructure, platform, and software collaboration, as shown in Figure 2.4. These levels of collaboration are further detailed below.

- **Infrastructure Collaboration**

Since each cloud server and edge server has different infrastructure resources, such as computing power, storage, network, and corresponding resource scheduling policies, unified cooperative management is necessary to achieve joint resource scheduling. So, infrastructure collaboration is essentially resource collaboration.

Resource Collaboration: The cloud provides centralized management of edge node resources. Edge nodes provide infrastructure resources, including computing, storage, network, and virtualization. They manage resources locally and receive and execute cloud-based resource scheduling and management strategies. The cloud provides a strategy for scheduling and managing resources for device, resource, and network connection management of the edge nodes.

In a scenario where the cloud server is burdened with multiple computing tasks, edge servers with available computing resources can act as offload destinations for these tasks. This allows them to sell their spare computing resources to generate revenue. Zhou et al. [511] utilized a computational offloading approach based on the Stackelberg game model to determine the optimal strategy for both cloud servers and edge servers through backward induction. Additionally, existing methods often overlook the competition among multiple edge service providers (ESPs). Wang et al. [394] addressed this by designing an intelligent tiered dynamic pricing mechanism at the edge. Cloud service providers will use edge computing services as the primary pricing basis to maximize their benefits. Each ESP at the edge node can participate in analyzing and predicting pricing, thus preventing a single adversarial disruption of the service pricing market.

The challenges of offloading computing and allocating resources in Cloud-Edge Computing are widely discussed. In their study, Zhou et al. [513] proposed a method called RACORAM, which utilizes a reverse auction mechanism to incentivize small base stations with edge servers to participate in the computing offload process of cloud service centers based on their resources and interests. Additionally, Thai et al. [359] introduced a framework that, for the first time, considers both vertical and horizontal offloading in a cloud-edge computing system. This framework can effectively handle different virtualized computing services on time and efficiently. It also allows for flexible selection of the offloading method based on various situations and needs, leading to improved system performance and reduced costs.

- **Platform Collaboration**

The platform layer generally includes data and application management. Data processing is a significant step, forming the basis for subsequent work. Platform collaboration is mainly between the platform services of edge servers and the cloud, including data collaboration, intelligent collaboration, application management collaboration, and business orchestration collaboration.

Data Collaboration: In a cloud-edge computing architecture, edge nodes are responsible for data collection, initial processing, and analysis. After this, the cloud server receives the relevant data and results and then stores, analyzes, and mines the critical data. Data collaboration facilitates the smooth data flow between the cloud and edge nodes, effectively reducing the consumption of storage, bandwidth, and computing resources. In addition, it efficiently manages the data life cycle and values mining.

Several studies have explored data collaboration. Li et al. [217] used deep learning methods and characteristics of cloud-edge computing to investigate the task unloading problem for accelerated automatic speech recognition (ASA). They designed a three-layer mobile network with an encoder-decoder model based on the convolutional neural network. In this model, the edge server is the encoder for feature extraction and compression of ASA audio data, while the cloud server acts as the decoder for complex computing tasks. Huang et al. [171] proposed a parallel control architecture called PSFlow for secure data sharing in mobile cloud-edge environments. This architecture achieved safe, finely grained data read and write control against malicious senders by introducing online and offline encryption and decryption modes, reducing computational overhead for both senders and receivers. PSFlow is specifically built on AOACE, and the purification process is accelerated by parallel computing of edge nodes.

Intelligence Collaboration: Traditional machine intelligence involves uploading data to cloud servers for training, optimization, and model inference. The inferred results are then sent to edge nodes. It's important to note that model inference requires far fewer resources than model training and optimization. Additionally, there are scenarios where new data is needed for real-time inference. In cloud-edge intelligence collaboration, deep learning (DL) models are centrally trained in the cloud. The trained models are then transferred to the edge server for DL model inference, enabling distributed intelligence. Data and inference results generated by the edge nodes are periodically uploaded to the cloud server for continuous model training and production analysis.

Some scholars have related research starting from the intelligent collaboration between the cloud server and the edge server. Ding et al. [86] proposed a cognitive service architecture that involves deploying a shallow Convolutional Neural Network model (ECNN) on the edge server and a deep Convolutional Neural Network model (CCNN) on the cloud server. CCNN aids in ECNN training by sharing its lower m levels. This approach allows ECNN to avoid starting training from scratch, save computational resources, and address the accuracy and overfitting issues of ECNN.

According to relevant research in academia, the rational use of computing power coordination can make edge training possible. Here's how the training process works: the cloud server trains a generic model by accessing the public data stored in the cloud and then sends it to the edge nodes. The edge server first reads the edge private data to obtain the edge model and then reads the incremental data and uses incremental learning to create an edge model that can be used for user-side reasoning. In conclusion, with the rapid development of IoT, achieving edge-side training is a significant trend for cloud-edge intelligence collaboration.

Application Management Collaboration: Edge nodes provide applications with distributed cloud computing capabilities and a low-latency, high-bandwidth, high-performance environment for application deployment and operation. They manage the lifecycle of several applications and schedule them. The cloud server mainly provides the environment for application development and testing and management policies for the application lifecycle. An application instance can be deployed simultaneously on the cloud center, edge node, or both. Under the cloud-edge computing framework, different types of applications and different edge locations where they are deployed will lead to varying choices of edge computing hardware deployment and provided environments. The cloud-native software architecture of microservices, containers, and virtualization technologies can reduce deployment, operation, and maintenance problems brought about by hardware infrastructure differentiation. These microservices-based applications can be easily deployed with container-based technologies on edge servers. Considering that IoT applications are distributed and often contain multiple computation-intensive tasks, the application's deployment and management must be done on each computing node.

Researchers have conducted several studies on deploying and scheduling edge computing applications among devices with different configurations to adapt to the cloud-edge computing distributed architecture. Xiong et al. [424] introduced an infrastructure in the edge environment called KubeEdge, which provides a network protocol and a runtime environment similar to that of the cloud server at the edge. This allows applications to seamlessly communicate with components on edge nodes and cloud servers, achieving container scheduling under the cloud-edge collaborative architecture.

Business Orchestration Collaboration: The cluster offers both cloud platform services and edge platform services through an Open API, which is accessible to both internal and external applications. The expansion interface is reserved for adding new functions to its services to facilitate business orchestration collaboration. Edge nodes host modular, microservices business instances, while the cloud primarily handles application business orchestration based on customer requirements. For example, the cloud application retrieves the state, processing results, logs, and other information from the edge application using the open API calls of the cloud-edge platform. It then performs additional processing in the cloud. This process allows for scheduling the edge application instances and storage resources based on policy.

The optimal state of collaborative business management involves a business application providing business functions by integrating services deployed in the cloud and the edge nodes. The business needs to consider how to invoke the services based on its requirements. For instance, it can directly call the cloud service or edge service to handle simple business tasks or integrate both to manage more complex business processes and fulfill service requests.

Based on the correlation between microservices, Qi et al. [291] designed a microservice deployment approach based on cloud-edge collaboration and modeled it as a service chain. The purpose is to reduce latency and achieve an equilibrium utilization ratio in edge clusters. Fu et al. [113] introduced a runtime system called Nautilus that ensures QoS for user-facing services based on microservices. At each node, the resource manager optimizes resource allocation for its microservices.

- **Software Collaboration**

The main component of software collaboration is service collaboration.

Service Collaboration: Service collaboration mainly discusses the synergy of service quality and service energy efficiency at the user application level between edge services and cloud services. The cloud server primarily provides cloud-based services, except for regulating SaaS service distribution. Edge nodes implement certain EC-SaaS services in compliance with the regulation and deliver customer-facing services as needed through collaboration between EC-SaaS and SaaS.

Combined with specific business characteristics (such as data security and delay requirements), edge nodes' workload and storage capacity are deployed and processed on demand according to the distribution of user requests and the data characteristics of end devices. Scheduling requests from edge nodes to homologous edge nodes precisely can enhance resource utilization and service experience and achieve the best balance between computing efficiency, user experience, and data security.

How to place edge nodes is a significant issue in cloud-edge computing systems. To optimize the performance of cloud-edge computing systems in mobile environments, Yuan et al. [470] studied a dynamic virtual edge node placement scheme that indicates the location of virtual edge nodes and resource allocation for each node according to a pay-as-you-go model and predictive information. Moreover, Cao et al. [37] proposed a method including offline and online phases. In the offline phase, the location of each heterogeneous edge server is optimized, and the offload policy from the base station to the server is calculated. In the online phase, a mobile sensing game theory method is proposed.

In the context of mobile computing, Yang et al. [452] developed an optimization method for joint computing offloading and data cache decision-making in a hybrid mobile cloud-edge computing system. This method addresses resource constraints and network delays, considering both the offloading of user-collected data and the caching of relevant databases. Additionally, deep reinforcement learning (DRL) is employed to promptly adjust scheduling algorithms in response to rapidly changing workloads and resources. Han et al. [145] introduced EdgeTuner, an online approach that utilizes DRL agents to select a suitable scheduling algorithm for cloud-edge tasks in real-time.

Table 2.3 provides a comparative summary of the literature discussed above. In summary, cloud-edge computing is a distributed open platform that integrates communication, computational power, data storage, and application services. It involves the collaboration of resources, data, intelligence, application management, business orchestration, and services. This creates a flexible and efficient cloud-edge computing environment, combining powerful resource capabilities with low-latency characteristics in a complementary manner, amplifying the value of both. It achieves collaborative optimization, where the edge supports the cloud application and the cloud helps meet the localized demands of the edge. Cloud-edge computing has gained wide attention and recognition from domestic and international academic circles and has a significant role in many scenarios. It is believed that cloud-edge computing will occupy a prominent place in the future Internet industry with the deeper progression of digitization.

TABLE 2.3

Summary of research related to cloud-edge computing.

Hierarchical Division	Ref.	Technology	Key Idea
Infrastructure Collaboration	[513]	Resource Collaboration	Adopt a computational offloading approach according to Stackelberg game.
	[511]	Resource Collaboration	Consider competition among multiple edge service providers.
	[394]	Resource Collaboration	Incentivize small base stations with edge servers to engage in the computing offload process.
	[359]	Resource Collaboration	Consider both vertical and horizontal offloading of computing tasks in the cloud-edge computing system.
Platform Collaboration	[217]	Data Collaboration	Design a CNN based encoder-decoder model.
	[171]	Data Collaboration	Introduce a framework that provides the network protocols at the edge and the same runtime environment as in a cloud server.
	[86]	Intelligent Collaboration	Deploy shallow models on edge servers and deep models on cloud servers.
	[424]	Application Management Collaboration	Introduce an infrastructure that provides a network protocol infrastructure on the edge and the same runtime environment as in the cloud.
	[291]	Business Orchestration Collaboration	Design a microservice deployment strategy based on could-edge collaboration and model it as a service function chain.
	[113]	Business Orchestration Collaboration	Propose a runtime system in which the resource manager optimizes resource allocation for each node's micro-services based on DRL.
Software Collaboration	[470]	Service Collaboration	Study a dynamic virtual edge node placement scheme according to predictive information and pay-as-you-go model.
	[37]	Service Collaboration	Design a method that determines the optimal deployment strategy in the offline phase and uses a mobile sensing game approach when online.
	[452]	Service Collaboration	Raise an optimization method of joint computing offloading and data cache decision making in hybrid mobile cloud-edge system.
	[145]	Service Collaboration	Utilize DRL agent to select a suitable scheduling algorithm online for edge-cloud tasks.

2.1.4 Conclusions and Discussions

With the continuous advancement of mobile Internet technology and the increasing number of smart devices, people's demands for data processing speed and real-time capabilities are constantly growing. The traditional cloud computing model focuses on processing data on the cloud server but faces challenges such as data transmission delays and network congestion. To address these issues, MCC brings cloud computing services closer to mobile devices, enabling faster data processing and service provision with lower latency. MEC extends computing resources to edge nodes like base stations and routers, reducing data processing delays. Cloud-edge computing combines the strengths of MCC and MEC to achieve more efficient data processing and service provision by leveraging both cloud and edge computing resources.

This chapter starts by discussing the general concepts of MCC, MEC, and cloud-edge computing. It emphasizes that the three concepts meet the requirements for data processing speed, real-time performance, and user experience in the mobile Internet era. The challenges of data security, resource consumption, and offload decisions for MCC are then discussed, along with potential solutions. In the meantime, we discuss how to enhance QoS further, reduce energy consumption, decrease time overhead, and carry out combined optimum task unloading and resource allocation design in MEC. Finally, we introduce cloud-edge computing collaboration at three levels and the corresponding six types of collaboration technologies. We also reveal that cloud-edge computing can better fulfill the demands of various application scenarios.

MCC, MEC, and cloud-edge computing provide solutions to many problems while also bringing some new challenges. Because the computing power of MCC is centrally deployed on the cloud server that is away from the device, this inevitably brings several relatively major challenges. Network congestion due to the massive number of connections can lead to unstable and unreliable connections. In addition, how to solve the problem of bandwidth waste and cost increase caused by redundant information contained in the data uploaded by the device is also worth pondering. When using MEC to deploy applications, it is necessary to consider the offloading problem and the competition for communication and computing resources when multiple devices are connected to the MEC system simultaneously. Only by properly solving the computing offloading problem and resource allocation strategy in MEC can the performance of MEC be better played. Hence, it needs to pay close attention and carry out in-depth research. Cloud-edge computing involves managing and scheduling resources across multiple nodes. Researchers should study more closely how to allocate tasks efficiently, optimize the use of resources, and reduce energy consumption. In addition, how to guarantee security and privacy protection when transferring and processing is worth considering. Moreover, achieving the consistency and synchronization of data from different nodes is also a problem that cannot be ignored, especially in distributed environments. Besides, designing appropriate computing architectures and models according to the specific needs of different application scenarios also brings new challenges to the development of cloud-edge computing. Nowadays, researchers are working on various solutions and technical means to achieve a more secure, efficient, and reliable cloud-edge computing system to accelerate the continuous innovation and development of mobile Internet technology.

2.2 Cloud-Edge-End Computing Concepts and Architectures

2.2.1 Understanding Cloud-Edge-End Computing

The cloud-edge-end computing integrates cloud servers, edge servers, and end devices, fully utilizing their heterogeneous computing and storage capabilities. This section will provide a detailed introduction to the cloud-edge-end computing from its origin, components, and data processing and storage mechanisms.

- **Origin of Cloud-Edge-End Computing**

Cloud-edge-end computing is a new architecture that extends cloud computing, storage, and network services from the traditional centralized cloud environment to edge nodes. In the conventional IoT model, the end devices have limited computing capabilities. Therefore, an effective solution is to offload computational-intensive tasks to the cloud, leveraging cloud servers' powerful computing and storage capabilities to meet the technical requirements of end devices. In recent years, the rapid development of IoT technology has led to an increasing number of end devices connecting to the network, resulting in a significant burden on the network due to the massive amount of generated data. Furthermore, cloud data centers are usually deployed far away from end devices, causing high response latency that cannot meet the needs of delay-sensitive tasks [306]. Therefore, in response to the shortcomings

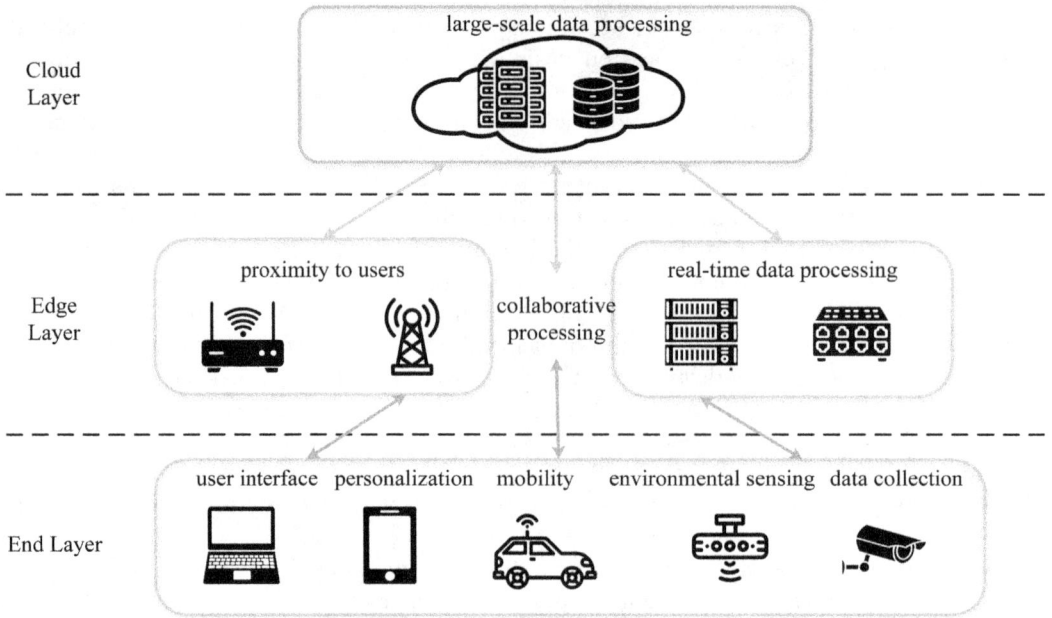

FIGURE 2.5
Components of cloud-edge-end computing.

of the cloud computing model, researchers have proposed adopting edge computing technology. This technology utilizes edge servers closer to the data source to provide services to end devices, effectively reducing data transmission latency and achieving low-latency application processing and data response. It should be noted that edge servers' computing and storage capabilities are relatively limited and cannot handle computational-intensive tasks. Therefore, the cloud-edge-end hierarchical computing architecture combining cloud platforms, edges, and end devices can integrate edge computing and cloud computing advantages. This approach effectively alleviates the data processing burden of the cloud center, dramatically reduces the data transmission delay, and thereby significantly improves service quality.

- **Components of Cloud-Edge-End Computing**

    With the rise in the number of IoT devices, the volume and velocity of data have also increased significantly. The traditional centralized data processing model, where the cloud serves as a core, can no longer meet massive data requirements for low latency, real-time data collaboration, intelligent decision-making, security, and privacy protection. Therefore, it is essential to utilize a distributed edge computing architecture, such as user computers, IoT devices, and edge nodes, among others, to bring computing as close as possible to the data source to reduce latency and bandwidth. As a result, the cloud-edge-end hierarchical computing architecture that combines cloud computing and edge computing has become an ideal option for addressing the needs of large-scale data processing, low-latency applications, and edge computing.

    As shown in Figure 2.5, the cloud-edge-end hierarchical computing architecture mainly consists of three parts, including core cloud, edge nodes, and end devices. The core cloud serves as the data center in the cloud-edge-end architecture. It usually has powerful computing and storage capabilities and is responsible for processing complex computing tasks and storing large amounts of data, providing highly scalable resources. Edge nodes are computing nodes deployed between the core cloud and end devices. They are usually located in the user environment and are responsible for collecting, processing, and transmitting various task data. Edge nodes have certain computing, storage, and network functions and can perform local data processing and decision-making and communicate with other devices or the cloud. The existence of edge nodes brings data processing and decision-making capabilities closer to the data source, reducing data transmission delays and network congestion problems. End devices are devices that communicate between users and the network and edge nodes, such as smartphones, tablets, sensors, etc. They collect and transmit data. A large amount of generated data can be processed and analyzed directly on the end device and then transmitted to edge nodes or the cloud for further processing. In addition, end devices must have security and privacy protection functions to ensure the confidentiality of user data. The core cloud, edge nodes, and end devices together form the infrastructure of cloud-edge-end computing. This distributed structure allows cloud-edge-end computing to achieve low latency, high availability, and flexibility while efficiently processing massive data and complex computing tasks.

- **Data Processing and Storage of Cloud-Edge-End Computing**

In the cloud-edge-end computing architecture, edge nodes need the cloud's powerful computing and storage capabilities to handle computing-intensive tasks, and the cloud also needs to rely on edge nodes to provide low-latency services to end devices. The cloud-edge-end computing architecture leverages the advantages of all three components, strategically plans and designs data processing and storage strategies, and dynamically adjusts task data according to the current status in the form of task migration to improve system performance and responsiveness and balance computing workload.

Data processing involves collecting data from end devices, analyzing the data, and making corresponding decisions based on application requirements. Part of the data processing can be completed on end devices, such as data preprocessing, real-time analysis, and basic decision-making logic. For more complex or meaningful data processing tasks, the data can be sent to edge nodes or the cloud for processing. Edge nodes have certain computing and storage capabilities to perform data processing and analysis, reducing dependence on cloud resources. Lightweight computing models and algorithms can be deployed on edge nodes for real-time data analysis, pattern recognition, anomaly detection, etc., providing instant decision-making and response to achieve low-latency data processing [55].

Data storage refers to saving the data generated and making it accessible at any time. On end devices, local storage can be used to save data generated by devices to meet immediate access needs. Edge nodes usually have a certain amount of storage capacity to temporarily store data that needs further processing, reducing communication frequency with the cloud. The cloud stores long-term, large-scale data that require in-depth analysis. The cloud typically features highly scalable storage systems such as distributed file systems or databases.

In addition, the cloud-edge-end computing architecture can also realize distributed computing, collaborative processing, and data aggregation functions. In this architecture, multiple edge nodes can work together to process data tasks. This distributed computing method can provide higher computing power and fault tolerance. On the other hand, different edge networks only store data related to their own business scenarios. When it comes to data sharing or the need to make global analyses or decisions, the cloud can aggregate the data of each edge node for further processing. In cloud-edge-end computing, data processing and storage not only involve the utilization efficiency of computing resources but also consider factors such as latency, bandwidth, privacy, and security. Proper planning and design of data processing and storage strategies can improve system performance and responsiveness while meeting the requirements of various application scenarios.

2.2.2   Enabling Technologies for Cloud-Edge-End Computing

The IoT has rapidly developed in recent years, resulting in a significant increase in data generated by IoT devices connected to the network. Processing this data only through cloud servers can cause considerable network pressure and may not meet the needs of applications that require low latency. While edge computing has obvious advantages in reducing latency, cloud computing is still necessary for computing and storage-intensive tasks. As a result, edge computing and cloud computing are often combined to establish a hierarchical distributed architecture that enables cloud-edge-end collaboration to meet diverse task requirements. However, due to the characteristics of networks and devices, the utilization of this architecture for task offloading and computation may encounter challenges related to data security leakage, limited application scenarios, and device battery depletion. Therefore, adopting key technologies in the cloud-edge-end computing architecture is essential to address these potential challenges. In the following sections, we will provide a detailed introduction to the relevant key technologies.

- **Blockchain**

    The cloud-edge-end computing architecture faces several challenges due to the diverse distribution and heterogeneity of mobile devices. These challenges include data security leaks, lack of trust among participants, and the management of system trustworthiness. However, blockchain technology can provide a solution to these problems. It offers a secure and transparent distributed ledger that prevents data tampering, enables traceability of historical data, and allows multiple participants to share data information while ensuring data integrity and immutability. As a distributed model that uses cryptography and multi-party consensus mechanisms to secure data and systems, blockchain can be integrated into the cloud-edge-end computing architecture to address security issues. Blockchain has several features that make it an ideal secure distributed storage system. These include data tamper-proofing, historical data traceability, and joint maintenance by multiple parties. Leveraging these features can guarantee data integrity and availability within the cloud-edge-end computing architecture. Additionally, blockchain can be an incentive platform for secure data sharing among multiple nodes within the cloud-edge-end architecture. Furthermore, blockchain smart contracts can establish a secure, trusted computing framework for the cloud-edge-end system. Blockchain's data security, transparency, and traceability characteristics can be leveraged to achieve security monitoring and auditing within the cloud-edge-end architecture. For a visual representation of the blockchain-based cloud-edge-end computing architecture, please refer to Figure 2.6.

    Scholars have combined blockchain technology with cloud-edge-end computing architecture to address the problem of data privacy protection. Xu et al. [431] proposed a secure and privacy-preserving decentralized learning system (SPDL) that employs differential privacy (DP) to protect data privacy. They integrated Byzantine fault-tolerant (BFT) consensus and BFT gradients aggregation rule (GAR) into the blockchain system to ensure efficient machine learning while maintaining BFT, transparency, and traceability. Yue et al. [471] developed a framework that uses blockchain technology within a distributed edge-cloud storage (ECS) environment to validate data integrity. The framework aims to solve the problem of incredibility in traditional verification mechanisms. They utilized a sampling approach to choose partial data slices for verification, balancing verification overhead and accuracy while ensuring high efficiency. Blockchain technology is used as an edge computing platform to integrate a distributed control system compliant with the IEC standard [333]. Blockchain participants exchange data, cooperate, and negotiate via smart contracts, effectively improving the system's security, reliability, and transparency. The authors designed a decoupled blockchain-based scheme in an internal healthcare monitoring ecosystem [23].

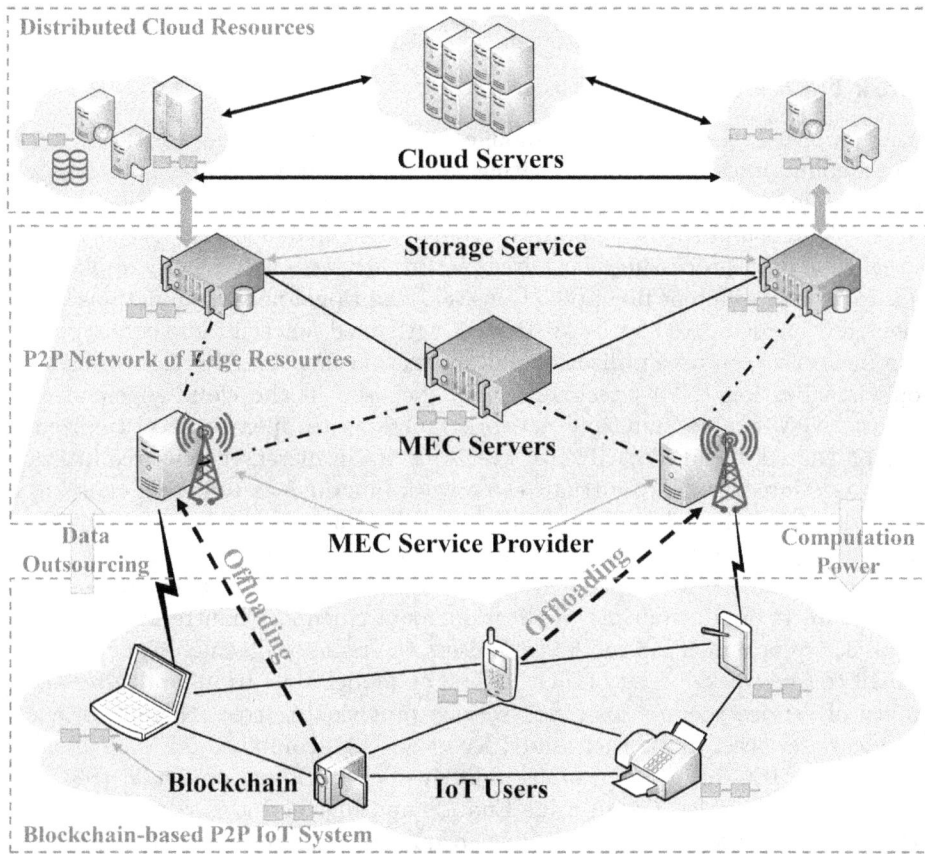

FIGURE 2.6
Blockchain-based cloud-edge-end computing architecture.

The scheme verifies and stores data from medical devices by decoupling complete blocks into block headers and block ledgers. Additionally, an incremental tensor training approach is used to reduce data duplication when transmitting large amounts of data in the healthcare network.

Utilizing blockchain technology in cloud-edge-end computing architecture prevents data leakage and ensures security and reliability during task scheduling. Huang et al. [173] proposed an edge blockchain system that can be used for resource allocation and mining consensus in an edge computing setting. The system achieves optimal data storage on resource-limited edge devices using a novel resource allocation strategy and block storage allocation scheme. Furthermore, the proposed proof-of-stake consensus algorithm decreases energy usage in block generation processes within edge environments. Tuli et al. [366] developed FogBus, a lightweight framework that integrates end devices, fog computing, and cloud computing infrastructures. FogBus aims to efficiently utilize edge and remote resources to meet the needs of IoT applications. Additionally, FogBus employs blockchain, authentication, and encryption techniques when transmitting confidential data, thus ensuring the integrity and confidentiality of sensitive data. Yao et al. [456] designed a blockchain-based task scheduling scheme in a cloud-edge-device computing system (BC-CED). BC-CED integrates task scheduling into the blockchain consensus mechanism and selects the output node with the optimum utility, thereby improving resource utilization. The task scheduling problem is also modeled as a partially observable Markov decision process, where blockchain participants address it through reinforcement learning.

- **Network Function Virtualization**

  In the era of IoT, a wide range of mobile devices connect to networks and generate an enormous amount of real-time data, creating various application demands. A highly efficient architecture called cloud-edge-end computing has been developed to address this challenge. It combines cloud computing with powerful computing capabilities and low-latency MEC to enable real-time data processing. This architecture has been successfully applied in various scenarios to achieve different functions. However, traditional network functions are typically implemented using dedicated hardware devices with fixed functions and configurations. This results in ineffective resource utilization and increased costs. To solve this problem, network function virtualization (NFV) technology can be used in the cloud-edge-end computing architecture. NFV enables multiple network functions to operate as virtualized network functions on shared versatile hardware, which results in increased resource utilization and reduced costs. Moreover, NFV virtualizes network functions as software, enabling them to be flexibly deployed, moved, and configured on different edge devices and cloud servers as needed to meet evolving business requirements and application scenarios.

  Numerous scholars have implemented NFV technology in the cloud-edge-end computing architecture to fulfill multiple application requirements. Chantre et al. [48] proposed a multi-objective optimization model to address the issue of placing edge devices in ultra-dense small cell networks based on 5G NFV. The model aims to minimize service latency, probability of service request loss, and service provisioning cost, thereby providing reliable broadcast services. In another study, Xu et al. [432] combined NFV technology, cloud computing, and MEC to design a multi-objective Service Function Chain (SFC) mapping model in the NFV-enabled Cloud-Edge-End Computing for Industrial Internet of Things (CECIIoT) architecture. This model aims to quantify different service requirements and specific network environments. Additionally, they proposed a Deep Q-learning-based SFC mapping algorithm (DQL-SFCM) to minimize resource consumption while guaranteeing a balanced workload in CECIIoT networks. To address the latency problem in Augmented Reality (AR) applications, Younis et al. [465] presented a new MEC framework for AR applications (MEC-AR). They proposed a mixed-integer linear optimization problem to find efficient application deployment schemes on the MEC-AR layer and reduce network latency.

- **Energy Harvesting**

  With the rapid advancements in IoT and 5G communication technologies, a new computing architecture has emerged that combines cloud, edge, and end computing. This architecture provides a more efficient way to run applications that require low latency and high computing power on end devices. Transferring these tasks to edge servers or cloud servers reduces the time it takes to execute a task, which significantly improves user experience. However, traditional end devices have limited battery energy and may face the challenge of battery depletion when performing task offloading, which can cause task execution interruptions. To address this issue, energy harvesting (EH) technology provides an effective solution by capturing and utilizing environmentally sustainable energy sources like solar power, wind power, and radiofrequency energy to power devices. With EH technology, mobile devices can efficiently meet applications' growing energy and computational demands. The devices acquire energy through EH technology and then offload computation-intensive and latency-sensitive workloads to edge servers or cloud servers for processing.

  Much research is being conducted to solve the task scheduling problem for mobile devices that can harness energy in the cloud-edge-end computing architecture. Sun et al. [344] proposed the MEC framework, which considers task interdependence and real-time energy harvesting capabilities. They developed two heuristic algorithms to reduce the time required to complete tasks. In addition, they created an adaptive energy harvesting approach that uses wireless power transfer (WPT) to assign flexible time slots for energy harvesting in end devices. In their paper, Liu et al. [236] discussed the issue of multi-user fairness in a fog-cloud computing network that relies on wireless technology. They proposed an optimization problem to maximize the minimum energy balance among all IoT devices. This problem is NP-hard and can be solved through a globally optimal solution method based on generalized Bender's decomposition (GBD) or a low-complexity suboptimal solution approach using penalized successive convex approximation (P-SCA). Zhao et al. [503] studied task scheduling in a multi-user edge computing scenario using EH devices. The study aims to minimize the total cost while maintaining system performance. The authors proposed an online dynamic offloading and resource scheduling algorithm (DORS) based on Lyapunov optimization theory to achieve this. In another study, Tian et al. [362] investigated dynamic scheduling problems in a green things-edge-cloud (TEC) system. They aimed to optimize the weighted average combination of delay and incomplete tasks over the long term while considering energy cost constraints. They introduced a streamlined algorithm grounded on Lyapunov optimization and a quasi-matrix mechanism to address this. This algorithm makes decisions based solely on the current system state without compromising long-term efficiency. Zhu et al. [526] studied the problem of maximizing the computation completion ratio in a wireless-powered mobile edge computing network (WP-MEC) with multiple edge devices. They designed the generalized allocation problem-based computation scheduling (GAP-CS) algorithm and the computation completion ratio maximization (CoCoRaM) algorithm to address this. These algorithms can effectively improve network performance.

Table 2.4 summarizes the literature on how blockchain, network function virtualization, and energy harvesting technologies are used in the cloud-edge-end architecture. These technologies strengthen the trust relationship between edge devices and cloud servers, provide dynamic network services and functions, and reduce battery reliance. As a result, the adoption and development of the cloud-edge-end computing architecture is driven forward.

2.2.3 Cloud-Edge-End Computing Architectures

Combining edge computing and cloud computing creates a distributed architecture with a collaboration between cloud, edge, and end devices. This integration allows applications with high computing power, large storage capacity, and low-latency services. The architecture can be categorized into two types depending on the extent of the user's visibility and access to the cloud when the end devices (EDs), edge servers (ESs), and cloud are merged into a cloud-edge-end computing system [89]. The two types of cloud-edge-end computing architectures are described as follows.

- **Hierarchical cloud-edge-end computing** is a three-tier service architecture, which is illustrated in Figure 2.7(a). When the ED generates tasks that need processing, it can use the ESs' resources by offloading the tasks to them. After deciding on resource allocation, the ESs may request additional resources from the cloud and transfer tasks for cloud-based processing if the ED's requirements are still unmet. This design hides the cloud from the ED, providing transparent services without direct user interaction. From the ED's perspective, the upper layer only includes the ESs.

- **Horizontal cloud-edge-end computing** is a service architecture that consists of two tiers. The EDs can access the cloud resources directly in this architecture. This means that when the EDs generate tasks to be processed, they can choose to offload them directly to either the cloud or the ESs. This eliminates the need

TABLE 2.4

Summary of enabling technologies for cloud-edge-end computing.

Technology	Ref.	System	Optimization Objective	Optimization Method	Year
BlockChain	[431]	SPDL	Strike a balance between privacy disclosure and convergence rate	Integrate blockchain, BFT, BFT GAR, and DP into SPDL	2023
	[471]	ECS	Reach a balance between verification overhead and precision	Use sample strategy and find the optimum sample size	2020
	[23]	Healthcare monitoring ecosystem	Ensure data security and privacy preservation/ Reduce data duplication	Use lightweight blockchain mechanism/ Use incremental tensor decomposition	2021
	[173]	Edge blockchain system	Minimize the weighted combination of fairness degree cost and range-distance cost	Use resource allocation strategy and block storage allocation scheme	2019
	[456]	BC-CED	Minimize the cost function of all tasks	Use reinforcement learning algorithm	2022
Network Function Virtualization	[48]	5G NFV-based cell network	Minimize service provisioning cost, service latency, and drop probability	Use two heuristic algorithms	2018
	[432]	NFV-enabled CECIIoT architecture	Minimize total resource consumption	Use DQL-SFCM algorithm	2022
	[465]	MEC-AR	Minimize system latency	Use branch-and-bound algorithm	2020
Energy Harvesting	[344]	MEC	Minimize task completion time	Use greedy algorithm and simulated annealing algorithm	2019
	[236]	Fog-cloud computing network	Maximize the minimum energy equilibrium across all devices	Use GBD-based method and P-SCA-based algorithm	2020
	[503]	MEC	Minimize average cost weighted sum	Use DORS algorithm based on Lyapunov	2021
	[362]	Green TEC system	Minimize long-term weighted average combination of latency and unfinished tasks	Use low-complexity algorithm based on Lyapunov and quasi-Matrix mechanism	2022
	[526]	WP-MEC	Maximize computation completion ratio	Use GAP-CS algorithm and CoCoRaM algorithm	2020

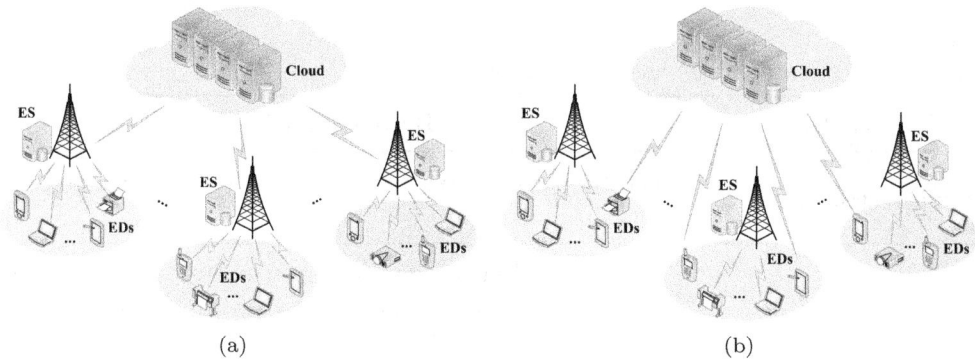

(a) (b)

FIGURE 2.7
Illustration of (a) hierarchical cloud-edge-end computing architecture and (b) horizontal cloud-edge-end computing architecture.

for intermediary handover by the ESs. In this architecture, the cloud is visible to the EDs, and both the cloud and ESs are perceived as valid processing options. Figure 2.7(b) shows a visual representation of this architecture.

Cloud-edge-end computing systems can potentially improve resource utilization and quality of experience. However, their performance is heavily influenced by their architecture [89]. Currently, most of the research on this topic focuses on hierarchical cloud-edge-end computing systems, with some studies attempting to integrate emerging technologies into the architecture. Wu et al. [408] proposed incorporating MEC into the smart IoT architecture to create an edge-driven IoT architecture. This would allow for cloud-edge-end collaboration, leading to efficient and timely processing of traffic data. They also analyzed the challenges of edge-driven IoT and presented a foresight for future work. Considering the feasibility and advantages of blending hierarchical cloud-edge-end computing systems architecture with blockchain technology to tackle the challenges of data security and multi-party mutual trust and interaction, Tong et al. [363] discussed the implementation scenarios of the three-layer architecture in terms of horizontal and vertical collaboration. They also analyzed the detailed technical issues and gave possible directions for future research. In [43], Ceselli et al. dealt with the problem of edge-cloud networks design for mobile access systems, which involves arranging all virtualized infrastructure components, starting from access nodes to cloudlets, and subsequently allocating users to cloudlets. They considered both static and dynamic network cases and presented heuristically supported link path formulas to compute the solution in a reasonable time. Researchers have also focused their efforts on implementing the hierarchical cloud-edge-end computing framework in the realm of data security. Given the inadequacy of current access control mechanisms for securing cloud data sharing against malicious actors, in [78], drawing inspiration from Access Control Encryption (ACE), Cui et al. developed a novel attribute-based data sharing framework. This innovative framework combines the cloud-edge-end architecture, which can carry out access control on both senders and receivers. Within this framework, the edge server assumes a central role in overseeing all communications, monitoring, and preventing unauthorized exchanges based on predefined access regulations. Experiments prove that this scheme has improved performance in terms of encryption and decryption compared with previous work. Further, in response to the constraints of current edge-cloud data sharing schemes, Huang et al. [171] introduced a fresh attribute-based outsourcing ACE approach (AOACE) that implements secure write control to confine the encryption capability of the senders and the

TABLE 2.5

Related works of two cloud-edge-end architectures.

Ref.	Architecture	Key Ideas
[408]	Hierarchical	Form a smart IoT architecture with cloud-edge-end collaboration
[363]	Hierarchical	Integrate cloud-edge-end architecture and blockchain for enhanced QoS and trusted computing
[78]	Hierarchical	Propose an attribute-based data sharing framework in cloud-edge-end architecture
[171]	Hierarchical	Develop a parallel secure flow control framework for private data sharing in cloud-edge-end computing architecture
[43]	Hierarchical	Propose heuristic link path formulations for cloud-edge-end network design
[89]	Hierarchical and Horizontal	Propose potential game-based algorithms for cloud-edge-end computation offloading optimization

decryption capability of the receivers. Building upon this method, a parallel secure flow control framework (PSFlow) for confidential data sharing within mobile edge-cloud environments was proposed, which expedites the purification process of AOACE by performing concurrent computation on edge nodes. Experimental results demonstrate that the PSFlow framework significantly outperforms existing schemes in terms of purification efficiency.

Unlike the research efforts mentioned above, Ding et al. [89] took a different approach to cloud-edge-end computing research by considering two types of architectures simultaneously. They developed a game model that specifically addresses the optimization problem of computation offloading strategies in this environment. The model allows each ED to minimize its revenue individually. The researchers proposed potential game-driven algorithms for both architectures, and experiments confirmed their effectiveness. They also comprehensively investigated the scalability and practicality of the two architectures across various scenarios.

As shown in Table 2.5, we have compared various cloud-edge-end architecture-related works regarding their application scenarios. It is important to note that the most suitable cloud-edge-end architecture depends on the specific application scenario and should be chosen based on the scale of users and task characteristics. Generally, hierarchical cloud-edge-end computing architecture is more appropriate for small-scale user scenarios, while horizontal cloud-edge-end computing architecture is preferable for large-scale user scenarios. Moreover, the hierarchical cloud-edge-end computing architecture is advantageous for tasks requiring high communication demands [89].

2.2.4 Conclusions and Discussions

In the previous sections, we have thoroughly explored the concept of cloud-edge-end computing and the technologies and models supporting it. As we come to the end of this chapter, it is essential to summarize our findings, discuss the outcomes of this computing model, and consider potential future developments in this dynamic field of study.

Cloud-edge-end computing is a major advancement in distributed computing. It overcomes the limitations of cloud-centric models by moving computational resources, data processing, and intelligence closer to data sources and end-users. This paradigm shift improves critical factors such as latency, bandwidth usage, privacy, and data sovereignty, essential for the next generation of IoT applications, real-time analytics, and smart environments. The development of enabling technologies such as 5G, IoT, AI, and blockchain has played a crucial role in realizing cloud-edge-end computing. Through various models such as fog computing, MEC, and cloudlets, we have seen how cloud-edge-end computing can address diverse application needs and scenarios. Each model offers unique advantages and addresses specific challenges, demonstrating the flexibility and adaptability of this computing paradigm.

Cloud-edge-end computing has the potential to solve many problems, but it also brings new challenges. The system's distributed architecture requires managing resources, ensuring security and privacy at all levels, and achieving seamless interoperability between different platforms and devices. These are significant obstacles that researchers and practitioners need to address. Several trends and developments will shape the evolution of cloud-edge-end computing in the future. Advancements in AI and machine learning will significantly boost the intelligence of edge nodes, empowering them with advanced data processing and decision-making capabilities at the edge. Furthermore, the widespread adoption of 5G and beyond 5G networks will provide the high-speed, low-latency connectivity required to manage the extensive scale of IoT devices and applications. Sustainability and energy efficiency will also become increasingly important considerations. Therefore, researchers will focus on green computing technologies and practices within the cloud-edge-end ecosystem. Additionally, the incorporation of emerging technologies like quantum computing and neuromorphic computing into cloud-edge-end architectures could present both new opportunities and challenges.

2.3 Cloud-Edge-End Computing Implementations

2.3.1 Server Placement in Cloud-Edge-End Architectures

Numerous studies nowadays distinguish between cloud computing and edge computing and focus on optimizing the architecture and system analysis. However, they often overlook the importance of collaboration among cloud, edge, and end. In recent years, the rapid growth of IoT and 5G has brought attention to the strengths and weaknesses of cloud computing and edge computing. This has led people to consider the value of combining their benefits, resulting in cloud-edge-end collaboration.

- **Cloud-Edge-End Collaboration Relationship**

Cloud centers are typically situated far away from end-users, leading to extended transmission delays in task completion. On the other hand, edge computing units often lack sufficient resources to process and store large amounts of data, which can negatively affect service quality and reliability. Therefore, cloud-edge-end architecture plays a crucial role in addressing these issues. The current architecture is mainly reflected in the following aspects:

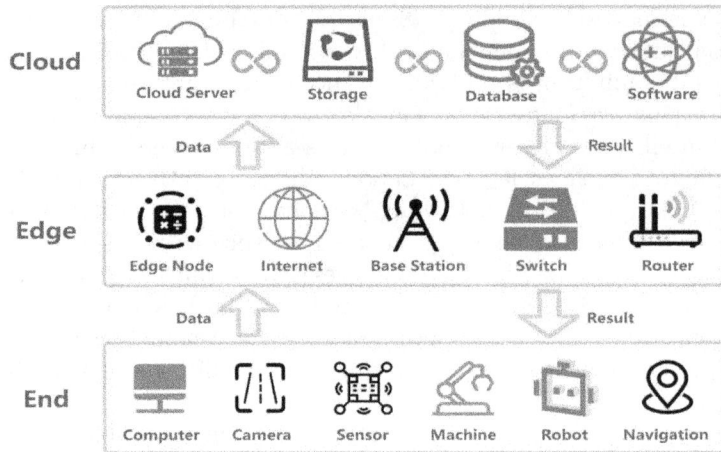

FIGURE 2.8
Cloud-edge-end architecture model diagram.

- **Unified Management:** Through the coordination of the complex and variable low-level resource management scheme, the perception of the business in the low-level details can be effectively decreased. The unified management of the cloud can guarantee basic business processing capability.

- **Task Scheduling and Collaboration:** The cloud-edge-end system normally consists of servers in different physical locations. Using optimization algorithms to distribute generated tasks among different servers intelligently can maximize the utilization rate of equipment resources to ensure the efficiency and stability of the system.

- **Data Processing and Storage:** The data produced by the end can be uploaded to the edge server via a wireless network. After the edge server pre-inference of the data, the results are gathered along the wireless network or optic-fiber network to the cloud server for analysis and training processing, after which the training results are returned step by step to the end, and this scheme can significantly improve the training efficiency.

- **Cloud-Edge-End Architecture Implementation**

Cloud-edge-end architecture is to realize the collaborative connection among end, edge, and cloud to achieve data worth jointly. The cloud-edge-end architecture model is shown in Figure 2.8.

So, how exactly is the cloud-edge-end architecture implemented? We have to mention edge computing, which plays an important role here. An important research field in edge computing is how to coordinate the resources of the edge nodes. Computing and storage resources ought to be distributed throughout the edge network to guarantee that all users can receive ubiquitous services. The deployment of computation and storage resources, in turn, depends on where the servers are placed. Therefore, mobile edge computing (MEC) servers should be arranged hierarchically and complement each other in terms of their physical locations. This will lead to the efficient utilization of resources while greatly meeting users' quality of service (QoS) and quality of experience (QoE) requirements. However, it is not as simple as it sounds since services may be incorrectly placed in crowded locations [374], which are farther away from users, resulting in more significant latency.

TABLE 2.6

Comparative summary of server placement and service placement. D: Delay, R: Resource, E: Energy Consumption, C: Cost. ✓ indicates the factor is considered in the literature.

Type	Ref.	Highlights	Approaches	Optimization Objectives			
				D	**R**	**E**	**C**
Server	[37]	Consider edge/cloud servers heterogeneity and fairness of base stations response time	Game Theory	✓			
Service	[127]	Propose weighted cost factor	Memetic Algorithm	✓		✓	
	[201]	Consider deadline satisfaction, community relationship of the devices and service availability	Graph Partition	✓			
	[438]	Formulate solution for shared and unshared resources among several network service providers	Game Theory		✓		
	[52]	Consider complete collaboration and strategy between small cells	Coalitional Game	✓		✓	✓
	[255]	Consider cooperation among multiple edge nodes	Queuing Theory	✓			
	[221]	Consider user deadline preference and edge clouds' strategic behaviors	Lyapunov			✓	✓
	[392]	Consider user mobility and the resulting dynamics	Lyapunov	✓			
	[29]	Consider dynamic nature of users' locations and network capabilities	Hungarian Algorithm			✓	
	[22]	Consider deadline for user's request	SDF, LFU, Hybrid, LRU	✓	✓		
	[393]	Involve placement across one or several levels of edges and cloud(s)	Approximation		✓		
	[259]	Prioritize application placement requests based on user expectations	Fuzzy Logic	✓	✓		
	[114]	Consider runtime dynamic migration of microservices	RL		✓		
	[374]	Consider current state of network and position of users and servers	ILP	✓			
Server & Service	[495]	Consider relationship between base stations and servers as well as service requests of servers	Clustering Algorithm	✓			✓

- **Studies for Cloud-Edge-End Architecture**

When it comes to placement in MEC systems, two important aspects to consider are server placement and service placement. The placement of services is dependent on the availability of servers, and the location of servers also affects the performance of service placement. Therefore, it is crucial to properly deploy servers and services in a way that is mutually beneficial [495]. Several research studies have investigated server placement, service placement, and joint placement strategies that take both aspects into account. For more details, please refer to Table 2.6.

Server Placement: In most existing edge computing studies, service providers often deploy a limited number of edge servers due to their expensive price and other environmental factors. However, the placement of edge servers has a critical effect on delay, but few studies have focused on the influence of server placement on edge services. Low-efficiency server placement will cause intolerable delay and unsatisfying QoE. In addition, there is an urgent need to achieve overall load balancing of edge servers to guarantee the efficiency of services and the reliability of the server system, avoiding overloading of some servers and underutilization of others. Consequently, an effective edge server placement strategy is necessary to guarantee the performance of edge services. Cao et al. [37] investigated the placement of heterogeneous edge servers to minimize response time. They proposed a method that includes both offline and online phases. In the offline phase, an integer linear programming approach is used to generate the best placement policy for servers. In the online phase, a strategy based on mobility-aware game theory is designed to handle the user's dynamic mobility characteristics. Their approach dramatically enhances the equity of base stations' response time and provides new ideas on the server placement problem.

Service Placement: Users can access the required services by approaching low-latency edge servers in practical applications. However, due to the uneven deployment of base stations, each edge server has different service requests impacting the server properties and service deployment schemes for all edge servers. In this situation, it becomes necessary to have a heterogeneous placement of edge servers and rational service placement to ensure better service delivery. For example, players at amusement parks can use AR devices to obtain better gaming experiences. Hence, deploying AR service on servers in this area to offer real-time service is an appropriate choice. Similarly, users tend to make more calls during the daytime. Therefore, it is recommended that the service provider supply abundant communication services at this particular time of day. The implementation of service placement depends on various factors such as resources, cost, energy consumption, and delay.

In the context of fog computing, Mahmud et al. [259] presented a QoE-aware method for placing application services based on user expectations and fog nodes' capabilities to optimize resource usage and minimize latency. Ascigil et al. [22] formulated a centralized approach to optimize resource allocation, which can improve service availability and QoS satisfaction. It has been experimentally shown that the proposed uncoordinated resource allocation algorithm can attain optimal behavior and eliminate transmission or coordination overhead. Wang et al. [393] developed online approximation algorithms with provable performance bounds for placing tree application graphs onto tree physical graphs, which enables online deployment of multi-component applications under optimal load balancing and resource consumption. Fu et al. [114] proposed a runtime system for efficiently delivering microservice-based user-facing services across the cloud edge continuum, ensuring users QoS while minimizing the required computational resources. Additionally, Xu et al. [438] developed different algorithms to solve the issue of sharing or not sharing resources for network service providers, which can significantly reduce the social cost of all players.

The articles [52, 221] discussed the issue of the cost incurred in service placement. Chen et al. [52] investigated the cooperative service placement of multiple small cell systems in the MEC environment. They proposed a collaborative service placement algorithm using parallel Gibbs sampling and exploiting graph coloring on small cell networks. Furthermore, using a coalitional game, they investigated the strategic behavior of selfish base stations (BSs). The proposed algorithm effectively reduces the operating cost of the whole system, whether for collaborative BSs or selfish BSs. Li et al. [221] developed a deadline-driven joint cooperative mechanism based on Lyapunov optimization. Their mechanism incorporates cooperative control, occasionally allowing workload migration among different edge clouds. They also designed a collaborative strategy based on auction to minimize the social cost, considering the selfishness of individual edge clouds.

The articles [127, 29] optimize system energy consumption. Regarding the service placement problem with weights, Goudarzi et al. [127] designed a new technology for batch placement decisions for concurrent IoT applications based on the Memetic algorithm to optimize energy consumption and execution time. Furthermore, the deployment issue of multi-component applications based on dynamic nature is also considered by Bahreini et al.

[29]. They designed an effective heuristic online approach and solved the problem by representing it as a mixed integer linear programming formulation, achieving minimized energy consumption.

The articles [201, 255, 392, 374] focused on reducing delay in their studied problems. In the paper [201], the authors proposed a strategy for placing relevant services in the device closest to the user. This service placement method considers community relations, deadlines, and application availability to maximize efficiency in complex networks. Ma et al. [255] formulated the problem of collaborative service caching and workload scheduling in MEC as a mixed integer nonlinear programming problem. This approach aims to minimize both the service response time and the amount of traffic sent to the central cloud. In [392], the authors developed a new method for solving a class of constrained Markov Decision Processes. This method takes into account user mobility and resulting dynamics. It focuses on minimizing execution time during dynamic service migration and workload scheduling. Finally, Velasquez et al. [374] proposed a service placement architecture that reduces latency. This architecture considers the current condition of the network and the position of users and servers.

Server & Service Placement: Many existing studies focus on either server placement or service placement alone, but the relative importance of these two factors varies widely across different works. Moreover, very few studies consider both factors together. Zhang et al. [495] addressed this gap by simultaneously considering service requests from edge servers and the links between these servers and base stations. They proposed a solution combining clustering algorithms and nonlinear programming to solve the coupled server and service placement challenge. The proposed solution aims to maximize the overall profitability of all edge servers.

2.3.2 Data Training in Cloud-Edge-End Architectures

Artificial Intelligence (AI) has made significant progress in various fields, such as computer vision, natural language processing, and big data analytics. To achieve this, deep neural network (DNN) models, convolutional neural network (CNN) models, and recurrent neural network (RNN) models have been proposed and are widely used, including in mobile applications. However, mobile devices have limited memory, battery capacity, and computational power, making it challenging to execute large-scale neural network models and data processing efficiently. One traditional solution is to send the collected data from mobile devices to cloud computing centers for processing. However, the increasing amount of data from terminals due to emerging technologies like IoT imposes an unaffordable transmission burden on the connection between the end and the cloud. Additionally, applications like autonomous driving require real-time feedback, and the transmission latency resulting from end-to-cloud communication can lead to incalculable losses. Edge computing technology can reduce transmission delays, but for some tasks with high computational and storage requirements, relying on the powerful computing capabilities of the cloud is still necessary. Therefore, the cloud-edge-end computing architecture based on CNN, as illustrated in Figure 2.9, is integrated to meet different model training requirements and task demands. In this architecture, small-scale neural network models are deployed at the end or edge to process simple data, while complex data is transmitted to powerful neural network models within the cloud for processing. However, there are risks of privacy leakage and data security during data transmission and storage. To address these risks, federated learning (FL) is used as a distributed machine learning method focused on privacy. It allows each participant to train models on the end devices or edge servers and only forwards model parameter updates to the central cloud server for merging without sharing the original data. This

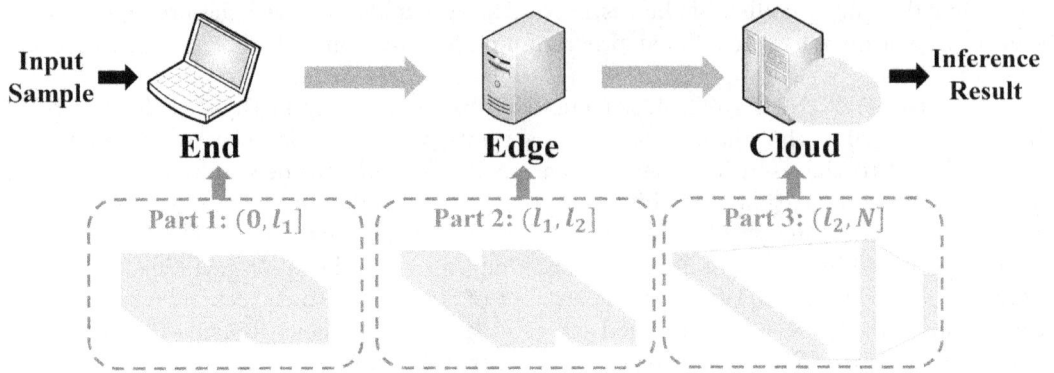

FIGURE 2.9
CNN-based cloud-edge-end computing architecture.

approach minimizes the risk of sensitive data exposure and provides better data privacy protection.

Numerous studies have analyzed the use of cloud-edge-end computing architecture for neural network training and inference. Pagliari et al. [283] proposed an input-dependent mapping method for RNN inference, which determines whether to execute inference locally or in the cloud according to factors such as the input sequence's length, neural network characteristics, and the current state of the system. The experiment results demonstrate that this approach reduces both total inference time and energy consumption compared to fully local or fully remote schemes. Long et al. [245] proposed a distributed system architecture comprising MEANet in the edge and CNN in the cloud for intelligent management and data processing. They also proposed a complexity-aware active training and inference strategy for the distributed AI system spanning the edge and cloud environments. This strategy aims to achieve intelligent distributed decision-making on resource-limited edge devices. Yang et al. [450] introduced a CNN inference method based on edge-edge-cloud collaboration called CNNPC. This method achieves rapid and precise CNN inference by employing joint model segmentation and compression tactics. A fast search algorithm is adopted to ascertain the optimal model separation and contraction strategies, enabling low-latency and high-accuracy inference. Li et al. [219] constructed Lasagna as an efficient DNN inference architecture for edge cloud supporting software guard extension. This architecture explores the hierarchical structure of DNN models, effectively balancing scarce enclave page cache resources and computational resources. In a collaborative framework between edge and cloud, Li et al. [209] presented an effective and precise approach named AppealNet for DNN-based inference. A balance between accuracy and cost is achieved by designing an edge neural network model with a dual-headed architecture and simultaneously improving the approximator and predictor during training.

Some researchers utilize the cloud-edge-end architecture to perform model training and inference for specific scenarios to achieve efficient task scheduling. Hao et al. [146] proposed an improved method for allocating AI workloads, which aims to optimize the performance of AI-centric workloads across all elements of the cloud-edge architecture. They have introduced two efficient workload allocation algorithms that reduce end-to-end latency in single-task and multi-task scenarios. Liu et al. [237] developed a cloud-edge-end cooperation framework for recommending cruising routes to vacant taxis. The framework employs federated learning to predict future waiting times for passengers and dispatch idle taxis between regions, thus improving the profitability of the taxi business.

TABLE 2.7
Summary of methods for data training in cloud-edge-end architecture.

Area	Ref.	System	Optimization Objective	Optimization Method	Year
RNN	[283]	Cloud-edge-end	Minimize the sum of total inference delay and energy usage	Use an input-dependent mapping method	2020
CNN	[245]	MEANet	-	-	2021
	[450]	CNNPC	Minimize latency and maximize accuracy	Use a fast search algorithm	2022
DNN	[219]	Lasagna	-	-	2022
	[209]	AppealNet	Minimize the overall expected loss	Conduct joint training for two-head small network	2021
Edge AI	[146]	Cloud-edge-end	Minimize the sum of transmission delay and inference processing delay	Use two workload allocation algorithms	2021
FL	[237]	CCF-CRR	Maximize the sum of all taxis' business profits and minimize the sum of pick-up distances	Conduct FL algorithm between vacant taxis and edge server	2023
	[406]	5G HetNet	-	-	2021
	[509]	Cloud-edge-end	-	-	2022

Some scholars have proposed combining FL technology with cloud-edge-end architecture to address the issue of data privacy protection. In a recent study, authors proposed an FL-empowered end-edge-cloud collaboration architecture for security in 5G HetNets [406]. This architecture involves distributing attack detection nodes at different levels of the 5G HetNet to enhance the accuracy and performance of attack detection. Another study by Zhong et al. [509] investigated user privacy protection within the cloud-edge-end framework. They designed a two-layer FL framework and proposed a semi-supervised learning method to effectively address three problems: hierarchical FL, heterogeneity of end device resources, and numerous unlabeled data under the cloud-edge-end framework.

Table 2.7 summarizes methods for data training in the cloud-edge-end computing architecture. Utilizing FL technology and neural networks such as DNN, RNN, and CNN in the cloud-edge-end architecture can provide efficient model training and inference capabilities and satisfy various requirements, including data privacy protection, real-time decision-making, and resource conservation. The advantages provided by these methods make the cloud-edge-end architecture an efficient prescription for processing large-scale data processing and implementing intelligent applications.

2.3.3 Resource Management in Cloud-Edge-End Architectures

Resource management involves two major aspects: the allocation of available resources and the control of system costs. Regarding allocation, tasks in the cloud-edge-end computing architecture can optionally be processed locally or delegated to the edge or cloud for execution over a wireless access network. This helps improve the speed of task processing. However, the finite resources of the edge servers mean that when tasks are transferred to the edge, the edge nodes need to allocate their computing and communication resources moderately to support task execution. This is known as the resource allocation problem. The range of resources discussed includes computation capability (typically assessed by

CPU frequency), the memory size of the edge server, wireless network bandwidth connecting the edge node and mobile device, and transmission power used by the mobile device, among others. Regarding controlling system costs, typical concerns include latency, energy consumption, reliability, security, network deployment, and operational overheads. Depending on the research issues, we compare some of the work related to resource management with respect to optimization objectives, constraints, and decision variables, as presented in Table 2.8.

Many studies have been conducted on resource allocation. Zhou et al. [523] proposed a cloud-edge resource allocation framework that utilizes a latency-conscious Lyapunov approach to address the performance-cost trade-off when allocating resources to various IoT applications. The strategy aims to achieve optimal long-term expenditure while meeting the hard/soft deadlines of the applications. To solve the computation resource management problems that arise in MEC and the cloud, Zhang et al. [500] presented an effective mobile edge-cloud computing network framework that allows the edge and cloud to collaborate on computation resources using wholesale and buyback transactions, thereby increasing utilization and profitability. Resource management is modeled as a profit optimization in [500], where the edge server decides on its wholesale and repurchase options and the cloud decides on its local resources, so both parties can benefit from sharing resources. Results show that the proposal can optimize the social welfare and profits of the edge servers and the cloud.

Resource management is closely connected with task scheduling and computation offloading policies in cloud-edge-end computing systems. These policies aim to achieve high efficiency and low consumption. In the domain of task scheduling, Yin et al. [463] proposed an efficient convergent firefly algorithm (ECFA) to find scheduling policies for tasks designated for execution either at the edge servers or the cloud data center. They also provided a detailed mathematical analysis to demonstrate the convergence of ECFA and suggested a range of parameter settings to avoid falling into local optima. The experiments show that ECFA has superior performance and the given parameter range is correct. In [348], Tan et al. suggested a generic model for the online assignment and scheduling of jobs in edge-cloud systems. They outlined a problem statement intended to minimize tasks' overall weighted response latency and designed an initial online job scheduling algorithm based on cloud-edge-end architecture. Simulation results show that their algorithm significantly reduces the total weighted response time compared to the heuristic approach. Liu et al. [239] investigated the inference task scheduling problem and developed a non-invasive performance characterization network (PCN) to accurately forecast the inference time of DNNs by leveraging the similarity in behaviors among comparable networks. They then proposed Sniper, a time-conscious and self-updating collaborative cloud-edge inference scheduling system. This system enables the flexible integration of PCN and time-based scheduling algorithms within its scheduling module. Experimental findings show that the system predicts the network inference time with low error and significantly reduces the waiting time for DNN scheduling while achieving a steady rise in throughput.

In recent research, several approaches have been proposed to solve the problem of task offloading. Zhang et al. [500] presented an energy-conserving offloading approach that optimizes both time and energy costs in smart IoT systems. They developed a system energy model for DNN-based systems, which takes into account the operation time and energy cost across all edge-cloud servers and IoT devices. By segmenting DNN layers, they reduced the execution time and energy consumption. Zhao et al. [504] presented the impact of task dependencies and service caching on task offloading strategies. They designed an algorithm that minimizes the task completion time by utilizing convex programming to reconfigure the issue as a convex optimization problem. The experimental results show that their strategies can significantly decrease the completion time of applications. To consider the environmental parameters in the DNN architecture design, Odema et al. [280]

TABLE 2.8

Summary of resource management optimizations in cloud-edge-end architectures. Cla.: Classifications, Opt_Obj_Rel_Cons: Optimization Objectives and Relevant Constraints, EC: Energy Consumption, L: Latency, R: Reliability, S: Security, C: Cost, T: Throughput, P: Profit, D: Dependency, CS: Cache Service, TES: Task Execution Sequence, OS: Offloading Strategy, CaS: Cache Strategy, CF: CPU Frequency, B: Bandwidth, TP: Transmission Power, TR: Transmission Rate. The terms "TS", "TO", "RA", and "OA" represent the task scheduling, task offloading, resource allocation, and task offloading and resource allocation problems, respectively. "✓" indicates that the factor is considered in the literature.

Cla.	Ref.	Opt_Obj_Rel_Cons										Decision Variables						
		EC	L	R	S	C	T	P	QoS	D	CS	TES	OS	CaS	CF	B	TP	TR
TS	[463]	✓	✓									✓						
	[348]		✓									✓						
	[239]		✓			✓						✓						
TO	[61]	✓	✓										✓					
	[504]		✓						✓	✓			✓					
	[93]	✓		✓									✓					
	[280]	✓	✓										✓					
	[162]	✓	✓										✓					
RA	[523]		✓			✓									✓	✓		
	[500]		✓					✓	✓				✓					
OA	[238]		✓										✓		✓	✓	✓	
	[57]	✓	✓										✓	✓	✓		✓	
	[187]		✓										✓		✓		✓	✓
	[42]		✓		✓	✓							✓	✓	✓	✓		
	[508]	✓	✓	✓									✓		✓			
	[512]		✓			✓							✓		✓			
	[62]	✓	✓										✓			✓		

developed a multi-objective neural architecture search design methodology. This methodology redefines the performance objectives to consider wireless communication parameters and searches for optimal deployment options for various architectures. Tests show that the method provides substantial improvements over traditional solutions on energy and latency metrics. Hu et al. [162] utilized the three-tier cloud-edge-end computing architecture to construct a multi-layer emotion-aware service architecture for emotion recognition applications. They defined the target as a delay-constrained energy minimization problem and realized an efficient application platform for emotion analysis. Dong et al. [93] designed a reliable-aware optimal computation offloading and allocation algorithm for resource-limited systems with multi-users and multi-servers. They constructed a combinatorial optimization problem and reduced the complexity by task aggregation. They also devised a novel shadow component design to meet the requirements for dependability. Test outcomes indicate that the suggested strategy can efficiently achieve the optimal solution.

There are several ongoing efforts to explore integrated approaches for task offloading and resource allocation. Chen et al. [62] addressed the problem of task offloading and channel assignment in a multi-channel wireless interference scenario. They introduced a decentralized computational offloading algorithm to achieve the Nash equilibrium in the computation offloading game among mobile devices. The algorithm demonstrates high computation efficiency and can scale well with increased user numbers. Casola et al. [42] investigated the task offloading and cloud resources allocation problem. They presented a new formulation of the cloud-edge resource allocation issue for Industrial IoT, which aims to characterize cloud-edge apps and their safety-conscious placement on cloud and local resources within specified performance, latency, and security constraints. They proposed a deterministic algorithm that can find suboptimal solutions in milliseconds. They can see the same solution much faster using the algorithm than a linear programming solver. Zhou et al. [512] addressed the task offloading and computing resources allocation problem. They introduced a new computation offloading and resource allocation mechanism based on reverse auction principles that formulate the MINLP problem in terms of minimizing the operational expenses of the cloud service center. They decoupled the initial problem and devised streamlined algorithms to tackle the optimization problems. Experiments show that the suggested method can remarkably reduce the cost of the cloud service center in different situations. Zheng et al. [508] developed a resource scheduling algorithm that utilizes multi-agent reinforcement learning. They incorporated a centralized training-decentralized processing framework and a deep deterministic policy gradient algorithm to construct and train a neural network for decision-making regarding the task offloading and computing resource allocation strategies in collaborative vehicle-road systems. Experiments prove that the new method can greatly enhance the utility values associated with latency and energy consumption.

Research is being conducted on multiple resource allocations in heterogeneous cloud-edge network environments. Several studies have been published [57, 187, 238]. Chen et al. [57] addressed computation offloading, caching selection, transmission power distribution, and computing resource allocation for multiple tasks. They proposed a two-layer alternation method framework that utilizes reinforcement learning and sequential quadratic programming to solve the problem. The simulation results show that this methodology delivers notable savings in overall costs compared to other benchmark techniques. Kai et al. [187] proposed a pipeline-based offloading strategy for collaborative task processing across user ends, edge nodes, and the cloud center. They focused on minimizing the total delay of all mobile device tasks, taking into account the task offloading, computation resources allocation, transmission rate, and transmission power allocation. The study shows that the presented strategy outperforms other offloading schemes. Liu et al. [238] formulated the combined task offloading, bandwidth allocation, transmission power allocation, and computation resource allocation problem as a mixed-integer optimization problem. They decomposed the problem using Lyapunov optimization and duality theory and developed an online semi-decentralized method to tackle it. The tests verify that the developed method outperformed the two baselines under various parameter settings.

It is not difficult to discern that the work on resource management is diverse in terms of optimization objectives, constraints, decision variables, and problem scenarios. These factors must be considered and amalgamated with the actual application requirements to construct the problem and further solve it effectively.

2.3.4 Conclusions and Discussions

Cloud-edge-end computing is a modern computing model that aims to bring computing resources and services closer to end-users and devices by integrating cloud computing, edge computing, and end devices. This creates a seamless computing environment that delivers faster and more efficient service. This chapter is structured around three key components that define the operational and strategic essence of this advanced computing model: server placement, data training, and resource management.

In the context of server placement in cloud-edge-end computing, the collaboration relationship, implementation strategies, and research efforts form a triad that drives the effectiveness and innovation of this distributed computing model. The implementation strategies focus on the practical execution of the architecture, detailing how servers are strategically placed at the edge to reduce latency and enhance performance. The research efforts address the ongoing efforts to refine server placement strategies and address the challenges associated with this architecture. Data training is a crucial part of data analytics and machine learning. It focuses on analyzing data and training models at the edge, where the data is generated, to make decision-making processes faster and more informed. Additionally, it reduces the amount of bandwidth required to transmit large datasets to the cloud. Resource management is critical for managing computational and network resources across the cloud-edge-end architecture. It involves strategies for allocating and monitoring resources to ensure system reliability, optimal performance, and energy efficiency.

Looking ahead, as technology continues to advance, AI and machine learning are expected to become increasingly prominent in optimizing server placement and resource allocation. In addition, the importance of security and privacy will require ongoing research in securing data at the edge and ensuring compliance with data protection regulations. In summary, cloud-edge-end computing represents a dynamic and evolving field with significant potential for growth and innovation. Continuous collaboration between researchers, industry practitioners, and policymakers is crucial to shaping the future of this technology and ensuring its successful integration into our digital society.

3

Cloud-Edge-End Orchestrated Systems

With the rapid advancement of science and technology, more research has been conducted on cloud-based coordination systems, and the cloud-edge-end orchestration system has made rapid progress. The cloud-edge-end orchestration system is an emerging computing architecture, which refers to a layered network computing architecture that integrates cloud resources, edge servers, and end devices to enhance the performance of Internet of Things (IoT) systems. This architecture consists of three layers. The cloud layer refers to a cloud computing center equipped with robust computing capabilities to manage and execute intricate computing tasks. The edge layer consists of devices with computing power, such as base stations, which can perform some of the computing tasks and ensure low latency. The end layer encompasses various devices, including smartphones, sensors, and others. These devices usually have limited computing power and storage and can only accomplish simple computing tasks.

The cloud-edge-end orchestration system accomplishes computation tasks together through collaboration between different layers. Its emergence has changed the current computation model and made computation more efficient. However, as an emerging computing paradigm, cloud-edge-end orchestration systems still have problems and challenges in some aspects that need to be optimized. This chapter aims to serve as a valuable resource for researchers and engineers, enhancing their comprehension and application of cloud-edge-end orchestration systems and fostering their continued evolution.

3.1 Collaborative Cloud-Edge-End Systems

In the context of rapid development in IoT devices and related technologies, cloud-edge-end technology offers efficient and reliable solutions for data processing and service deployment as an integrated technical system combining cloud computing, edge computing, and end devices. Cloud-edge-end technology finds applications in various domains, such as smart cities, smart homes, and industrial automation. However, with the progress of time and technological advancements, cloud-edge-end technology faces different challenges in meeting the growing demands for data processing and services and adapting to complex application scenarios. This section will delve into cloud-edge-end technology's challenges regarding latency, reliability, security, and energy consumption.

3.1.1 The Motivation of Latency Optimization

A cloud-edge-end system is an integrated system built between cloud computing, edge computing, and end devices. This integration leverages the resources, computing capabilities, and data from the three-layer architecture model of cloud computing, edge computing, and end devices, enabling comprehensive data processing, storage, and distribution. The advantages of a cloud-edge-end system lie in its ability to leverage the resources from the

DOI: 10.1201/9781003540281-3

three-layer architecture model of cloud computing, edge computing, and end devices. This enables efficient data processing and distribution and enhances user experiences and service quality. It can be applied across diverse domains, including smart factories [295], the Industrial Internet of Things (IIoT) [391], the Internet of Medical Things [97], etc., offering powerful computing and data processing capabilities for various application scenarios. However, with the widespread adoption of cloud-edge-end systems, the issue of latency in related applications has become a growing concern. The latency in cloud-edge-end systems has gradually been exposed. There are still many challenges in addressing and improving the latency issues in cloud edge systems.

Latency optimization of cloud-edge-end systems is critical. Cloud-edge-end systems consist of three parts. The cloud is responsible for providing high-performance computing and storage resources for processing large-scale data and complex computational tasks. The edge refers to the computing resources and service nodes located between the cloud infrastructure and end devices. The primary purpose of edge computing is to bring computing and storage capabilities close to users and devices. This enables low-latency services. End devices include smartphones, sensors, IoT devices, etc. They serve as the end-users or data sources of the cloud-edge-end computing system and interact with the edge and cloud servers to access computing and services support.

Collaborative computing tasks in cloud-edge-end systems necessitate extensive data exchange between the cloud, edge, and end devices. It is worth noting that in small-scale computing tasks, the issues caused by latency may be minimal and acceptable. However, if collaborative computing tasks in cloud-edge-end systems require high real-time performance and precision, the issues caused by high latency can hinder the development in related fields. For example, driverless cars, smart factories, VR technology, etc. However, latency issues involve many aspects of cloud-edge-end systems. To guarantee efficient data transmission, the internal network within the cloud-edge-end system must be optimized, and techniques such as load balancing, traffic scheduling, and congestion control must be utilized to enhance network efficiency and minimize latency. Methods such as data compression, data preprocessing, and resource management can also improve network communication efficiency. While significant progress has been made in latency optimization in cloud-edge-end systems, including data offloading [417] and intelligent resource management [453], there is still room for improvement.

The latency of cloud-edge-end systems still faces huge challenges, mainly in three aspects: computing resources, resource allocation, and data collection.

- **Challenge in Task Offloading**

As mentioned above, with the rapid growth of IoT devices, the limited resources of devices make it challenging to satisfy the requests of increasingly complex applications [352]. Furthermore, assuming the constraints posed by insufficient resources on the end devices in the cloud-edge-end system, they may be unable to compute the required results in real-time, leading to significant accidents [97]. However, although current technologies can partially compensate for these limitations, they have certain drawbacks. For example, efficient task offloading can be achieved through artificial intelligence (AI) algorithms [108], but there are still challenges in cost control and other aspects. Therefore, task-offloading technology still holds potential for further development.

- **Challenge of Resource Allocation**

In cloud-edge-end systems, the design of computational resource allocation strategies inevitably faces a challenging problem known as latency-sensitive dual-queue optimization. More specifically, the coupling of edge processing queues and cloud processing queues introduces difficulties in ensuring end-to-end latency requirements. However, existing methods need to address this problem better. For example, current approaches treat the two queues as serial queues and cannot handle multiple queues simultaneously [446]. Additionally, network resource allocation is also challenging. It requires flexible resource allocation based on different user network requirements while providing the lowest possible latency.

- **Low-Latency Collection of Heterogeneous Device Data**

    With the development of IoT technology, industrial IoT has emerged and started to integrate with cloud-edge-end systems to improve productivity and efficiency. Most current technologies rely on the conventional pyramid approach for data collection[391]. However, with the proliferation of heterogeneous devices, interoperability and latency become challenging and open problems, posing difficulties in data collection.

3.1.2   The Motivation of Energy Optimization

Technological advancements have driven the widespread adoption of smart devices and the Internet of Things for the past few years. More and more smart devices and environments have given the cloud-edge-end system more attention. The cloud-edge-end collaboration systems represent a paradigm shift in data processing and management. Their goal is to shorten response latency and conserve bandwidth by bringing data storage and computation closer to the location where they are needed. Cloud-edge-end collaboration systems can achieve more computing power and accomplish more complex tasks by coordinating cloud centers, edge nodes, and end devices. However, this will involve the issue of reducing energy consumption, and it is interesting to think about how to balance computing power and energy consumption. Some studies have been conducted to reduce energy consumption, which is enough to show the importance of energy in the cloud-edge-end orchestrated system. However, there are still some problems and challenges in the current studies.

Energy consumption in the cloud-edge-end system is an essential metric for evaluating the system. In cloud-edge-end systems, data transmission can be significantly reduced as data processing occurs closer to the data source, thus reducing energy consumption. This reduction in energy consumption can be significant for large-scale data processing tasks. Consequently, there is still a need to research and improve energy efficiency.

Currently, the cloud-edge-end collaboration system is developing rapidly, with an increasing number of devices, such as sensors, smartphones, cameras, and others, gaining access to it. The devices often rely on battery power. In some scenarios, the increased mobility of services and the expanded data processing capacity impose greater demands on node processing performance, which will cause an increase in energy consumption [244]. However, high energy consumption can lead to shorter device endurance, and users commonly fear battery depletion, particularly when the battery level is already low [469]. As a result, high energy consumption affects the device's ability to keep working and the user experience. In addition, edge computing nodes may be deployed in cloud-edge-end collaboration systems on base stations, enterprise servers, or other network nodes. The energy consumption of these computing nodes directly affects operators' power costs. Reducing energy consumption can reduce operational costs and improve economic efficiency. Optimizing energy management is also conducive to improving efficiency. For example, an energy management system can provide real-time data and reports that enable users to better understand energy usage, optimize equipment use and maintenance, and improve efficiency.

It is well known that in the 21st century, energy and the environment have received more attention, and conservation of resources and protection of the environment are often advocated. The growing integration of renewable energy sources underscores the importance of resource management [481]. In cloud-edge-end orchestrated systems, high energy expenditure becomes a limiting factor for the development of the system, and energy increases operational expenses and can even hurt our environment [11]. Moreover, many devices are deployed in resource-constrained environments, and reducing system energy consumption can help conserve resources such as electricity, making the system more economical and sustainable. Otherwise, the number and size of cloud-edge-end collaboration systems are growing with the development of technology and the popularization of applications. The system's total energy consumption has an increasing influence on the environment. Reducing energy expenditure helps to reduce the carbon footprint and has a positive impact on environmental protection. In conclusion, the energy consumption of cloud-edge-end collaboration systems is critically important, and there is still ample space for research. However, many challenges remain. There are still enormous challenges for energy consumption in cloud-edge-end collaboration systems, with three main challenges: limited resources, flawed resource management, and limited scenarios.

- **Challenge in Limited Resources**

In a cloud-edge-end collaboration system, many end devices and edge nodes often have a certain amount of computational power capable of handling a portion of the task. However, due to the lightweight nature of many end nodes, they have limited computational power and energy resources. It is difficult for these IoT nodes to perform large or complex data tasks independently [208]. In addition, although edge servers have more computational resources than end devices, the limitations of edge smart device capabilities and the relatively weak computational processes and storage capabilities of edge servers [364] make them still resource-constrained. Meanwhile, the limited resources will significantly impact the cloud-edge-end collaboration system, which may cause significant delays or even cause it to fail to accomplish specific tasks. For example, task offloading may not be able to execute all offloaded tasks due to constrained computational and communication resources [468]. Therefore, decreasing energy consumption in resource-limited equipment and environments is essential, and reducing energy expenditure in resource-limited environments becomes a new challenge.

- **Challenge in Resource Management**

The cloud-edge-end collaboration system comprises three layers: end devices, edge servers, and cloud servers. Therefore, how to synergize the communication and computational resources between different layers cannot be ignored. Task offloading is a resource coordination technique in edge computing that enables resource-limited devices to offload tasks to edge servers and cloud servers for execution [401]. This aims to reduce the computation load on end devices, minimize the latency of task completion, improve energy efficiency, and enhance user experience. Task offloading will inevitably result in energy loss, and balancing energy consumption and other metrics poses a significant challenge. For instance, the trade-off between offloading delay and energy consumption is challenging [436]. There is also a temptation to ignore security and privacy while reducing energy consumption. In addition, since edge devices are usually distributed over a wide range of geographic locations, it is also challenging to manage and optimize energy consumption effectively.

- **Challenge in Limited Scenarios**

Cloud-edge-end collaboration systems are evolving faster and faster, and many studies target energy consumption. They may involve different application scenarios or handle different computational tasks, which brings great convenience to people's lives. However, current research on energy consumption has limitations. Existing studies are often conducted for some specific scenarios, while in others, the studies could be more extensive and sufficient [380]. In addition, some studies simplify the modeling in cloud-edge-end collaborative systems [200], but in real-world environments, it tends to be more complex [356]. Therefore, the current research still needs to be more idealized. Better application of optimized energy consumption in more practical scenarios is still a significant challenge.

3.1.3 The Motivation of Security and Privacy Optimization

The cloud-edge-end system is an emerging computational paradigm that integrates cloud servers, edge devices, and end equipment, ushering in a new era of flexible and efficient computing. With the rapid advancement of information technology, the convergence of cloud servers, edge devices, and end equipment has become a key driver of digital transformation. Cloud-edge-end systems provide more flexible and efficient solutions for data processing and computation and inject new vitality into the innovation and development of various industries, so more and more scenarios begin to use cloud-edge-end orchestrated systems, such as Power Internet of Things (PIoT) [492], Artificial Intelligence of Things (AIoT) [241], smart grid [105], among others. However, as user privacy and system security have received more attention, privacy and security issues in cloud-edge-end systems have been gradually exposed. There are still many challenges in solving and improving the privacy and security issues in cloud-edge-end systems.

Security and privacy of cloud-edge-end orchestrated systems are critical. The cloud-edge-end orchestrated system consists of three parts. The end layer is always composed of a large number of sensors, actuators, or even smartphones, computers, and similar devices, which are responsible for generating a massive volume of data and information, saving a certain quantity of data, and completing some simple calculations; the edge layer is typically consisted of edge servers, base stations, and others, which are responsible for storing data and carrying out a large number of calculations; the cloud layer is accountable for providing large-scale computing and storage resources. It is evident that each hierarchical structure retains a certain amount of data, making effective data management crucial to ensuring the security and privacy of data on devices. If there is a data management problem, it will be easy to cause a data breach. Furthermore, cloud-edge-end systems collaboratively perform computational tasks, which means there is a constant flow of information between the cloud, edge, and end layers. At the same time, data exchange or transfer can occur among devices within the same layer [242]. For example, task offloading, information search, resource allocation, and others require information exchange or sharing. It is worth noting that the data transmission process may be affected by attacks or channel quality. In addition, data transmission often involves senders and receivers, and the integrity and correctness of the data between them cannot be ignored. This involves the privacy of the systems and may impact the system's performance. Meanwhile, it may be challenging to recognize tampered or forged data at the edge side or cloud side [109]. Therefore, ensuring the security and privacy of data transfer and device-to-device sharing is particularly significant. Moreover, as technology undergoes rapid development, the means of cyberattacks have evolved and increased. Attackers employ sophisticated and advanced techniques, such as SQL injection attacks, false data injection attacks, and others, to meet their different objectives, including obtaining sensitive information, disrupting services, and extortion. Although some progress has been made in the field of security and privacy protection in cloud-edge-end orchestrated systems, which the privacy protection of cloud-edge-end systems mainly includes data privacy, key negotiation, and identity verification [105], there are still things that could be improved in security and privacy protection. The data of the cloud-edge-end systems often come from the end layer. However, due to all the reasons mentioned above, the users of the end layer are not willing to share their data, which impedes the progress of the cloud-edge-end systems and makes it easier to produce the phenomenon of "data island" [243]. To sum up, the security and privacy of cloud-edge-end orchestrated systems are essential, and there is still much room for research, but there are still many challenges.

There are still enormous challenges to the security and privacy of cloud-edge-end orchestrated systems, and there are four main challenges: malicious behavior and security and privacy protection, trust issues, encryption cost and performance, and incentive mechanisms.

- **Challenge in Malicious Behavior and Security and Privacy Protection**

As mentioned above, there is a wide variety of malicious behaviors at this stage, yet security and privacy protection technologies remain deficient. In reality, there can be malicious devices in the cloud-edge-end system, which leads to data leakage or tampering that can lead to severe consequences [497]. Furthermore, suppose the cloud-edge-end system is contaminated or bugged by an attacker. In that case, it can lead to bad decisions due to false information or even many confidential technologies being compromised [109]. However, many technologies available at this stage do not compensate for these shortcomings well. For example, federated learning techniques can protect privacy to a certain extent but cannot provide comprehensive privacy protection [198]. In addition, recognizing malicious devices in some specific tasks requires manual operation and human judgment [398]. Therefore, there is still much space for improvement in security and privacy technologies.

- **Challenge in Trust Issues**

In cloud-edge-end systems, the issue of trust is an essential factor that affects system security and privacy. The issue of trust is not only related to trust between devices but also to the environment. Firstly, nonconfidence between devices hinders cloud-edge-end systems' security, reliability, and efficiency [294]. However, existing methods need to solve the problem better. For example, the existing approach fails to address the trust issues arising from constructing a black box model using conventional methods [478]. Moreover, cloud servers and edge devices are often viewed as semi-trusted entities, possibly because of internal malfunctions or external security breaches [53]. Secondly, untrustworthy environments can quickly raise security and privacy concerns, such as offloading application tasks to servers in unreliable environments, posing security risks for mobile users [410]. Hence, ensuring trust between devices and that cloud-edge-end systems operate in a trusted environment remains challenging.

- **Challenge in Encryption Cost and Performance**

Encryption has recently garnered significant focus as the demand for security and privacy continues to rise, making it a prevalent method to enhance system security and safeguard individual privacy. Encryption is often used to encrypt sensitive data in cloud-edge-end systems to guarantee the security and privacy of transmission and data and avoid illegal access to data. However, while encryption provides a certain level of security and privacy, it sacrifices a sizeable computational cost and the ability to process data [105, 455]. Consequently, how to balance the limited performance and the computational cost of cryptography is also a question worth exploring.

- **Challenge in Incentive Mechanisms**

In recent years, security and privacy have been protected to a certain extent, and data has become increasingly available. However, given the potential implications for data privacy and security, data owners may hesitate to share their information and hand over their data to third parties [105, 243]. The above situation makes the system prone to a "data island", which can hamper the optimization of the system. In addition, in cloud-edge-end systems, multiple devices often collaborate to accomplish a particular computational task, but some users are reluctant to consume their resources. For example, advances in federated learning are constrained by users' unwillingness to expend the resources to upload models [243]. Most of the above situations occur because of the absence of effective and rational incentive mechanisms. Only by developing appropriate incentive mechanisms can users or devices better participate in collaborating with cloud-edge-end systems. Therefore, developing incentives is also a significant challenge in cloud-edge-end systems.

3.1.4 The Motivation of Reliability Optimization

The cloud-edge-end system is an architecture that integrates the cloud servers, edge servers, and end devices to optimize data processing, storage, and service delivery. Computation, storage, and services are as close to data sources and consumers as possible to provide faster, more efficient, and responsive data processing and services. It provides the advantages of low latency, data shredding, privacy and security, flexibility, and scalability, which are suitable for a variety of application scenarios. The cloud-edge-end system can improve system performance and users' quality of experience.

The cloud-edge-end system consists of three parts:

- The cloud computing center provides highly scalable computing and storage resources where large-scale data processing and storage operations are performed efficiently.

- Edge devices and servers are closer to data sources. It can provide low-latency data processing. Execution tasks at edge servers can reduce the delay in data transmission to the cloud.

- User devices, including smartphones, sensors, and so on, can collect data and transmit it to the cloud or edge servers for data processing. They also receive the processing results from the cloud or edge servers and display or use them.

In the cloud-edge-end system, the ability of the system to continue operating and provide stable services is critical. The reliability of the cloud-edge-end system has the following aspects:

- **Sustainability:** Regardless of failure caused by network failure, hardware failure, or software failure, the system can quickly recover and continue to provide services.

- **Availability:** Even under particular circumstances, such as high load, network congestion, and equipment failure, the system can still provide available services when users need them.

- **Security:** The system can protect user data's integrity, confidentiality, and availability, ensuring the security of data and systems.

Reliability is a goal that needs to be considered and pursued during the design and implementation of cloud-edge-end systems to ensure they can always remain efficient and reliable in the face of various challenges and uncertainties. Currently, the problems faced by cloud-edge-end systems include:

- **Network latency and instability:** Since data transmission between edge devices and the cloud may pass through multiple network nodes, network latency and fluctuations may lead to uncertainty in data transmission and service instability.

- **Single point of failure:** A single edge node or edge server may cause service interruption or performance degradation of the entire system.

- **System scalability and resource management:** As business needs grow, data transmission volume is also growing explosively. If the system cannot effectively expand and manage resources, it may lead to system performance degradation, system overload, resource waste, and other problems.

- **Complexity and integration issues:** Cloud-edge-end systems usually comprise multiple components and technologies, including cloud computing centers, edge computing, network communications, etc. These components' complexity and integration issues can lead to configuration errors, compatibility issues, and system failures. Therefore, proper system design, testing, and verification are required to ensure coordination and stability among the various components.

To improve the reliability of cloud-edge-end systems, these issues need to be fully considered and resolved during system design, implementation, operation, and maintenance.

3.1.5 Conclusions and Discussions

This chapter summarizes the challenges cloud-edge-end technology faces in terms of latency, energy consumption, security, and reliability based on a review of relevant literature. Regarding latency, challenges include task offloading delay and low-latency data collection from heterogeneous devices. Energy consumption is a challenge as end devices typically possess computational capabilities. Additionally, security challenges involve defending against malicious behavior and ensuring data encryption, which is crucial for protecting user privacy in practical applications of cloud-edge-end technology. Lastly, reliability is challenging due to growing demands on data processing and services. Issues such as network latency-induced instability and the complexity arising from many heterogeneous devices pose obstacles to better developing cloud-edge-device technology. Effectively utilizing end devices in resource-constrained and complex scenarios is also a significant challenge.

3.2 Performance Modeling for Cloud-Edge-End Systems

Cloud-edge-end orchestrated systems usually involve a combination of cloud computational resources, edge computational nodes, and end devices, each with its role in processing and delivering content and services. Performance modeling of cloud-edge-end systems is an important research area that facilitates managing and coordinating computational resources of the cloud layer, edge layer, and end devices by modeling the performance of cloud-edge-end orchestrated systems, including latency, energy consumption, security and privacy, and reliability. Hence, it is critical to optimize the performance of the system.

In cloud-edge-end systems, performance is modeled in various ways and may be modeled differently in different scenarios. These performance models provide more intuitive and feasible optimization goals for the system. In the subsequent discussion, several different performance models will be explored.

3.2.1 Latency of Cloud-Edge-End Systems

Cloud-edge-end computing presents a hopeful solution for computation-intensive and time-sensitive tasks. To address these challenges, it leverages the abundant computing resources of cloud data centers and the low-latency services delivered by edge computing servers. In a cloud-edge-end environment, servers and networks exhibit more significant heterogeneity, which makes traditional solutions for cloud computing and mobile edge computing (MEC) inadequate. This heterogeneity results in differences in compute speeds and access latency among servers. Considering the distinct characteristics of fifth-generation (5G) heterogeneous networks within the cloud-edge-end environment, task offloading with latency optimization remains an area that receives relatively little attention. Thus, Xiao et al. [417] focused on the problem of collaborative task offloading in MEC-enabled small-cell networks within the cloud-edge-end computing paradigm.

- **Delay Computing Models for Processing Tasks on Local, ES, and CDC in Cloud Edge Architecture**

Xiao et al. [417] first established a wireless communication model. They assumed that regardless of whether user devices (UDs) choose to execute tasks on edge servers (ES) or the cloud data center (CDC), the tasks must be uploaded to their associated small cells first.

They defined two parameters: $s_{mn\iota}^{\text{SC}}$ and $p_{mn\iota}$, $\iota \in \{1,2\}$. $\iota = 1$ means m-th UD offloading its task to ES n, while $\iota = 2$ means m-th UD offloading its task to CDC. Note that because only one CDC, when $\iota = 2$ satisfies, n equals 0. $s_{mn\iota}^{\text{SC}}$ represents the proportion of bandwidth allocated for m-th UD transmitted to n-th small cell. $p_{mn\iota}$ represents the fraction of power assigned for the same transmission. The uploaded rate achievable by m-th UD connected to n-th small cell is

$$r_{mn\iota}^{\text{SC}} = s_{mn\iota}^{\text{SC}} B \log_2 \left(1 + \frac{p_{mn\iota} g_{mn}}{s_{mn\iota}^{\text{SC}} B N_0} \right). \tag{3.1}$$

Here, the total bandwidth of the radio spectrum is denoted by B, N_0 represents the spectral density of radio noise, and the channel gain between m-th UD and n-th small cell is denoted by $g_{mn} = |g_0|^2 d_{mn}^{-\nu}$. In the formula, g_0 represents the Rayleigh fading channel coefficient. The distance between m-th UD and n-th small cell is d_{mn}. ν represents the path loss exponent. Because UDs have limited mobility during the brief offloading period, they assumed that the value of g_{mn} remains constant.

During the offloading process of UD tasks to the CDC and subsequent forwarding to the gateway, the variable s_{mn}^{GW} is assigned to represent the proportion of bandwidth allocated to the n-th small cell for delivering the m-th UD's tasks to the gateway. The rate at which data forwarding from the n-th small cell to the gateway for m-th UD is

$$r_{mn}^{\text{GW}} = s_{mn}^{\text{GW}} B \log_2 \left(1 + \frac{P_n \bar{g}_n}{s_{mn}^{\text{GW}} B N_0} \right). \tag{3.2}$$

Here, $\bar{g}_n$ is the channel gain between n-th small cell to the gateway. The transmit power of n-th small cell is expressed by P_n.

Based on the wireless communication model, they proposed the latency and energy consumption models for executing tasks on CDC, ES, and UD. This section mainly introduces the calculation equations related to the latency model. The task executed by m-th UD can be characterized by the tuple (D_m, F_m, T_m). D_m symbolizes task input size, F_m is the number of CPU cycles required to complete the task, and T_m is maximum tolerable latency. The local execution latency for m-th UD in n-th small cell is

$$T_{mn}^l = \frac{F_m}{f_m^l}. \tag{3.3}$$

Here, f_m^l represents the frequency of the m-th UD.

Considering the uploaded rate model described in Equation (3.1), the time for m-th UD to upload its task to n-th small cell is

$$T_{mn\iota}^{SCt} = \frac{D_m}{r_{mn\iota}^{SC}}, \tag{3.4}$$

where D_m symbolizes the input data size of the task of m-th UD. Let f_m^e represent the CPU frequency allocated to m-th UD by n-th small cell. The execution latency of the related task on n-th small cell is

$$T_{mn}^{ec} = \frac{F_m}{f_{mn}^e}. \tag{3.5}$$

Consequently, the response latency for m-th UD when its task is executed on the respective ES is

$$T_{mn}^e = T_{mn1}^{SCt} + T_{mn}^{ec}. \tag{3.6}$$

As stated in Equation (3.2), the transmission latency for the task of m-th UD from n-th small cell to the gateway can be performed as

$$T_{mn}^{GWt} = \begin{cases} 0, & j = 0 \\ \frac{D_m}{r_{mn}^{GW}}, & \text{otherwise} \end{cases} . \tag{3.7}$$

Considering both the transmission latency and propagation latency, the combined effects are considered during the upstream transmission from the gateway to the CDC. Hence, by representing f_{mn}^c as frequency allocated to m-th UD by the CDC, the response latency for the task of the m-th UD to be executed at CDC is

$$T_{mn}^c = T_{mn2}^{SCt} + T_{mn}^{GWt} + \frac{D_m}{C} + \chi + \frac{F_m}{f_{mn}^c}. \tag{3.8}$$

Here, χ represents the propagation delay, while C represents the bandwidth of the fiber network.

- **Calculation Model for Task Offloading and Transmission Delay of Data Flow in Devices, CAP, and Cloud Servers**

  Edge computing exhibits low latency and high bandwidth characteristics, making it well-suited for handling multi-source data streams produced by IoT devices. Wu et al. [412] suggested a three-tier collaborative cloud-edge-end architecture encompassing the cloud, edge, and IoT devices. This architecture facilitates efficient cloud-edge-end computing by integrating M IoT devices, N computing evaluation points, and the centralized cloud server with robust computing capabilities.

  Initially, they developed a transmission model in which wireless upstream channels support the flow of offloaded data from the m-th ($1 \le m \le M$) device to the n-th ($1 \le n \le N$) CAP and subsequently from the n-th CAP towards the cloud server. They assumed these channels are independent and identically distributed, representing them in their model as Rayleigh channels. $h_{m,n}$ and $g_{m,n}$ represent the channel gain between the m-th device and the n-th CAP, as well as between the n-th CAP and the cloud, individually. Based on Shannon's theory, denoting p_m^{trans} as the transmission power of the m-th device and p_n^{trans} as the transmission power of the n-th CAP, the corresponding transmission rates can be calculated as

$$r_{m,n}^{CAP} = w_{m,n}^{CAP} \log_2 \left(1 + \frac{p_m^{trans} |h_{m,n}|^2}{\sigma^2} \right), \tag{3.9}$$

$$r_{m,n}^c = w_{m,n}^c \log_2 \left(1 + \frac{p_n^{\text{trans}} |g_{m,n}|^2}{\sigma^2} \right). \tag{3.10}$$

Here, σ^2 signifies the variance of additive white Gaussian noise.

Let $w_{m,n}^{CAP}$ and $w_{m,n}^c$ represent the bandwidth allocated to the m-th device for wireless from the n-th CAP and the cloud, respectively. They must satisfy the following constraint

$$\sum_{m=1}^{M} w_{m,n}^{CAP} \leq W_n^{CAP}, \tag{3.11}$$

$$\sum_{m=1}^{M} w_{m,n}^c \leq W^c. \tag{3.12}$$

Here, W_n^{CAP} and W^c represent the overall bandwidth owned by the n-th CAP and the cloud server, respectively.

Based on the wireless transmission model, they proposed a computational model. Let α_m denote the proportion of the data stream transferred from the m-th device to CAP, and β_m represents the proportion further uploaded from CAP to the cloud. Here, it is trivial to agree that $\alpha_m \in [0, 1]$ and $\beta_m \in [0, 1]$. For the m-th device, the size of the data stream l_m it is assigned to for local, CAP and cloud computing can be expressed as

$$l_m^{\text{local}} = l_m (1 - \alpha_m), \tag{3.13}$$

$$l_{m,n}^{CAP} = l_m \alpha_m (1 - \beta_m), \tag{3.14}$$

$$l_{m,n}^c = l_m \alpha_m \beta_m. \tag{3.15}$$

Consequently, for data stream l_m, the computing delay for the local device, CAP, and cloud are given by

$$T_m^{\text{local}} = \frac{l_m^{\text{local}} \omega \kappa}{f_m^l}, \tag{3.16}$$

$$T_{m,n}^{CAP} = \frac{l_{m,n}^{CAP} \omega \kappa}{f_{m,n}^{CAP}}, \tag{3.17}$$

$$T_{m,n}^c = \frac{l_{m,n}^{\text{local}} \omega \kappa}{f_m^c}. \tag{3.18}$$

Here, ω represents the processing density, measured by the CPU cycles required to process each bit of the data stream. κ denotes the coefficient factor per unit scale from Mbps to bits, $f_{m,n}^{CAP}$ and f_m^c respectively represent the CPU frequencies allocated to l_m by the n-th CAP and the cloud, while f_m^l represents computation capacity of the m-th device.

Based on Equation (3.9) and Equation (3.10), the transmission delay between the m-th device and the n-th CAP and between the n-th CAP and the cloud can be separately expressed as:

$$T_{m,n}^{\text{trans},CAP} = \frac{l_m \alpha_m \kappa}{r_{m,n}^{CAP}}, \tag{3.19}$$

$$T_{m,n}^{\text{trans},c} = \frac{l_m \alpha_m \beta_m \kappa}{r_{m,n}^c}. \tag{3.20}$$

Given the assumption that each CAP is equipped with both a transmission unit and a computation unit that can function concurrently, we neglect the return delay owing to the small size of the resulting feedback. It is reasonable to presume that data flow l_m can be managed and transmitted in parallel, resulting in the processing delay of l_m calculated as:

$$T_m^{\text{total}} = \max\left\{T_m^{\text{local}}, T_{m,n}^{\text{trans, CAP}} + T_{m,n}^{\text{CAP}}, T_{m,n}^{\text{trans, CAP}} + T_{m,n}^{\text{trans, c}} + T_{m,n}^{\text{c}}\right\}. \tag{3.21}$$

Considering the scenario where either the CAP or the cloud server generates a virtual machine for each device, enabling the concurrent execution of all data flows from M devices, the aggregate of processing delay for all devices can be computed in the following manner:

$$T^{\text{total}} = \max_{m \in \mathcal{M}} T_m^{\text{total}}. \tag{3.22}$$

- **A Communication and Task Execution Delay Model for Heterogeneous Server Base Stations**

The response time optimization can be achieved by deploying edge servers in a mobile edge-cloud computing architecture. However, existing methods may severely reduce the service quality offered to mobile users by neglecting two crucial aspects: cloud servers/the heterogeneity of edge and fairness in the base station (BS) response time. Cao et al. [37] focused on deploying heterogeneous edge servers to optimize the overall and individual BS's expected response time. They proposed a response time model. For communication delay, they required that each BS transmit the tasks of mobile users to the distributed server (i.e., edge/cloud server), which then performs these tasks. In a given scenario, multiple tasks from mobile users are received by each BS B_m. Supposing these tasks obey a Poisson process, the average arrival rate γ_m (CPU cycles/s) denotes the rate at which the task reaches each BS B_m. Additionally, it is assumed that a link exists between BS B_m and distributed server S_n, with a transmission capacity denoted as ϑ_{mn}.

Let $d(B_m, S_n)$ represent a non-negative variable denoting the Euclidean distance from BS B_m to the distributed server S_n. This model follows the approach described in reference [188]. Tasks' average delay of communication transmitted from BS B_m toward the distributed server S_n can be calculated as follows

$$T_{m,n}^{\text{tr}} = \frac{d(B_m, S_n)}{\varrho} + \frac{D_m}{\vartheta_{m,n}}. \tag{3.23}$$

Here, ϱ is a constant representing the propagation rate of electromagnetic waves, while D_m is a nonnegative variable denoting the average input data size for the tasks generated by BS B_m.

To model the task execution delay during distributed server S_n task processing, they utilize the well-known M/G/1 queue model [117]. Incoming external tasks are queued and wait for execution at a rate of p_n (CPU cycles/s). The task execution delay on the distributed server S_n within the M/G/1 queueing framework can adhere to a diverse range of usual distribution, where variance $\alpha_n^2 > 0$ and the mean is denoted as $\beta_n > 0$.

$\mathcal{B}_n$ represents the collection of BSs associated with the distributed server S_n when it is attached to multiple BSs. According to [117], the average task execution delay, which includes accounting for the time to wait at the task queue, for the BS $\mathcal{B}_n$ on the distributed server S_n can be represented as

$$T_{m,n}^{\text{ex}} = \frac{\gamma_m}{p_n} + \frac{(\alpha_n^2 + \beta_n^2)(\gamma_m + \Phi_n)}{2(p_n - \gamma_m + \Phi_n)}. \tag{3.24}$$

Here, $\Phi_n = \sum_{B_k \in \{\mathcal{B}_n - \{B_m\}\}} \gamma_k$.

A BS's anticipated response delay comprises both task execution delay and communication delay. By considering Equation (3.23) and Equation (3.24), BS B_m's anticipated response delay can be illustrated as follows

$$
\begin{aligned}
T_m^{\text{total}} = \sum_{j=1}^{M} \sum_{n=0}^{N} \Bigg(& \mu_{j,n,m} \left(\frac{d\left(B_m, S_n\right)}{\varrho} + \frac{D_m}{\vartheta_{m,n}} \right) \\
& + \mu_{j,n,m} \left(\frac{\gamma_m}{p_n} + \frac{\left(\alpha_n^2 + \beta_n^2\right)\left(\gamma_m + \Phi_n\right)}{2\left(p_n - \gamma_m + \Phi_n\right)} \right) \Bigg),
\end{aligned}
\tag{3.25}
$$

where $\mu_{j,n,m}$ denotes a binary indicator. $\mu_{j,n,m} = 1$ refers to the following conditions: (1) S_n serves as an edge server colocated with BS B_j that belongs to the set B_n and BS B_m is associated with S_n, or (2) S_n functions as a cloud server, and BS B_m is connected to S_n with a fixed value of j equal to 1. Otherwise, $\mu_{j,n,m} = 0$.

The system's anticipated response delay can be calculated as the average response time of M BSs, which can be formulated as

$$
T^{\text{avg}} = \frac{1}{M} \sum_{m=1}^{M} T_m^{\text{total}}.
\tag{3.26}
$$

3.2.2 Energy of Cloud-Edge-End Systems

In this section, we introduce the energy consumption model of the cloud-edge-end system. The energy consumption model can be divided into the global energy consumption model and the end energy consumption model.

- **Global Energy Consumption Model**

  The primary energy-saving measure is to reduce the energy consumed. The energy expenditure accounted for here primarily encompasses the energy required for data transmission and computational processes. However, energy utilized during task idle periods awaiting execution is not considered. The primary reference for the global energy consumption model section is [364]. The following will calculate the energy used by tasks at different computational stages and throughout the entire system.

  Local Computing: The local computational model alone takes into account the energy consumed during the calculation of a given task. Let d_i and c_i denote the data size and the number of CPU cycles needed to execute unit data of task i. f_u represents the CPU frequency associated with user equipment (UE) u. The energy consumption of UE u executing task i is as follow[364]:

$$
E_i = \epsilon d_i c_i \left(f_u\right)^2,
\tag{3.27}
$$

where ϵ symbolizes the efficient switching capacitance within the core.

  Edge Computing: When the task is offloaded to the MEC server for execution, the energy consumption includes the transmission energy for the device and the execution energy for the server. Transmission energy is related to the rate of transmission. The rate of transmission from UE u toward BS n can be calculated as follows [358, 260]:

$$
r_{u,n} = W \log_2 \left(1 + \frac{p_u g_{u,n}}{N_0 W} \right),
\tag{3.28}
$$

In this context, W represents the communication bandwidth, and $g_{u,n}$ represents the channel gain. Moreover, p_u refers to the transmission power of UE u. N_0 is the BS's noise power spectral density. The energy consumption for task i executing on server m at BS n is given by

$$E_i = p_u \frac{d_i}{r_{u,n}} + d_i q_{n_m}, \tag{3.29}$$

where q_{n_m} indicates the energy used by server m at BS n to compute a single bit of task data.

Cloud Computing: If the task is offloaded to the cloud for execution, the energy consumption encompasses three main components: input data transmission energy, execution energy, and result transmission energy. The energy consumption for offloading task i from UE u to the cloud server is:

$$E_i = p_u \frac{d_i}{r_{u,n}} + p_n \frac{d_i}{r_{n,c}} + d_i q_c + p_n \frac{d_{i,r}}{r_{n,c}}. \tag{3.30}$$

The variable p_n represents the transmission power between the cloud computing center and base station n in this scenario. On the other hand, the variable q_c indicates the energy needed for the cloud server to compute a single piece of data. Based on the above formula, the evaluation indicator for the system's energy usage is based on the aggregate energy used by all tasks, and this metric is established in the following manner:

$$E_{\text{total}} = \sum_{i=1}^{n} E_i. \tag{3.31}$$

- **End Energy Consumption Model**

The computing and communication models are crucial for determining the energy consumption and time delay. Next, the communication and computing models in task offloading are described in detail concerning [244].

Communication Model: The wireless access communication model is presented. Each mobile device is equipped to transmit requests to an edge node, which can offer service and network coverage. Every edge node is equipped with a distinct base station responsible for receiving transmission requests. The decision to offload the computational task $T_{i,s}$ from mobile device i is represented as $x_{i,s} \in \{0, 1, \ldots, n, \ldots, N, N+1\}$, where $i \in \boldsymbol{m-th}$ $(1 \leq m \leq M)$, $s \in \boldsymbol{L}$. The offloading decision vector X for all devices' computing tasks is recognized as $\boldsymbol{X} = \{x_{1,1}, x_{1,s}, \ldots, x_{1,l}, x_{2,1}, \ldots, x_{i,s}, \ldots, x_{m,l}\}$. The uplink rate at which a task is offloaded over the wireless channel to the edge node is determined using the Shannon formula [231].

Suppose the processing of the computational task $T_{i,s}$ can either take place at the edge node v or it can be sent to the cloud for processing; the transmission rate for computing task t can be expressed as follows:

$$r_{i,s} = B \log_2 \left(1 + \frac{p_i g_{i,v}}{\omega_0}\right), \tag{3.32}$$

where B and p_i denote the bandwidth and transmission power of the wireless channel used for transmission by mobile device i, respectively. $g_{i,v}$ is the channel gain between edge node v and mobile device i, while ω_0 stands for the background noise. Besides, $g_{i,v}$ is defined as $R_{i,v}^{-\beta}$, where $R_{i,v}$ denotes the distance separating mobile device i from edge node v, and β represents the exponent of path loss.

Computing Model: This paragraph tells the computational model utilized for the computational problem. Several mobile devices (MD), each with various computing tasks, are considered in this model. Every task can be offloaded to a covering edge node for execution on the device or transferred to a cloud data center. Mobile device i assigns and submits the computing task $T_{i,s} = \{S_{i,s}, C_{i,s}, D_{i,s}\}$, and the collection of tasks is denoted as $T_i = \{T_{i,1}, T_{i,2}, \ldots, T_{i,l}\}$. In this context, $S_{i,s}$ denotes the input data size associated with computing task $T_{i,s}$. $C_{i,s}$ denotes the overall count of CPU cycles necessary to finish the computing task $T_{i,s}$, while $D_{i,s}$ denotes the latest deadline for the completion of computing task $T_{i,s}$. Subsequently, the methods for calculating computing tasks' energy usage and time delay are discussed under a cloud-edge-end orchestrated environment.

(1) Local Computing: In the scenario where mobile devices process computing tasks locally, the power usage for local computing offered by mobile devices varies for each mobile device. Then, the energy expenditure for processing computing task $T_{i,s}$ locally is as follows [244]:

$$e_{i,s}^L = C_{i,s} \times e_i, \tag{3.33}$$

where e_i is the energy expenditure for each CPU cycle on mobile device i.

(2) Edge Computing: Edge computing requires the mobile device to establish wireless communication with the host edge node, which assumes control of the computational operations previously performed by the mobile device. Consequently, The factors to be considered include the delay in transmission and energy expenditure for computing tasks, along with the delay in execution and energy usage for tasks processed at the edge nodes. Based on the size of the input data associated with the computing task $T_{i,s}$ and the transfer rate in the communication model, the delay in data transmission can be characterized as follows:

$$t_{i,s,tran}^E = \frac{S_{i,s}}{r_{i,s}}. \tag{3.34}$$

Simultaneously, the energy consumption associated with data transmission can be expressed as

$$e_{i,s,tran}^E = p_i \times t_{i,s,tran}^E. \tag{3.35}$$

Furthermore, different edge nodes are considered to have varying total computing capacities. The computing capacities designated by edge node v for processing the computing task $T_{i,s}$ are denoted as $f_v^{i,s}$. In contrast, the overall computing capacity offered by edge node v is represented as F_v, measured in the CPU frequency. The delay in executing task $T_{i,s}$ is as follows:

$$t_{i,s,exec}^E = \frac{C_{i,s}}{f_v^{i,s}}. \tag{3.36}$$

Subsequently, as the edge node performs computational work instead of the mobile device, the mobile device experiences energy consumption during periods of inactivity. Therefore, the energy expenditure during execution is derived as

$$e_{i,s,exec}^E = p_i^{idle} \times t_{i,s,exec}^E, \tag{3.37}$$

where p_i^{idle} indicates the idle energy consumption of mobile device i.

(3) Cloud Computing: When a mobile device transfers its computing work to the cloud for processing, it first transmits its computing tasks to the edge nodes wirelessly. Subsequently, the edge node sends the task to the cloud server to be executed using a physical circuit. Consequently, when evaluating the offloaded computing work transmitted to the edge node, the delay in data transmission and energy consumption associated with the input data must be considered. Additionally, it is significant to consider the delay in transmission. Suppose W_{EC} represents the bandwidth of the wired connection linking the edge server to the cloud center. When the computing task $T_{i,s}$ is transferred to the cloud, the latency can be expressed as follows:

$$t_{i,s,tran}^C = \frac{S_{i,s}}{r_{i,s}} + \frac{S_{i,s}}{W_{EC}}. \tag{3.38}$$

The energy required for data to be sent from the edge to the cloud layer is calculated by multiplying the time taken to transmit by the energy consumed when the device is idle. Consequently, the energy used for transmission is described as follows:

$$e_{i,s,tran}^C = p_i \times \frac{S_{i,s}}{r_{i,s}} + p_i^{idle} \times \frac{S_{i,s}}{W_{EC}}. \tag{3.39}$$

In addition, the computing capability from the cloud center available for the task $T_{i,s}$ is denoted by q_c. It is established that $N_{i,s}$ indicates the overall number of CPU cycles needed to execute the computational task $T_{i,s}$. Thus, the time that is needed to execute the computational task $T_{i,s}$ within the cloud center C is specified as

$$t_{i,s,exec}^C = \frac{N_{i,s}}{q_c}. \tag{3.40}$$

Under similar circumstances, mobile devices demonstrate idle energy consumption while computational tasks are performed at edge nodes. Therefore, the energy usage during execution is given by

$$e_{i,s,exec}^C = p_i^{idle} \times t_{i,s,exec}^C. \tag{3.41}$$

Following the approach in [400], it is observed that the volume of output data following the processing of a computational task is considerably lower than the input data. Therefore, the delay in downlink transmission is overlooked when the completed offloading computing task results are returned to the devices. In accordance with the offloading decision for the computing task and the calculation above formula, the energy usage for the computing task $T_{i,s}$ is formulated as follows:

$$e_{i,s} = \begin{cases} e_{i,s}^L, & \text{if } x_{i,s} = 0, \\ e_{i,s,tran}^E + e_{i,s,exec}^E, & \text{if } x_{i,s} = 1, 2, \ldots, N. \\ e_{i,s,tran}^C + e_{i,s,exec}^C, & \text{if } x_{i,s} = N + 1 \end{cases} \tag{3.42}$$

The aggregate energy consumption of a mobile device is determined by the cumulative sum of its various computing operations.

3.2.3 Security and Privacy of Cloud-Edge-End Systems

This section will introduce the security privacy model of cloud-edge-end systems from the aspects of trustworthiness evaluation, differential privacy, threat model, security score of resource deployment, and security definition of cross-domain data sharing.

- **Trustworthiness Evaluation**

In the cloud-edge-end orchestrated vehicular network system, formulating a resource allocation scheme requires methods for evaluating the security of the project. The trustworthiness of a node is an essential metric for assessing its security. A higher trustworthiness of a node typically implies that it is more secure and reliable. Thus, a method of evaluating the node's trustworthiness is proposed by [497]. Node trustworthiness is assessed using the nodes' static security and activity level. The situation varies across different nodes, each with distinct behaviors and security levels. Generally, the more active node is considered more trustworthy and secure. In addition, the static security of a node plays a crucial role in determining the trustworthiness of nodes. Static security is determined by two factors: the security level of the individual node and its adjacent nodes. The static security S_1 is expressed as [497]

$$S_1\left(N_m^P\right) = \beta \times SL\left(N_m^P\right) + \gamma \times \frac{\sum_{a=1}^{|N^n|} SL\left(N_a^P\right)}{|N^n|}, \tag{3.43}$$

where the N_m^P denotes the m-th node in the physical network, and the function SL represents the security level of the node. Moreover, N^n represents the set of adjacent nodes of N_m^P. β and γ, respectively, represent the weights of the security level of the node itself and its adjacent nodes.

Afterward, the activity level is calculated by how often it engages in virtual network request embedding and its communication condition. Let S_2 indicate the activity level.

$$S_2\left(N_m^P\right) = \frac{\sum_{i=1}^{|N^P|} \tau\left(N_m^P, N_i^P\right)}{|N^P|} + \frac{fre\left(N_m^P\right)}{Succ\left(G^V\right)}, \tag{3.44}$$

where $fre\left(N_m^P\right)$ represents how often the node N_m^P engages in virtual network request embedding. $Succ\left(G^V\right)$ indicates the count of virtual network requests successfully embedded, and G^V is a virtual network request. Besides, $\tau\left(N_m^P, N_i^P\right)$ denotes whether there is communication interaction between N_m^P and N_i^P; use one if it exists; otherwise, use zero.

The node's trustworthiness is composed of static security and activity level. Let α denotes the weight; the equation of trustworthiness is given by

$$Tru\left(N_m^P\right) = \alpha \times S_1 + (1 - \alpha) \times S_2. \tag{3.45}$$

- **Differential Privacy**

Differential privacy is a cryptographic method for ensuring and proving the security of a cloud-edge-end system. Based on a robust data theory foundation, differential privacy technology can enhance quantitative analysis, proof, and so on while ensuring good scalability. Therefore, differential privacy technology is always used to quantify privacy, protect privacy, and defend against attacks. Let DS and DS$'$ represent two different data sets. If DS and DS$'$ are different in one entry, then DS, DS$'$ is referred to as neighboring data sets. An algorithm $\mathcal{F}$ with Ψ as its range and Θ as its domain is given. In the case of a subset of output $\psi \in \Psi$ and any two neighboring data sets $\theta, \theta' \in \Theta$, the inequality is given by [132]

$$\Pr\left[\mathcal{F}\left(\theta\right) \in \psi\right] \leq e^{\varepsilon} \times \Pr\left[\mathcal{F}\left(\theta'\right) \in \psi\right]. \tag{3.46}$$

It indicates that if the inequality holds, the algorithm $\mathcal{F}$ meets ε-differential privacy. ε represents the privacy budget of differential privacy, also called privacy loss. If the privacy budget ε is smaller, then the privacy protection level is higher, and the data availability is lower. Furthermore, another differential privacy is introduced for the practicality of the algorithm. The inequality can be given by [198, 138]

$$\Pr\left[\mathcal{F}\left(\theta\right) \in \psi\right] \leq e^{\varepsilon} \times \Pr\left[\mathcal{F}\left(\theta'\right) \in \psi\right] + \varrho. \tag{3.47}$$

If established, it demonstrates that the algorithm $\mathcal{F}$ can achieve (ε, ϱ)-differential privacy. ϱ indicates the relaxation factor and the smaller ϱ that the attacker is able to distinguish if the individual target record exists within the data set. Differential privacy technology is always utilized in the cloud-edge-end orchestrated system to quantify the privacy constraint using the value of ε. Besides, security is often demonstrated in systems that employ differential privacy techniques by proving that the system satisfies differential privacy.

- **Threat Model**

    In the cloud-edge-end orchestrated system, a vast amount of data is uploaded to the cloud server for storage. Additionally, the cloud-edge-end orchestrated system is more susceptible to attacks and security threats due to the substantial computing tasks executed on the edge nodes. In such an environment, searchable encryption schemes must satisfy keyword privacy, file privacy, and trapdoor security to ensure data security [105]. Establishing a security model for evaluation is required to assess the schemes' security and privacy. Consequently, initially assuming the cloud and edge nodes are considered semi-trusted, three privacy protection threat models in the cloud-edge-end orchestrated system will be put forward [105]. The first is known as a ciphertext attack, in which the attackers acquire encrypted data exchanged between the server and the data owner through communication interception. Based on this, the attackers produce extra retrieval requests by gathering valid search requests. The second is known as a background attack, in which the attackers use background information related to the dataset. Hence, the attackers are able to infer keywords by analyzing the background information and frequency of captured keywords. The final threat model is an adaptive chosen keyword attack, in which the attackers cannot deduce extra information from the service process except for the content previously known from the query outcomes or other sources. In the cloud-edge-end system, the goal is to ensure the security of the system and user data by establishing a threat model to identify and evaluate potential security threats.

- **Security Score of Resource Deployment**

    The security in resource allocation cannot be overlooked in the cloud-edge-end orchestrated system of the Industrial IoT. To deploy applications in the cloud-edge-end orchestrated systems and allocate application components reasonably to various computing resources while ensuring security, a security score is proposed by [42]. Let set $C = \{c_1, ..., c_n\}$ represent the Industrial IoT application specification, including n interconnected software components, and set $O = \{o_1, ..., o_m\}$ indicate available deployment resources consisting of m various computational resources, known as offerings. Specifically, offerings can be local edge devices on-premises or cloud-based IaaS instances. It is worth mentioning that the local area network is considered secure because it encompasses both edge devices and end devices. Security features are associated with each offering and component. Besides, assuming that common security controls represent the security attributes of an offering within a reference architecture. Next, security controls can be implemented at different levels, employing various mechanisms and technologies, thereby producing different effects. Let $SC = \{sc_1, ..., sc_S\}$ and $LEV = \{0, l_1, ..., l_L\}$ denote the set of S security controls under consideration and the set of L potential implementation levels, respectively. It is worth mentioning that LEV is a subset of $\mathbb{N}_0$.

A function denoted as $providedLevel : SC \times O \mapsto LEV$, associates offering and security control with the corresponding implementation offered by o_j for the security control sc_t. If $providedLevel(sc_t, o_j) = 0$, the security control sc is not implemented at all. Then, the comprehensive security profile of o_j is expressed as [42]

$$SecSLA(o_j) = \{(sc_t, providedLevel(sc_t, o_j)) : sc_t \in SC\}. \tag{3.48}$$

In addition, a component's security profile can be expressed as the minimum required implementation level for each control under consideration. Let $requiredLevel : SC \times C \mapsto LEV$ denote a function that calculates the minimal needed implementation level for c_i by mapping security control and components. If $requiredLevel(sc_t, c_i) = 0$, it represents that security control sc_t is unnecessary for c_i. The comprehensive security profile of c_i can be expressed as indicated below:

$$SecReq(c_i) = \{(sc_t, requiredLevel(sc_t, c_i)) : sc_t \in SC\}. \tag{3.49}$$

In terms of security, the security score is defined by considering the additional security that a common solution can provide, contingent upon fulfilling the security requirements. All considered security controls' implementation level is crucial in determining the additional security. Specifically, it depends on the variance between the required and offered level for each security control. The security score is given by

$$sec\_score(s) = \sum_{p_l \in P} \sum_{c_i \in p_l} \sum_{sc_t \in SC}$$
$$(providedLevel(sc_t, \lambda(p_l)) - requiredLevel(sc_t, c_i)), \tag{3.50}$$

where $P = \{p_1, ..., p_k\}$, a partition of C, is a potential solution to the allocation issue. Element p_l belongs to P, and p_l denotes a subset of components intended for deployment on the same resource. Furthermore, $\lambda : P \mapsto O$ is a function mapping p_l to a particular offering within the set O, and the pair $s = \langle P, \lambda \rangle$ denotes a generic solution for resolving the allocation challenge.

- **Security Definition of Cross-Domain Data Sharing**

  In cloud-edge-end orchestrated systems, data sharing is always necessary, and security is also an important consideration in data sharing. For cross-domain data sharing based on blockchain, the security definition is provided by [242], consisting of four parts: data sharing liveness, policy registration liveness, cross-domain no conflict, and intra-domain consistency. Assuming Π is a scheme for sharing data across domains. Consider $T_{PR}, T_{PT-CDDS}$, $T_{ZT-CDDS}$ as liveness parameters. If the above properties are satisfied, it can be stated that Π is secure. In addition, let C_e^I represent the I-th domain committee in epoch e, composed of numerous nodes represented by p. Each domain needs to undergo periodic reconfiguration. Therefore, the time interval between reconfigurations, starting from one reconfiguration to the next, is called an epoch for each domain. Because each domain implements a Byzantine Fault Tolerance (BFT) protocol, the block generated by the I-th domain in the r-th round is denoted as $B_r^I := (Hash(B_{r-1}), mr_{II} \| \cdots \| mr_{IN})$, where r refers to a voting round in the BFT protocol and mr represents the root of cross-domain and intra-domain policy Merkle tree mt. Furthermore, the protocol necessitates that each committee utilizes the BLS threshold signature mechanism as a foundational element of the BFT protocol. Each member within the domain produces a signature σ share by employing its secret key.

  For policy at a particular time, a valid policy will be guaranteed to be accepted and recorded on the blockchains of corresponding domains after a particular time. The definition of policy registration liveness is as follows:

Definition 3.1 (Policy Registration Liveness) *For $\forall e \in \mathbb{N}^*, \forall I, J \in [1, N]$, if a valid policy P regarding C^I and C^J is submitted by an object at time t, then at time $t + T_{PR}$, all honest members of C^I and C^J must output $\langle B_r^I \rangle_{\sigma^I}$ and $\langle B_{r'}^J \rangle_{\sigma^J}$ where:*

$$B_r^I = (str, mr_{IA} \| \cdots \| mr_{IJ} \| \cdots \| mr_{IN}), P \in mt_{IJ}.leaf$$
$$B_{r'}^J = (str', mr_{JA} \| \cdots \| mr_{JI} \| \cdots \| mr_{JN}), P \in mt_{JI}.leaf$$

Data sharing liveness refers to the need to execute a valid access request within a specified timeframe in accordance with predetermined policies. The concept of policy registration liveness is as stated:

Definition 3.2 (Data Sharing Liveness) *For $\forall e \in \mathbb{N}^*, \forall I, J \in [1, N]$, for a valid policy $P \in mt_{IJ}.leaf \wedge mr_{IJ} \in \langle B_r^I \rangle_{\sigma^I}$ with some r where $mr_{IJ}.height = \max(\{mt_{IJ}.height\})$, if a valid AccessR is submitted following P at time t by a subject, then at time $t + T_{PT-CDDS}$, AccessCert must be generated and data sharing can be done. Otherwise, the arbitration process must be done at time $t + T_{ZT-CDDS}$.*

The concept of intra-domain consistency implies that the perspectives of any two honest members in the same domain regarding the execution outcomes and policy registration should be consistent. Note that "$\Rightarrow$" signifies the output of information by a node. The definition of intra-domain consistency is as follows:

Definition 3.3 (Intra-Domain Consistency) *For $\forall e \in \mathbb{N}^*, \forall i, j \in [1, n]$, $\forall I \in [1, N], \forall P_i, P_j \in C_e^I.honest, \forall r \in \mathbb{N}^*$, if:*

$$P_i \Rightarrow \langle B_r^I \rangle_{\sigma^I}, \ B_r^I = (str, mr_{IA} \| \cdots \| mr_{IN})$$
$$P_j \Rightarrow \langle \hat{B}_r^I \rangle_{\sigma^I}, \hat{B}_r^I = (s\hat{t}r, \hat{mr}_{IA} \| \cdots \| \hat{mr}_{IN})$$

then it must hold that $B_r^I = \hat{B}_r^I$.

The meaning of cross-domain no conflict is that the perspectives of two trustworthy individuals in distinct domains regarding the access control policies related to those domains should be in agreement. The concept of cross-domain no conflict is as stated:

Definition 3.4 (Cross-Domain No Conflict) *For $\forall e \in \mathbb{N}^*, \forall i, j \in [1, n]$, $\forall I, J \in [1, N], \forall P_i \in C_e^I.honest, \forall P_j \in C_e^J.honest$, if $m_{IJ}.height = mr_{JI}.height$ and:*

$$P_i \Rightarrow \langle B_r^I \rangle_{\sigma^I}, \ B_r^I = (str, mr_{IA} \| \cdots \| mr_{IJ} \| \cdots \| mr_{IN})$$
$$P_j \Rightarrow \langle B_{r'}^J \rangle_{\sigma^J}, \ B_{r'}^J = (str', mr_{JA} \| \cdots \| mr_{JI} \| \cdots \| mr_{JN})$$

then it must hold that $mr_{IJ} = mr_{JI}$.

3.2.4 Reliability of Cloud-Edge-End Systems

This section introduces the reliability model for large-scale cyber-physical systems (CPSs), sustainable CPSs, and software-defined networks (SDN).

- **Reliability Model for Large-Scale CPSs**

The continuous advancement of information technology has enabled a profound integration of the physical and software components within systems, leading to the emergence of entities referred to as CPSs. Within this, large-scale CPS is a significant subcategory in the expansive field of CPS, encompassing various domains such as manufacturing and daily life.

Cao et al. [36] proposed an edge-intelligent solution that optimizes both service latency and system lifecycle, considering reliability and energy consumption. They developed a comprehensive system model through theoretical analysis and experimentation. Within the system tasks, they identified two distinct categories of transient faults: (1) bit errors that emerge during task handovers and (2) soft errors that arise during task processing. The occurrence of error codes is often modeled using the exponential distribution. Let $\mathcal{B}_{\text{biterr}}^{m,n}$ represent the average bit error rate of the connection between the user group G_{end}^m and edge server E_{serv}^n, based on which the delivery reliability can be evaluated as [39]

$$\mathcal{R}_{\text{com}}^{m,n} = \exp\left\{-\mathcal{B}_{\text{biterr}}^{m,n} \times \mathcal{L}_{\text{com}}^{m,n}\right\}. \tag{3.51}$$

Afterward, by letting $\mathcal{B}_{\text{softerr}}^n$ represent the average incidence of soft errors on CPU $\mathcal{P}_{\text{serv}}^n$, we can obtain $\mathcal{B}_{\text{softerr}}^n = C_n \times \exp\left\{-\varpi_n \times F_{\text{serv}}^m\right\}$, with C_n and ϖ_n being constants related to the hardware architecture [518, 40]. Indeed, the exponential distribution is frequently employed to emulate the incidence of soft errors. Assuming that the exponential distribution holds, the reliability of CPU $\mathcal{P}_{\text{serv}}^n$ performing task $\mathcal{T}_m$ in G_{end}^m is denoted by [518]

$$\mathcal{R}_{\text{exe}}^{m,n} = \exp\left\{-\mathcal{B}_{\text{softerr}}^n \times \lambda_m \times T_{\text{slot}} / V_{\text{serv}}^n\right\}. \tag{3.52}$$

Indeed, the dependable processing of tasks relies on both the task transmission and executing processes. Thus, the task reliability of G_{end}^m associated with E_{serv}^n is the multiplier of $\mathcal{R}_{\text{com}}^{m,n}$ and $\mathcal{R}_{\text{exe}}^{m,n}$. They employed task replication redundancy techniques to achieve the necessary level of reliability in the face of transient failures. These techniques tolerate both bit errors and soft errors. With the help of $\tau_{m,n}$ duplications (including the original task itself), the augmented transient fault reliability can be formulated as

$$\mathcal{R}_{\text{relia}}^{\text{task}}(\tau_{m,n}) = 1 - (1 - \mathcal{R}_{\text{com}}^{m,n} \times \mathcal{R}_{\text{exe}}^{m,n})^{\tau_{m,n}}. \tag{3.53}$$

System reliability is characterized as the multiplier of the transient fault reliability of all tasks, which can be given as

$$\mathcal{R}_{\text{relia}}^{\text{sys}} = \prod_{m=1}^{M}\left(\sum_{m=1}^{M}\sum_{n=1}^{N} A_{m,n} \times \mathcal{R}_{\text{relia}}^{\text{tas}}\right). \tag{3.54}$$

- **Reliable Edge-Cloud Computing Model for Sustainable CPSs**

The existing edge cloud computing based on delay awareness can not consider energy budget and reliability requirements simultaneously, which may significantly diminish the feasibility of CPS applications. Cao et al. [38] proposed a two-stage reliable service delay optimization composed of static and dynamic. In the paper, they established a system model. The reliability of base station tasks pertains to the likelihood of successfully transmitting these tasks to the designated distributed server, free from bit errors, and their subsequent flawless performing by the target server without encountering any software errors. During the digital transmission process, bit errors are primarily caused by various factors such as environmental noise, interference, distortion, or bit synchronization errors occurring on the link. Let $\Theta_{\text{bit}}^{j,m}$ represent the fixed bit error rate for the link connecting $\mathcal{S}_{\text{base}}^j$ to the distributed server S_{server}^m. Subsequently, the transmission reliability can be characterized as [206]

$$\Upsilon_{\text{com}}^{j,m} = \exp\left(-\Theta_{\text{bit}}^{j,m} \times \mathcal{L}_{\text{com}}^{j,m}\right). \tag{3.55}$$

In contrast to bit errors, software errors are mainly triggered due to transient faults caused by cosmic radiation or electromagnetic interference. The mean failure rate of distributed server S_{server}^m can be formulated as [40, 207]

$$\Theta_{\text{soft}}^m = C_m \times \exp\left(-\varpi_m \times \mathcal{P}_{\text{server}}^m\right), \tag{3.56}$$

where C_m and ϖ_m are constants that are contingent upon the hardware architecture. Under the assumption of the exponential distribution, execution reliability is given by

$$\Upsilon_{\text{exe}}^{j,m} = \exp\left(-\Theta_{\text{soft}}^m \times \frac{\mathcal{R}_{\text{base}}^j}{\mathcal{P}_{\text{server}}^m}\right). \tag{3.57}$$

To meet the system's reliability requirements, they employed robust backup technology that tolerates both bit and soft errors. Furthermore, acceptance tests are conducted after each backup is performed on a distributed server to verify the successful processing of tasks. If the acceptance test shows no errors, the output of the present backup is received; if not, it will simply be discarded. When $\mathcal{I}_{\text{back}}^{j,m}$ backups are reserved for BS $\mathcal{S}_{\text{base}}^j$, the reliability can be defined as

$$\Upsilon_{\text{back}}^{j,m} = 1 - \left(1 - \Upsilon_{\text{com}}^{j,m} \times \Upsilon_{\text{exe}}^{j,m}\right)^{\mathcal{I}_{\text{back}}^{j,m}}. \tag{3.58}$$

The system reliability is defined as the multiplication of all BSs' reliabilities in the system, i.e.

$$\Upsilon_{\text{back}}^{\text{sys}} = \prod_{j=1}^{\mathcal{J}} \sum_{m=0}^{\mathcal{K}} \mathcal{A}_{j,m} \times \Upsilon_{\text{back}}^{j,m}. \tag{3.59}$$

- **Reliability Model for SDN**

Mobile applications are progressively gaining significance in our everyday lives. However, due to the hardware conditions of equipment, it is difficult to adapt to the increasingly high demand in life. Mobile cloud-edge computing networks (MECCNs) can offer services with low latency and plenty of computing resources. Nevertheless, the uneven distribution of users within MECCNs results in an imbalanced distribution of network demand and workload. In large-scale multi-control networks, the control plane at the edges can handle network requests in time, but the fixed control plane cannot dynamically allocate network traffic. Xu et al. [425] proposed a deep reinforcement learning-based methodology to place load-aware controllers dynamically for SDNs-supported MECCNs.

Xu et al. established a system model for SDN-enabled MECCN in their paper. A stable and reliable connection between the controller and the managed equipment is essential to guarantee the continuous and steady operation of the control plane in a cloud-edge collaborative computing network. The controller exhibits high reliability and can keep a consistent connection with the collected equipment, enabling it to control the network effectively even under complex network conditions. The link for transmitting both control and data flow is the same. This means that the control reliability of the controller is affected by the reliability of the nodes and links on the connection between the controller and the equipment [159, 12].

Multiple paths may exist between controller $\mathcal{C}$ and its controlled equipment e. Each path can be divided into $\mathcal{N}^l$ sub-links. r_i^l denotes the i-th sub-link reliability of this path. There are $\mathcal{N}^n$ nodes between the controller $\mathcal{C}$ and the controlled equipment e. r_j^n denotes the reliability of node j. The reliability of k-th path between $\mathcal{C}$ and e can be calculated by multiplying link and node reliability. The equation is as follows:

$$R_{\mathcal{C},e}^k = \prod_{i=1}^{\mathcal{N}^l} \left(r_i^l\right) \times \prod_{j=1}^{\mathcal{N}^n} \left(r_j^n\right). \tag{3.60}$$

Controllers and switches use the wired network to communicate. Communication links may fail in the wired network, resulting in disconnected links between nodes. Due to the transmission of SDN control information and data through the same network, a heavy transmission load may cause delays in the arrival of control information. The failure possibility of i-th sub-link due to physical factors is r_i^p. Then the i-th sub-link reliability is

$$r_i^l = (1 - r_i^p) \times \text{sigmoid}\left(\frac{B_i}{f_i}\right), \tag{3.61}$$

where B_i and f_i, respectively, represent the bandwidth and traffic of the i-th sub-link. It is clear that the reliability decreases as the traffic increases. A monotonically increasing function, sigmoid function, is utilized to model the relationship between them.

They assume there are $\mathcal{N}^p$ paths between $\mathcal{C}$ and e to guarantee the timing constraint. The control reliability between $\mathcal{C}$ and e can be calculated as

$$R_{\mathcal{C},e} = \sum_{k=1}^{\mathcal{N}^p} R_{\mathcal{C},e}^k. \tag{3.62}$$

The controller $\mathcal{C}$ controls N devices. The control reliability of $\mathcal{C}$ is expressed as

$$R_{\mathcal{C}} = \frac{\sum_{e=1}^{N} R_{\mathcal{C},e}}{N}. \tag{3.63}$$

3.2.5   Conclusions and Discussions

This chapter surveys the literature on cloud-edge-end systems from a performance modeling perspective. These works discuss four critical aspects of optimizing cloud-edge-end systems: latency, energy expenditure, security and privacy, and reliability. Mathematical models are available for latency modeling to calculate the latency of tasks performed by end equipment, edge nodes, and cloud processing data centers. Then, in terms of energy consumption modeling, more typical energy consumption models are provided: the global energy consumption model and end energy consumption model. In addition, performance modeling through trustworthiness evaluation schemes, differential privacy, threat models, and security scores are covered in the security and privacy modeling section. Finally, this chapter discusses reliability modeling and provides feasible reliability modeling approaches. The cloud-edge-end system performance modeling covered in this section can provide some reference for optimizing the system.

3.3   Performance Optimization for Cloud-Edge-End Systems

In daily life, smart homes that connect and work with home devices and sensors with the cloud and edge nodes and smart cities that use cloud-edge-end systems to monitor and manage urban infrastructure are increasingly gaining popularity. The cloud-edge-end system concept aims to provide computing capabilities closer to data sources and end devices to meet the needs of scenarios with high real-time requirements, large data volumes, and high data security requirements. With the rapid popularization of cloud-edge-end systems, cloud-edge-end systems with better performance can better meet users' needs. This section explores the performance optimization of cloud-edge-end systems in depth by examining the optimization of cloud-edge-end systems in terms of latency, energy, security, and reliability.

3.3.1   Optimization for Latency of Cloud-Edge-End Systems

In this section, we review the works on latency optimization in cloud-edge-end collaborative systems. As shown in Table 3.1, we summarize these works and compare them in terms of optimization targets, energy consumption, latency performance optimization, and benchmark schemes.

Jin et al. [186] investigated the resource allocation problem of network slices in edge computing. They observed that 5G network slicing technology divides the network into multiple independent networks. As a result, it becomes challenging to devise an efficient resource allocation strategy in edge computing. Therefore, they have optimized the cloud-edge collaboration hybrid computing (CECHC) model and proposed a novel network-slicing algorithm designed explicitly for CECHC. Firstly, they analyzed the latency requirements for three network slices. Subsequently, it was proposed that the deployment of Cu and Du be optimized, the CECHC model be optimized, and storage and computing capabilities be improved. Finally, a network slicing algorithm will be designed to enable flexible resource allocation, considering the distinct latency requirements of various network slices. Based on experimental results, it has been proven that the new algorithm and model have the ability to reduce latency significantly.

Fang et al. [108] investigated the energy-saving issues brought about by the growth of mobile internet traffic. They observed that although cloud edge networks currently attain energy conservation, their computing power is limited due to resource heterogeneity and limitations. Therefore, they proposed a content task offloading solution based on DRL for cloud-edge-end environments, which minimizes power consumption and improves computing power. Firstly, they transformed the energy conservation issue into a power consumption minimization model to solve the network request caching problem. Finally, they proposed a new DRL algorithm that maximizes power consumption by utilizing coordinated caching and optimized computing power allocation strategies. The results demonstrate that compared to other power reduction strategies, the approach suggested in the literature achieves improved content response and power savings, particularly in scenarios with increasing request loads.

Wang et al. [391] conducted a study on an interoperable and flat IIoT with a focus on low-latency data collection in manufacturing systems. They analyzed traditional pyramid structures and found that data isolation issues result in higher data collection delays. Addressing these challenges solely with OPC UA technology results in compatibility issues with legacy devices and leads to expensive equipment replacement costs. Therefore, they proposed an IIoT architecture that combines OPC UA and Software-Defined Networking to attain interoperability and flatness. First, they conducted experiments to demonstrate the issues related to interoperability and latency in the traditional pyramid structure. Subsequently, they introduced an edge information layer in the proposed IIoT architecture to address the compatibility issues of OPC UA with legacy devices. Finally, they tested the feasibility of the architecture on an actual production platform. The experiments show that the proposed architecture enables low-latency data collection and exhibits good interoperability.

Zhang et al. [492] focused on a secure cloud-edge collaborative PIoT architecture to address the data security and complexity issues in PIoT architectures. They observed that existing research primarily focuses on improving performance but lacks adequate solutions for data security and complexity issues. First, they illustrated the application scenarios of PIoT architecture and summarized related issues. Therefore, they proposed a secure cloud-edge collaborative PIoT architecture based on AI and blockchain. Subsequently, to solve the problem of data security and low-latency computation offloading, they proposed a blockchain-based federated deep learning algorithm, and the algorithm separated long-term

TABLE 3.1

A Comparison Summary of Latency Optimization from Multiple Aspects, Such as Optimization Target.Power/Energy Savings, Latency Performance Optimization, Benchmarks Schemes.

Ref.	Optimization Target	Power/Energy	Latency	Benchmarks
[295]	Latency	*	34.1%↓	PCRO
[391]	Latency	*	*	OPC UA and SDN
[97]	Latency and energy	93.4%↓	*	OMA
[417]	Latency and energy	59.02%-255.68%↓	*	Random
[453]	Energy	*	*	GRU and PPO
[352]	Latency and energy	94.29%-183.33%↓	70.08%-192.51%↓	OLN and CFO
[108]	Energy	6.59%-73.05%↓	*	DRL
[412]	Latency	*	62.91%-70.79%↓	PPO and DRL
[37]	Latency	*	47.37%↓	GT, ILP and M/G/1
[186]	Latency	*	46.15%-63.12%↓	FC and MAEC
[492]	Latency	*	18.99%-50.17%↓	DAC
[225]	Latency and energy	16.15%-16.79%-18.43%↓	51.29%-71.11%-86.63%↓	DTO, FTO and EMM
[125]	Latency	*	198.01%-200.00%↓	MOEA/D
[187]	Latency	*	28.79%-29.58%↓	PS and SCA
[426]	Latency	*	806.11%↓	MARL
[166]	Latency	*	74.93%↓	KM and MaOEA

*: No relevant experiments and data are mentioned in the literature;

↓: Compared to the benchmark scheme, the method proposed in the article saves energy or reduces latency;

PCRO: priority-based chemical reaction optimization algorithm; OPC: OLE(object linking and embedding)for process control; UA: unified architecture; OMA: cloud-first offloading strategy; GRU: gate recurrent unit; PPO: proximal policy optimization; OLN: offloading nothing strategy; CFO: cloud-first offloading strategy; GT: game theory; ILP: integer linear programming; FC: fog computing; MAEC: multi-access edge computing; DAC: deep actor-critic-based algorithm; DTO: distributed deep ac-based online computing offloading; FTO: federated deep ac-based task offloading algorithm; EMM: energy-aware mobility management; MOEA/D: multi-objective evolutionary algorithm based on decomposition; MARL: multi-agent reinforcement learning; PS: pipeline strategy; SCA: successive convex approximation; KM: knowledge mining; MaOEA:many-objective evolutionary algorithm.

security constraints from short-term queuing delay optimization by utilizing Lyapunov optimization. The results obtained from the experiments confirm the algorithm's outstanding performance in terms of consensus delay and overall queueing delay.

Liao et al. [225] investigated the multi-layer resource allocation and timescales and dimension issues faced by AGI-PIoT to realize intelligent cloud-edge-end collaboration. Therefore, they proposed a multi-layer, multi-time-scale, and multi-dimensional resource allocation algorithm based on federated deep reinforcement learning. To tackle the overall optimization problem, they decomposed it into manageable subproblems. First, they proposed a joint semi-distributed algorithm based on actor-critic to address the task offloading and power control problems and strike a balance between the cost of learning and performance. Then, they solved the admission control through quadratic programming. Finally, the resource allocation problem is solved through Lagrange dual decomposition and smooth approximation. Through experiments, it can be proved that the algorithm suggested in this article performs better than others in terms of delay, power consumption, and other concerns.

Yang et al. [449] investigated the challenges faced by existing methods in ensuring efficient real-time interaction and content delivery in VR technology. These challenges include high caching and computational costs and difficulties in ensuring real-time responsiveness. Firstly, they proposed a cloud-edge-device collaborative service architecture that supports AI-assisted content generation. This architecture enables independent encoding and distribution of virtual reality backgrounds and interactive content. Subsequently, a GNN-based cloud-edge collaborative caching strategy is proposed. Finally, a GNN-aided-based request prediction minimum cost update algorithm is proposed to achieve efficient caching and updating of background content for optimal performance. Based on the experimental results, the algorithm proposed in the paper demonstrates superior performance in terms of low power consumption, low latency, and high user satisfaction compared to other algorithms.

Liu et al. [240] investigated the existing drawbacks of inefficient, unstable, and high-power-consumption methods for wind turbine (WT) blade surface damage detection. Therefore, they proposed a cloud-edge-end collaborative approach to lightweight deep learning-based networks for real-time detection of surface damage on WT generator blades. The method involves analyzing real-time images through edge computing to reduce latency and save bandwidth. Firstly, they optimized the YOLOv3 model and trained a lightweight model on a cloud server, which is then deployed on edge devices. Next, they proposed using unmanned aerial vehicles (UAVs) to capture blade images for damage detection on edge devices. Finally, the blade damage reports can be uploaded to the cloud server for immediate inspection and analysis. Experimental results validate that the proposed method achieves an accuracy rate exceeding 90% while providing rapid detection speed and minimal bandwidth consumption.

Yang et al. [453] investigated the optimal scheduling of renewable resources and the problem of data containing missing values. Therefore, they proposed a novel cloud-edge collaborative computing scheme consisting of two hierarchical levels and designed two algorithms to address the missing value problem and the optimal adjustment of resources. Initially, they introduced a data restoration and prediction algorithm based on GRU to repair missing values within the dataset and forecast future data. The repaired dataset is utilized as input for the prediction layer. Lastly, they proposed a PPO-based algorithm. This algorithm introduces rotational spare capacity to address renewable energy variability, enhance grid flexibility, and enhance the accommodation efficiency of renewable energy. Experimental results demonstrate that the proposed approach in the paper can achieve optimal results in accommodating renewable energy resources.

Tang et al. [352] focused on the task dependency relationships of IoT devices in complex applications and networks. They employed DAG to model task dependencies and formalize offloading as a multi-objective mixed-integer optimization issue. The objective was to minimize IoT devices' average energy consumption and latency. Therefore, they suggested a task-offloading algorithm based on task priority and DRL. First, the decision of task offloading is formulated as a Markov decision process. Finally, an optimized DRL method based on task priority with an action mask is proposed to utilize server computing resources to determine the optimal computing offloading strategy. Experimental results demonstrate that the algorithm proposed in this article surpasses others in minimizing energy consumption and reducing time delays.

Du et al. [97] investigated the challenge of efficiently and low-latency meeting real-time demands in the current healthcare Internet of Things context. They observed that the current wireless communication rates based on OMA can no longer meet the growing mobile communication demands. Therefore, they proposed a novel network communication model that combines cloud-edge computing technology with NOMA. First, using a hierarchical structure, they decomposed the joint optimization problem of task offloading decisions and transmission time into two subproblems. Second, they proposed a deep learning algorithm based on a DNN to address the problem of optimal offloading decisions. Lastly, they introduced a linear search-based algorithm to identify the optimal transmission time. The experimental results demonstrate that the algorithm proposed in the paper can consistently converge to the optimal value and reduce high costs compared to other approaches.

Xiao et al. [417] focused on task offloading in the cloud-edge environment of 5G heterogeneous networks. They observed and analyzed the influence of heterogeneity in servers and network environments within the cloud-edge-end architecture on task offloading. Therefore, they proposed a method for distributed wireless backhaul-supported collaborative cloud-edge-end task offloading to reduce power consumption. Firstly, they proposed a joint optimization approach that aims to minimize the energy consumption of UDs. This is achieved through optimizing task offloading decisions, the transmission power of UDs, computing resources, and spectrum allocation. Finally, they decoupled the problem above into three subproblems, applied various convex optimization techniques to derive solutions for each subproblem, and proposed an iterative methodology to obtain the solution for the original problem iteratively. The experimental results indicate that the presented strategy outperforms other approaches in reducing the energy consumption of all user devices.

Kai et al. [187] introduced the problem of limited computational and communication resources in mobile edge computing networks. They observed that the limited battery life of mobile devices, combined with the increasing number of latency-sensitive applications, has led to increased costs associated with task offloading. Therefore, they developed a cloud-edge-end-based collaborative computing framework and proposed a pipeline-based offloading scheme. Subsequently, they proposed a non-convex optimization problem that minimizes the total latency of all MDs by considering factors such as offloading strategies and computational resources. Finally, they introduced the SCA method to convert the problem above into a convex optimization problem. They also proposed a joint offloading decision-making and power allocation scheme. The experimental results affirm that the paper substantiates the superior performance of the proposed cooperative offloading scheme over other offloading schemes.

Yuan et al. [468] presented the problem of minimizing the cost associated with task-dependent user association and partial offloading. They observed that in mobile edge computing, task computation and task offloading can be performed simultaneously, and quick response time can be achieved by adjusting task partitioning. Therefore, they considered the three aspects of task offloading, task partitioning, and user association and formulated them as a unified optimization problem. Subsequently, a hybrid particle swarm optimization algorithm based on genetic operations and simulated annealing, incorporating the Metropolis criterion, was proposed to address the joint optimization problem. The experimental results validate that the proposed genetic simulated annealing-based particle swarm optimization (GSPSO) algorithm surpasses other methods in terms of overall cost while effectively meeting the latency constraints of mobile devices.

Gong et al. [125] focused on the problem of task offloading in mobile edge computing. They observed that current task-offloading strategies often overlook economic costs, making achieving sustainable cloud-edge-end collaborative computing challenging. Therefore, they optimized the problem of collaborative task offloading in a multi-user and multi-server environment encompassing cloud-edge-end interactions. Firstly, they modeled and formalized the multi-objective optimization problem to optimize offloading delay and execution rewards. Finally, they presented an efficient multi-objective evolutionary optimization algorithm based on dependency constraint decomposition. This algorithm effectively addresses the challenges of latency and rewards optimization in mobile edge computing task offloading, achieving the goals of minimizing latency and maximizing rewards simultaneously. The experimental simulation results confirm the significant improvement achieved by the algorithm proposed in the paper, particularly regarding the user's offloading utilization rate measured by the Q-value. It surpasses the greedy algorithm by 34% and demonstrates a remarkable 3.8-fold superiority over the random algorithm in terms of Q-value.

Yang et al. [446] developed the problem of computing resource allocation in mobile edge cloud computing networks. In the context of MECC networks, they observed the coupling between edge computing and cloud computing queues, giving rise to challenges in resource allocation and end-to-end latency guarantees. Therefore, they adopted a simple computing speed controller to reflect resource allocation, optimizing resource allocation by adjusting the computing speed of edge servers and cloud servers. Firstly, they modeled the MECC network as a tandem computing queue consisting of multiple servers. Subsequently, they introduced a DRL algorithm to learn a computing speed adjustment strategy and seek the optimum solution for the computing controller. The experimental results demonstrate that the proposed method can achieve good end-to-end latency guarantees in dynamic network environments while avoiding excessive computing resource use.

Qu et al. [295] focused on the emergency task offloading strategy in smart factories. They observed that in the cloud-edge-device three-tier architecture, the computing capacity of any single layer is insufficient to meet the computational requirements of emergency strategies. Therefore, they proposed an emergency task offloading strategy that leverages the collaborative capabilities of cloud-edge-end systems. Firstly, they formulated the overall task execution latency and critical task execution latency as a single objective function. Next, they introduce the fast chemical reaction optimization (Fast-CRO) algorithm, which prioritizes offloading critical tasks in emergencies. Finally, the Fast-CRO algorithm is employed to achieve the optimal solution of the objective function. The experimental results demonstrate that the Fast-CRO algorithm outperforms other algorithms regarding average execution time and overall latency.

Wu et al. [412] investigated the issue of handling multi-source data streams from IoT devices. They observed that the computational and communication resources in mobile edge computing are limited and can be enhanced by the collaborative work of cloud servers and edge servers. Therefore, they suggested a cloud-edge-end collaborative computing architecture in dynamic network environments. Firstly, they considered the characteristics of wireless channels and data streams in a multi-source environment to be time-varying. Then, they modeled and decomposed the optimization problem using Markov Decision Processes, dividing it into subproblems involving allocating data stream offloading ratios and resources. Finally, they proposed a new method combining the integration of the proximal policy optimization scheme and the convex optimization scheme for data flow offload and resource allocation, respectively. The experimental results demonstrate that the proposed cloud-edge-end architecture can achieve lower overall system latency, thus proving its superiority.

Hao et al. [147] designed task-offloading schemes for intelligent devices. They observed that existing edge computing research primarily focuses on addressing energy efficiency issues but overlooks the need for personalized and fine-grained task offloading specifically for intelligent applications. Therefore, they proposed the iTaskOffloading architecture, which combines a cognitive engine and a conventional cloud-edge collaborative system. Firstly, the information collection of intelligent devices is achieved through the local device layer. Then, computational tasks and information storage are processed through the edge and remote cloud layers. Finally, they proposed personalized task offloading strategies by integrating the cognitive engine with SDN technology. The experimental results demonstrate that the iTaskOffloading architecture outperforms conventional cloud computing by improving performance by 10%. In high-load scenarios, it can achieve performance improvements of up to 30%.

Xu et al. [426] proposed low-latency transmission issues in millimeter-wave networks. They observed that the network's latency explosion is due to the extensive signaling exchange required for coordinated multi-point processing and centralized processing. Therefore, they proposed a distributed transmission scheme based on MARL. First, they formulated the aforementioned problem as a CMDP, with the aim of achieving two objectives: minimizing network latency and ensuring quality of service (QoS). Next, they proposed a hybrid learning framework that combines decentralized execution and centralized training. They extended the actor-criticism model to achieve this framework. Lastly, they designed a reward function using the Lagrangian method. This reward function guides the learning process and optimizes the trade-off between minimizing network latency and meeting QoS requirements. The experimental results validate the proposed algorithm, demonstrating its effectiveness and performance advantages.

Fan et al. [105] focused on the security concerns associated with data retrieval in the CEEO intelligent power grid. They observed that in order to address the issue of data silos, it is necessary to encrypt relevant data to ensure security. However, traditional encryption techniques can compromise the retrievability of the data. Therefore, they proposed a searchable encryption algorithm called MSIAP. Firstly, they improved the Apriori algorithm to mine the correlations in electric power grid data and construct a multi-level index structure. As a result, they efficiently retrieved subsets of multiple keywords and dynamic updates. Finally, they used bilinear pairing and hash functions to apply encryption to the keywords, resulting in the accomplishment of trapdoor untraceability. This approach enhances retrieval efficiency and ensures the confidentiality of the encryption key is not compromised. The experimental results demonstrate that the MSIAP algorithm significantly reduces time complexity and achieves efficient data retrieval.

Hu et al. [166] introduced the optimization of network metrics for cloud-edge-end collaborative caching in the context of the Internet of Things. They observed that existing cache strategies based on popularity prediction can only address two or three network metrics using multi-objective evolutionary algorithms. Therefore, they proposed a many-objective optimization-based popularity prediction for cooperative caching (MaOPPC-Caching) framework to popularity prediction. Firstly, the MaOPPC-Caching framework integrates three predictive algorithms to forecast content popularity. Based on the prediction results, it generates cache collaboration decisions in both horizontal and vertical forms, thereby achieving joint optimization of multiple network metrics. Finally, they defined the MaOPPC-Caching problem as a multi-objective mixed-integer nonlinear programming problem. To address this problem, they developed a many-objective evolutionary caching algorithm, which leverages knowledge-mining techniques. The experimental findings indicate that the proposed framework in the paper outperforms existing prediction algorithms in terms of latency, offloading traffic, prediction accuracy, and load balancing.

This section provides a literature review of cloud-edge systems from the perspective of latency optimization. These works indicate that task offloading and deep learning algorithms are promising solutions for latency in cloud-edge systems. Some works focus on single-objective optimizations, specifically on either task offloading or deep learning approaches for latency optimization. However, a few works have combined task offloading and deep learning to optimize latency further while minimizing energy consumption.

3.3.2 Optimization for Energy of Cloud-Edge-End Systems

This section reviews efforts focused on optimizing energy consumption in cloud-edge-end collaboration systems. As shown in Table 3.2, we summarize these works and compare them in terms of optimization targets, energy consumption, performance, and benchmark schemes.

Aiming at optimizing the cloud-edge-end collaboration performance in 6G space-air-ground integrated power internet of things, Wang et al. [401] proposed a hybrid and hierarchical paradigm for orchestrated resource management in cloud-edge-end systems, with artificial intelligence as its foundation. The framework includes admission control, task splitting, task offloading, task processing, and result feedback, which can adapt to network dynamics and resource heterogeneity with multiple dimensions. Based on the framework, they proposed a task offloading algorithm called queue-aware deep actor-critic (Q-DAC). This algorithm aims to identify suitable servers for task offloading in order to minimize energy consumption. The algorithm takes into account constraints related to task offload queuing delay, server-side buffer queuing delay, and result feedback queuing delay. Experimental results show that compared with energy-aware mobility management [345] and DAC-based task offloading [66], Q-DAC can effectively reduce energy consumption by 33.52% and 41.48% respectively, and reduce end-to-end queuing latency by 44.56% and 30.98%.

To prevent the unauthorized dissemination of confidential data during the offloading service procedure in cloud edge computing, Xu et al. [436] proposed an offloading method based on locality sensitive hash (LSH) to preserve privacy while improving service utility in edge cloud computing environments. The authors first calculated the waiting time for the Internet of Multimedia Things services by constructing an M/M/S queuing model based on queuing theory. Then, the cloud edge computing paradigm is used to model the transmission latency and execution consumption during service offloading and perform data processing on external resources. Finally, they proposed a scheduling algorithm LOM based on LSH to extract appropriate services from massive data, achieve collaborative offloading, and protect privacy. Simulation experiments show that compared with time-greedy, energy-greedy, and all-cloud algorithms, LOM can have less offloading delay than energy-greedy and all-cloud with the least energy consumption.

Long et al. [244] introduced a task offloading algorithm called "TO-EEC" that is based on the "end-edge-cloud" framework, which reduces response delay and energy consumption in cloud-edge collaboration. The authors approached the task offloading problem as a multi-objective optimization problem, whereby the latency, energy consumption, and load balance of edge nodes are considered within the framework of the "end-edge-cloud". Load balancing in multi-objective optimization helps prevent latency increases due to edge nodes overloading and energy wastage from idle edge nodes. The authors suggest a TO-EEC offloading algorithm that enhances AR-MOEA's initialization approach, mutation probability, and crossover approach. This improvement aims to mitigate the algorithm's susceptibility to local optimality and expedite its convergence toward the global optimal solution. Furthermore, the algorithm does not require specifying the weights between the various objectives,

TABLE 3.2

A Comparison Summary of Energy Optimization Methods from Multiple Aspects, Such as Optimization Target, Power/Energy Savings, Performance Improvement, Benchmarks Schemes.

Ref.	Optimization Target	Power/Energy	Performance	Benchmarks
[401]	Energy and delay	33.52%↓-41.48%↓	44.56%↓-30.98%↓	EMM and DAC
[436]	Energy and privacy	21.2%↓	21.79%↓	ATC
[244]	Energy and delay	10%↓-9%↓	14%↓-11%↓	AR-MOEA and NSGA-III
[336]	Energy and privacy	65.52%↓-45.02%↓	64.97%↓-44.69%↓	SCC and SCC-Random
[208]	Energy and convergence	86.07%↓	*	RND
[380]	Energy and social welfare	*	2x↑	GA
[364]	Energy and response time	18.51%↓	*	RND
[335]	Energy	45.23%↓	*	OPT-edge
[200]	Energy and task computation time	23.71%↓	36.97%↓	BAT-SAA
[81]	Energy	5.59%↓	*	PSO
[419]	Energy and delay	20.13%↓-18.63%↑	589.02%↑-50.63%↓	DO and EO
[94]	Energy	5.8%↓-11.0%↓	*	CBPSO
[469]	Battery life and latency	20.14%↑	12.49%↓	DRA, DRL-E2D, and MUDRL
[356]	Energy	37.18%↓-45.63%↓	*	NBS and RBS
[452]	Energy and delay	9.50%↓-13.37%↓-18.55%↓	11.01%↓-16.07%↓-21.39%↓	OMCWC, OOCWC and TCO
[485]	Energy	97.86%	*	SFP
[481]	Energy and frequency regulation	36.3%↓	*	ACT
[494]	Energy	53.0%↓	*	FSRP
[489]	Revenue-cost ratio and revenue	13.0%↑	21.0%↑	TDM, CLF, and DMRT-SL
[463]	Energy and runtime	82.52%↓	*	FAB
[59]	Energy	24.79%↓-11.57%↓-14.74%↓	*	DQN,DDPG and A3C
[107]	Energy	14.36%↓-28.74%↓	*	POP and LRU
[468]	Energy and latency	36.66%↓-3.9%↓-16.53%↓	*	SA, GA, and SAPSO
[11]	Energy	13.57%↓	*	GB

*: No relevant experiments and data are mentioned in the literature;

↑: Compared to the benchmark scheme, the methods proposed in the article do not save energy or reduce latency but may also be performance improvements such as welfare and revenue;

↓: Compared to the benchmark scheme, the method proposed in the article saves energy or reduces latency;

ATC: all-to-cloud; AR-MOEA: an indicator-based multiobjective evolutionary algorithm with reference point adaptation; NSGA-III: non-dominated sorting genetic algorithm; SCC: single cloud computing; SCC-Random: single cloud computing with multiple edge computing units based on random task allocations; RND: random; GA: greedy algorithm; OPT-edge: only minimizing edge energy; BAT-SAA: bat-based service allocation algorithm; PSO: particle swarm optimization algorithm; DO: delay optimal; EO: energy optimal; CBPSO: chaotic binary particle swarm optimization; DRA: deep reinforcement approach; DRL-E2D: a DRL-based Model-free task offloading approach; MUDRL: multi-update deep reinforcement learning; NBS: nearest baseline scheme; RBS: random baseline scheme; OMCWC: optimal offloading to APs or cloud servers without caching; OOCWC: optimal offloading and caching at AP without offloading to cloud servers; TCO: task caching and offloading; SFP: sequential fractional programming algorithm ; ACT: actual; FSRP: fixed service risk probability; TDM: time delay mapping; CLF: closed-loop feedback algorithm; DMRT-SL: dynamic minimum response time considering the same level; FAB: firefly algorithm-based; DQN: deep Q-network; DDPG: deep deterministic policy gradient; A3C: asynchronous advantage actor-critic; POP: popularity; LRU: least recently used; SA: simulated annealing; GA: genetic algorithm; SAPSO: simulated annealing-based particle swarm optimizer; GB: greedy-based.

which allows the user to be non-professional and reduces the user's workload. The evaluation and simulations demonstrate that the suggested approach can decrease service delay and conserve energy expenditures compared to solutions based on reservations.

Federated learning (FL) facilitates the efficient utilization of digital medical data, hence fostering the advancement of intelligent healthcare systems (IHS). Nevertheless, this also encounters obstacles regarding computer and storage resources, along with a heightened susceptibility to privacy breaches. To tackle this problem, Su et al. [336] proposed a cloud-edge collaboration (CEC) architecture for IHS that combines blockchain and FL, which aims to mitigate the computational and storage demands placed on the cloud master station while simultaneously addressing concerns about data privacy and the establishment of mutual confidence between users and service providers. The CEC framework comprises four layers: cloud, management, edge, and end. These layers are designed to extract data value, including execution latency, processing complexity, and electricity consumption, enabling timely responses from IHS applications. In addition, the framework creates a secure, dependable, intelligent, and confidential system for accessing and sharing data, effectively dealing with the potential dangers related to healthcare data sharing and data privacy in IHS. Finally, a two-level IHS scheduling model is designed to reduce management costs and harmonize the interests of various stakeholders, including distributed generation, controllable load, and energy storage. Simulation results validate the effectiveness of the proposed CEC framework in minimizing the time it takes to execute tasks and use energy while enhancing the coordination of interests across several stakeholders.

Mobile edge computing can significantly enhance the computing power of mobile devices, thereby reducing energy consumption and service latency for Internet of Things applications. However, MEC faces challenges such as constrained network and computer resources, suboptimal or imprudent resource management, and disregarded security and reliability concerns. Thus, Li et al. [208] proposed a cloud-edge coordinated computing offloading architecture for blockchain-enabled Internet of Things systems over 6G networks to support security and reliability while improving system performance. The proposed framework is formulated to minimize system consumption overheads by concurrently considering the offloading decision, transmission power, and block interval. Subsequently, the authors established the parameters of the action space, reward function, and state space inside the structure of the Markov Decision Process (MDP). A cumulative reward learning approach is devised to address the MDP's optimization problem, thereby guaranteeing the system's optimal operation. In the context of shared training experiences, blockchain technology guarantees the system's security and dependability, consequently augmenting the overall user experience. Experiments show that the suggested structure has superior properties regarding convergence and efficacy compared to other current techniques.

Wang et al. [380] examined the difficulties associated with the management of emergency demand response (EDR) in cloud-edge orchestrated systems, particularly in the context of increasing artificial intelligence workloads, which leads to difficult compromises between energy usage and model precision, model degradation, limited training deadlines, and non-deterministic task arrivals. To address these issues, an auction-based approach is proposed specifically for federated learning. Firstly, the authors articulated the issue as a long-term social welfare maximization problem and consider several aspects of FL training duration, model quantization, and computation/communication patterns while guaranteeing the correctness of the prescribed model and complying with the EDR's energy restrictions and resource limitations. The problem is NP-hard since no assumptions are made about the EDR dynamics and federated learning tasks, and it requires compatibility with both soft and hard deadlines. Thus, a polynomial-time online approximation method is proposed for generating candidate training plans, reformulating the original problem into an equivalent but relatively easy plan selection problem. Then, the authors designed an online algorithm based on primal duality to acquire feasible solutions without knowing the future FL tasks. The experiments conducted in this study utilized EDR events, real-world training data, electricity prices, and varying quantities of FL tasks that were dynamically received over 168 hours. They showed that the proposed approach achieves approximately 2x the social welfare of two greedy bid scheduling algorithms and is more advantageous as the penalties for violating the deadline log increase.

Tong et al. [364] proposed a self-learning algorithm called SLTRA to minimize energy consumption and task response latency in mobile cloud edge collaborative computing while considering system reliability and resources. The task offloading problem is initially represented as a Markov decision process, and a deep reinforcement learning algorithm is developed to solve the MDP and optimize performance adaptively. The DRL algorithm under consideration incorporates the evaluation of task reaction time and total system energy expenditure to determine appropriate states and incentives while also taking into account the preferences of users and service providers. Therefore, the DRL algorithm can effectively choose suitable computing nodes for certain work, minimizing task reaction time and energy consumption while considering the limitations imposed by dependability and system resources. The findings from the simulation demonstrate that the SLTRA algorithm, as proposed, significantly enhances user experience quality and significantly reduces system energy consumption.

In order to reduce system energy expenditure and meet user experience requirements in a heterogeneous cloud-edge orchestrated environment, Su et al. [335] designed a cloud-edge orchestrated computing offloading method for near-real-time decision-making. They formulated a standard cloud edge computing offloading model in multi-ES and multi-MT heterogeneous scenarios by using multi-dimensional resource-constrained integer linear programming, taking into account minimizing the energy consumption of computing and transmission. The computational offloading problem is then proven to be NP-hard based on the generalized allocation problem. In order to solve this NP-hard problem, the authors designed a novel prima-dual algorithm based on the prima-dual theory and concepts in economics for near-real-time computational offloading. The algorithm can make offloading decisions in a near-real-time manner by sequentially considering task requests that arrive sequentially. Simulation experiments demonstrate that the suggested offloading approach outperforms the best solution found by IBM CPLEX [176] and the approximate results of two different baselines. It achieves a high level of performance that is near the optimal solution, even when there is uncertainty about future tasks.

The performance of task execution is greatly influenced by the cloud-edge collaboration mode, such as the pre-trained artificial intelligence models required for tasks assigned to edge servers and the data preprocessing or model rendering from edge servers as necessary for tasks assigned to cloud servers. Without considering the cloud-edge collaboration mode, the task scheduling solution may cause additional communication overhead or uneven resource utilization, resulting in task delays and high energy consumption. Thus, Laili et al. [200] established a task scheduling model of cloud-edge computing task scheduling, considering the communication overhead between cloud servers, edge servers, and end devices under two cloud-edge collaboration modes. Subsequently, a parallel combinatorial merger evolutionary algorithm (PGMEA) is put forward for the online scheduling of large-scale tasks to minimize energy consumption. The algorithm categorizes jobs and identifies sub-solutions for each category based on the dynamic status of edge nodes and cloud servers. Then, a heuristic crossover algorithm is used to merge sub-solutions to generate a comprehensive solution. The findings from simulation trials demonstrate that PGMEA exhibits superior solution quality, reduced job computing time, and decreased energy usage compared to existing scheduling algorithms.

To reduce the energy consumption of the chilled water system, Deng et al. [81] proposed a coal-water slurry optimization control strategy in a cloud-edge-end platform. They first designed an ensemble learning method for cooling load forecasting based on error compensation and conditional fuzzy matching. This method can be applied to cooling load prediction under different working conditions with unbalanced sample distribution. Based on the prediction results, a new control strategy is developed to optimize the selection of process control inputs, ensuring that the cooling load demand is met with lower energy consumption. Furthermore, the authors also used cloud edge terminals to train optimal control strategies in real-time, which can be used for big data modeling to improve the effectiveness of system response. Experimental results show that the proposed control strategy reduces energy consumption by 5.59% and increases the coefficient of performance from 5.64 to 6.14.

In distant places without roadside infrastructure deployment or network coverage, single-vehicle intelligence is susceptible to complete or partial functional failure. The high costs associated with deployment and service fees also lead to low user acceptance and penetration of edge servers. Therefore, Xiao et al. [419] proposed a new vehicular mobile-edge-platooning cloud (MEPC) platform for task scheduling, resource sharing, and autonomous traffic. MEPC utilizes platoon resources' aggregation, sharing, and economy to provide aggregated computing resources, storage, communication, and intelligence. The authors then integrated sophisticated technologies, such as blockchain and artificial intelligence, to achieve intelligent, decentralized, and scalable applications. Blockchain technology can provide a decentralized architecture while ensuring the adequacy and security of training data. Therefore, the proposed MEPC can solve the problem of distributed intelligent scheduling, meet application delay requirements, and ensure data security and anti-tampering. Experimental results demonstrate that MEPC can effectively balance energy consumption and execution delay, resulting in improved delay and energy consumption performance.

Dong et al. [94] proposed an economical energy dispatch framework based on cloud edge computing architecture aimed at minimizing the long-term operating costs of microgrids. The proposed scheduling framework employs machine learning to optimize energy scheduling sequences and make real-time inference decisions. Historical operations are used as samples for supervised training, and advanced machine learning models can achieve complex mapping of input and output spaces. This approach avoids predicting random variables and reduces the difficulty of designing adjustment strategies or reward strategy functions for real-time scheduling. Furthermore, the proposed energy scheduling solution enables real-time decision-making through local reasoning, significantly reducing energy consumption and increasing deployment flexibility. Experimental results indicate that the suggested solution can accomplish a cost close to optimal scheduling, even lower than the algorithm, without uncertainty in some cases.

Distributed energy resources (DER) are an essential component of power systems, especially those that support the large-scale integration of renewable resources. However, differences in capabilities and characteristics between DERs may lead to undesirable characteristics such as slow overall response and severe overshoot. In order to attain comprehensive coordination among DERs, Han et al. [144] proposed a coordination scheme based on cloud-hosted and edge-hosted DER digital twins (DT). The solution includes a centralized, coordinated control algorithm and a distributed, coordinated control algorithm for different control systems. The centralized coordinated control algorithm uses a DT hosted in the cloud to calculate the DER's real-time output, which is then utilized by a controller also in the cloud to coordinate the DER's response. The distributed coordinated control algorithm hosts DTs and controllers at the edge sites of individual DERs to estimate the DER's real-time output and minimize the need for real-time communication. Experiments have proven that both centralized and distributed algorithms can significantly reduce the need for and dependence on real-time communications while concurrently offering efficient assistance to the power grid in emergencies.

User preference is a crucial factor in influencing task offloading strategies to enhance the battery's longevity in intelligent mobile devices. For example, fear of low battery anxiety may lead users to accept higher latency to extend battery life. Therefore, Yuan et al. [469] put forward a user preference-based offloading approach (UPOA), which takes into account user preferences during the offloading decision-making process. In UPOA, the authors first establish a user preference rule to define the user's offloading preference based on the user's battery energy status. Subsequently, a task offloading model is constructed to represent the task assignment of each node in the offloading link, simplifying complex task transfer between different devices. Based on the model, they solved the uncertainty issue of offloading tasks through the task prediction algorithm of the long short-term memory (LSTM) neural network model. Finally, the authors design an online offloading algorithm based on particle swarm optimization by combining task prediction algorithms and user preference rules to provide the best long-term offloading strategy. The experimental findings demonstrate that the proposed UPOA outperforms the three leading offloading methods, namely A Double-Q Deep Reinforcement Learning Approach [496], DRL-E2D [223], and MUDRL. The UPOA is capable of generating efficient policies that align with user preferences, resulting in a 12.49% reduction in average latency when battery energy is adequate and a 20.14% extension of battery life when battery energy is low.

To enhance the efficiency of cloud-edge collaboration, Tang et al. [356] suggested an LEO-assisted terrestrial-satellite network (TSN) framework for energy-efficient task offloading. The cloud-edge orchestrated optimization problem is first formulated to minimize the energy consumption of the entire TSN under limitations on the quality of service, which has proved to be a mixed integer programming problem. Further, the optimization problem is decomposed into two optimization sub-problems of integer variables and continuous variables based on the optimization variables. Then, successive convex approximations and deep neural networks are used to solve these two optimization sub-problems, respectively. DNN is used to obtain user association patterns and formulate task offloading strategies, and the SCA algorithm is used to solve the remaining subproblems based on the DNN output. The experimental findings demonstrate that the suggested cloud-edge orchestrated computing framework is capable of significantly mitigating the energy expenditure associated with cloud-edge collaboration purposes.

Aiming to reduce the total computing latency of mobile users, Yang et al. [452] put forward a hybrid mobile cloud/edge computing system that implements computing offloading and data caching. The authors initially formulated the challenge as the simultaneous optimization of compute offloading and data caching solutions, with the objective of minimizing the overall execution delay experienced by the mobile user. Considering the limited resources of mobile edge computing servers and mobile battery capacity, joint optimization must also meet the constraints of each user's maximum tolerable energy usage and each MEC server's computing power and cache. Since the computational offloading and data caching decision variables are highly coupled binary values, the constructed optimization problem is proven to be a discrete bilinear non-convex optimization problem. To simplify the problem, the authors used the McCormick envelope method and introduced auxiliary variables to decouple the bilinear discrete variables, thereby converting the problem into a linear programming problem. Then, an alternating-direction multipliers algorithm with low computational cost was developed to achieve near-optimal computational offloading and data caching decisions. The experimental findings demonstrate that the method presented as a solution significantly mitigates the overall processing delay experienced by mobile users while simultaneously maintaining individual users' performance.

The presence of a heterogeneous network, known as Het-Net, which consists of several base stations (BS) operating within the same frequency band, can lead to significant mutual interference and a decrease in global energy efficiency (GEE) due to spectrum sharing between distinct BS-user links. To address this problem, Zhang et al. [485] proposed a cloud-edge collaboration-assisted intelligent power control architecture to improve GEE and enhance the transmit power of each BS. The proposed framework establishes an edge deep neural network at each edge base station for formulating the transmission power policy while establishing a shadow DNN with the same structure for each edge DNN in the cloud. Subsequently, all edge BSs coordinate the local data with the shadow DNN in the cloud, enabling the cloud to comprehend the overall information of the whole Het-Net. Based on global information, a DRL-based energy-saving power control algorithm is suggested to train shadow DNN to optimize transmission power and enhance GEE. With this approach, each edge BS can perform real-time transmit power configuration based only on local observations by combining local edge DNN and associated shadow DNN parameters. The simulation results demonstrate that the proposed approach exhibits significantly reduced time complexity in comparison to the SFP algorithm [473] in order to attain equivalent performance in terms of GEE in both static user and mobile user scenarios.

Variable renewable energy sources are an important part of the power system. However, the output power of these energy sources is uncertain and cannot be directly connected to the power system, which increases the need for resources related to power system frequency regulation. Considering that distributed energy resources have good potential to provide frequency support, Zhang et al. [481] proposed wireless communication coordination architecture and a DER-based cloud edge collaboration to achieve frequency regulation. The framework aims to optimize and collaborate tasks and resources, taking into account the balance between the level of control uncertainty and the expenses associated with computation and transmission. In this framework, the cloud is used to balance uncertainty power with edge controllability as well as computational burden and wired and wireless communication loads, and the edge is used to adjust the power output of the storage for achieving frequency regulation. Furthermore, an automatic gain control signal scheduling algorithm is proposed to minimize the impact of photovoltaic uncertainty and maximize resource utilization. Experiments prove that the proposed framework maintains an average of 95.5% of the theoretical maximum frequency modulation adjustable power capability while reducing the computational cost by 36.3%.

Energy efficiency optimization of cloud/edge computing systems is an important issue, and the technology known as dynamic voltage and frequency scaling (DVFS) has the capability to adjust its computation frequency in response to demand, which helps reduce energy costs when processing uncertain workloads. Therefore, Zhang et al. [494] focused on the energy-reducing adaptive workload allocation and computing frequency configuration issues in distributed cloud/edge computing systems. First, they defined the service risk probability (SRP) as the probability that a virtual machine will fail to handle the incoming workload during the current time slot. Subsequently, a stochastic energy optimization model that comprehensively considers energy cost and SRP is proposed for virtual machines supporting DVFS. Based on the model, they derived the closed form of achieving minimum energy cost and optimal SRP when the workload does not follow any distribution and follows the Gaussian distribution, respectively. For workloads that do not assume any distribution, they proposed a heuristic algorithm for finding the optimal workload allocation vector (WAV). For workloads that follow a Gaussian distribution, an alternative optimization algorithm is proposed to solve this fixed computing power WAV optimization problem. The performance of the proposed technique is evaluated using Monte Carlo simulations and real-world virtual machine workload trace data. Empirical findings demonstrate that the suggested algorithm can attain negligible energy consumption compared to algorithms employing a fixed SRP.

Focusing on the high cost of providing precise positioning services in collaborative intelligent transportation systems, Zhang et al. [489] proposed a DRL-based framework that integrates both cloud and edge network infrastructures to reduce resource consumption. They modeled service optimization as a service function chain (SFC) embedding problem from the viewpoint of resource virtualization, and a DRL-based SFC embedding algorithm is proposed for a cloud-side vehicle collaboration network (CEVCN). The DRL-based SFC algorithm perceives the environment of CEVCN and derives the optimal node selection strategy by constructing a multi-layer policy network. Then, they used a breadth-first search to complete the embedding of virtual links. Simulation results show that compared with CLF [343], TDM [488], and DMRT-SL [342], the algorithm under consideration demonstrates a 31% rise in long-term average revenue, a 13% increase in the long-term average revenue-cost ratio, and an 8% increase in embedding rate.

To achieve efficient scheduling of cloud edge servers and cloud data centers, Yin et al. [463] proposed an efficient convergent firefly algorithm (ECFA), which designed a probability-based mapping operator and a low-complexity location update approach. The presented mapping operator relates the firefly space to the solution space by converting individual fireflies into scheduled solutions, while the approach of low-complexity position update improves computational efficiency by reducing the burden when updating firefly positions. Subsequently, the proposed ECFA is demonstrated through a series of theoretical analyses to possess the capacity to converge to the global optimal individual within the firefly space. To mitigate the occurrence of local optimality in ECFA, the authors subsequently present the notion of the boundary trap, which is used to analyze the trajectory of firefly movement and adjust parameter values accordingly. The usefulness and efficiency of the proposed ECFA are demonstrated by simulation studies that compare it with four other metaheuristic-based methods and three FA-based methods.

Focusing on the limitations of static edge computing technology in dynamic environment applications, Chen et al. [59] put forward a DRL-based cloud-edge collaborative mobile computing offloading (DRL-CCMCO) mechanism to reduce energy usage costs in dynamic environments. The problem is framed as a collaborative optimization problem of offloaded task execution latency and energy consumption, considering mobility, task dependencies, offloading strategies, and network resources. Mobility is addressed by implementing the digital twin mobility solution and cloud-edge collaboration with powerful computing capabilities. This solution significantly reduces the costs incurred in offloading computing tasks to edge nodes due to user devices relocation. Then, a distributed task resource network is employed to solve user task dependencies so that tasks with relevant characteristics can be quickly processed and returned regardless of the movement of the user devices. Furthermore, the proposed algorithm leverages cloud-edge collaboration to enhance the parameters of the neural network in a distributed manner, thereby enabling swift decision-making for optimal offloading. The simulation results demonstrate that the DRL-CCMCO algorithm, as described, is capable of achieving optimal offloading decisions while maintaining the lowest overall cost when compared to alternative approaches such as local computing, offloaded computing, deep deterministic policy gradient, deep network, and A3C algorithm. Comparison with three other representative deep reinforcement learning methods also demonstrates the proposed DRL-CCMCO algorithm's superiority in terms of stability and convergence speed.

The increase in mobile device users and rich media content services has significantly increased the demand for network resources and energy. Fang et al. [107] presented a cloud-edge joint task offloading and resource allocation algorithm (TORA-DRL) based on deep reinforcement learning to address the heterogeneous network resource allocation problem, which arises when the spatiotemporal distribution of mobile user requests can undergo significant changes. The initial formulation of the resource allocation problem involves minimizing energy usage while considering the limitations imposed by restricted computer resources, communication resources, and network cache. Implementing request aggregation and in-network caching techniques mitigates redundant transfers of network content, enhancing energy efficiency. Subsequently, the proposed TORA-DRL algorithm utilizes historical request data and current network resources to make optimal resource allocation and task offloading decisions. The approach enables the algorithm to effectively adapt to constant fluctuations in network status and user demand. Compared with existing popular algorithms, simulation results indicate that the suggested TORA-DRL algorithm can significantly alleviate the duplicate content distribution problem and reduce network power usage, thereby achieving higher energy efficiency in cloud-edge collaboration environments.

Edge nodes in mobile edge computing execute computing tasks offloaded from users' mobile devices. However, edge nodes cannot perform all tasks in time due to constrained communication and computing resources, requiring cloud data centers to provide low-latency services. Thus, Yuan et al. [468] investigated the optimization problem of task offloading, task segmentation, and user association in a cloud-edge system. This objective is to minimize the overall cost of the system while ensuring that user latency limitations are carefully met. Given the ability to divide the application into many subtasks with interdependencies, it has been demonstrated that the problem can be classified as a mixed integer nonlinear program. Next, a hybrid algorithm based on a genetically simulated annealing particle swarm optimizer generates a near-optimal task allocation strategy. The algorithm integrates the Metropolis acceptance criteria of genetic operations and simulated annealing into the particle swarm optimizer. The experimental findings demonstrate that, compared to the baseline approaches, the proposed GSPSO reduces the cloud-edge system's total energy consumption cost while strictly meeting user task completion time requirements.

Alnoman et al. [11] proposed a dormancy mechanism for edge devices and small base stations (SBS) to reduce the energy usage of heterogeneous cloud radio access networks under the constraints of instantaneous task completion time and long-term statistical response time. First, the authors consider a cloud-edge orchestrated system where workloads are not shareable, and cloud or edge devices can only provide computing resources independently. The problem is defined as a 0-1 knapsack problem that seeks to minimize the quantity of active SBSs. The weight of the SBS utilization is determined by the number of incoming computational jobs, while the number of incoming computational tasks determines the value of the SBS. Furthermore, SBSs with low service loads are assigned higher values than other SBSs. This problem is solved by a centralized SBS hibernation solution based on dynamic programming. The objective is to identify the most suitable subset of hibernation SBSs, taking into account both cloud and user limitations. Then, a new shared cloud edge computing architecture coordinated with cellular infrastructure is introduced to share workloads at the cloud edge. Based on the shared architecture, an exhaustive search method is utilized to optimize energy savings under response time constraints. Experimental results prove that the suggested sleep algorithm has the potential to substantially decrease the overall energy usage of cloud-edge systems while shared computing systems perform better in terms of response time.

The explosive growth of computing-intensive and latency-sensitive applications, such as autonomous driving, has created huge challenges for the computing power of the mobile edge. Focusing on leveraging cloud-edge systems to reduce application latency, Bachoumis et al. [26] put forward a collaborative cloud-edge-end offloading solution to accommodate these smart mobile applications. This solution combines mobile devices, edge servers, and the central cloud to form a multi-layer heterogeneous architecture that better utilizes available computing, storage, and network resources and achieves a higher level of QoS.

Considering multi-component applications with component dependencies, a multi-objective mixed integer linear program (MILP) is designed to minimize application latency and system energy consumption. Based on LP relaxation and rounding, the authors propose an efficient approximation algorithm to solve the MILP in polynomial time. Simulation experiments show the superiority of the proposed solution in terms of application request acceptance rate, system energy consumption, and latency.

This section reviews the literature on cloud-edge-end collaboration systems in terms of optimizing energy management and reducing energy consumption. These works show that there is still much room for improvement and enhancement of energy consumption in cloud-edge-end collaboration systems. Many works focus on minimizing the system's energy consumption while considering the constraints of performance metrics such as execution latency and response time. They often consider other performance metrics, such as security, reliability, and so on, while optimizing the energy consumption to ensure the system's performance. Many works have adopted different measures and approaches for optimization, such as deep reinforcement learning-based approaches and digital twin-based approaches, to minimize energy consumption, and all of them have achieved good results.

3.3.3 Optimization for Security and Privacy of Cloud-Edge-End Systems

This section discusses the efforts made toward optimizing security and privacy in cloud-edge-end collaboration systems. As listed in Table 3.3, we summarize these works and compare them regarding optimization goals, methods used to protect security, performance, and benchmarking schemes.

Zhang et al. [478] focused on the security and reliability of data generated by industrial devices within trusted AI-enabled IIoT. They observed that the complicacy and heterogeneity of AI-driven IIoT systems result in a lack of trust among devices, consequently impacting data security within the network. Moreover, the current trust mechanisms must be better suited for distributed networks. Therefore, they introduced a cloud-edge-end framework built upon blockchain technology and devised a trust mechanism grounded in blockchain consensus. Firstly, they leveraged proof of replication (PoRep) based on BLS as its consensus mechanism to achieve mutual trust among devices. Ultimately, a verifiable delay function employing secret sharing is implemented to deter the server from dynamically computing replicas within the PoRep scheme and enhance efficiency. Experimental results indicate that compared to several representative schemes, this scheme offers the advantages of lower computing, storage, and communication costs while ensuring security and reliability.

Li et al. [210] investigated a method of exploring human factors when designing partitioning algorithms for orchestrating cloud-edge-end systems within a secure and reliable environment. They discovered that current cloud-edge-end orchestrations mainly focus on technical metrics but need to consider the influence of user interaction habits on component usage. Hence, they first designed a system consisting of the profiling and blockchain modules. The profiling module collects user behavior data and provides data support for cloud-edge-end orchestration while preserving privacy. Furthermore, the blockchain module utilizes IPFS and blockchain technology to enable secure large data transmission. Finally, to observe user-driven relationships among method-level components, they suggested a clustering algorithm framework for analyzing the data of each subject. After conducting a case study, it has been demonstrated that user behavior varies, and the findings additionally confirm the viability of integrating human factors into the partitioning algorithm.

Souri et al. [330] investigated a way to manage the resources in cloud-edge computing within the environment of S-IoT. Thus, enhancing the performance of the S-IoT system through the optimal selection of processing locations is proposed for task execution while

TABLE 3.3

A Comparison Summary of Security and Privacy Optimization Methods from Multiple Aspects, Such as Optimization Target, Security Method, Performance Improvement, Benchmarks Schemes.

Ref.	Optimization Target	Method	Performance	Benchmarks
[478]	SP and cost	Blockchain; BLS-based PoRep mechanism	38.10%↓-80.82%↓	RSAB and BLSPRFB
[210]	SP and QoE	Blockchain; IPFS	*	*
[330]	Trustworthiness and efficiency	Bayesian deduction; Trust-based framework	*	*
[68]	SP and consumption	Blockchain	*	*
[132]	SP	Differential privacy	*	*
[492]	SP and delay	Blockchain	50.17%↓	DAC
[398]	SP and accuracy	WEMI detection	92.4%	*
[53]	SP and overhead	Blockchain; IPFS	92.13%↓-50.01%↑-91.06%↓	CLPAEKS, PPCBEKS, and BSPEFB
[198]	SP and accuracy	Federated learning; Differential privacy	3.84%↑	PPFLEC
[410]	SP, latency, and energy	Blockchain	1.32%↑-2.07%↓	LARAC
[235]	SP and cost	Attribute-based searchable encryption	66.38%↓-9.14%↓-40.67%↓	MK-ABSE, VMK-ABSE, ABSE-UR
[204]	SP and accuracy	Federated learning; False data injection attack detection	3.72%↑-1.26%↑	GFDIA and CFDIA
[293]	SP and precision	Hierarchical federated split learning	40.47%↑-9.34%↑	Hier-FAVG and Fedsplit
[455]	SP and precision	Quadtree indexing; Geocoding; LDP	*	BR
[419]	SP and latency	MEPC; Blockchain	589.02%↑-50.63%↓	DO and EO
[105]	SP and delay	Improved Apriori; Bilinear pairing; Hash function	9.2%↓-4.62%↓-52.41%↓	CLAEKS, CL-SPKAE, and Tc-PEDCKS
[294]	SP and latency	Blockchain	15.24%↓	DQL
[243]	SP and utility	Federated learning; Blockchain	332.69%↑-43.89%↓	SMC and SDC
[227]	SP and overhead	Paillier homomorphic encryption; Additive secret-sharing scheme	62.13%↓	BAS
[138]	SP and cost	Differential privacy	8.58%↓-59.35%↓-43.61%↓	ItAdap, FLGDP, and FEEL
[116]	SP	Privacy-preserving average-consensus algorithm	*	CAC

(Continued on next page)

TABLE 3.3
(*Continued*)

Ref.	Optimization Target	Method	Performance	Benchmarks
[241]	Efficiency	Federated learning	*	AsyncFL
[242]	SP, scalability, and fairness	Plaintext checkable encryption scheme; Blockchain	*	*
[42]	SP and cost	Security score	*	LPS
[497]	SP and revenue	Dynamic trust evaluation	12.69%↑-15.22%↑-26.55%↑	SA-VNE, RL-VNE, and STEC
[109]	SP and performance	GS-SNC	*	DP-SNC and SPOC
[402]	SP and accuracy	EM-Based Federated Control Scheme	48.38%↓-31.18%↓-17.68%↓	LS, LSST, and SPA

*: No relevant experiments and data are mentioned in the literature;

†: Compared to the benchmark scheme, the methods proposed in the article do not save energy or reduce latency but may also be performance improvements such as revenue, accuracy, and precision;

↓: Compared to the benchmark scheme, the method proposed in the article saves energy or reduces latency;

SP: security and privacy; RSAB: RSA-based; BLSPRFB: BLS and PRF-based; IPFS: interplanetary file system; WEMI: weak electromagnetic interference; CLPAEKS: certificateless public key authenticated encryption with keyword search scheme; PPCBEKS: pairing-free and privacy-preserving certificate-based encryption with keyword search scheme; BSPEFB: blockchain-based searchable public-key encryption scheme with forward and backward privacy; PPFLEC: privacy protection scheme for federated learning under edge computing; LARAC: lagrangian relaxation-based aggregated cost scheme; MK-ABSE: multi-keyword attribute-based searchable encryption; VMK-ABSE: verifiable and multi-keyword attribute-based searchable encryption; ABSE-UR: attribute-based searchable encryption with user revocation; GFDIA: GNN-based false data injection attack detector; CFDIA: CNN-based false data injection attack detector; Hier-FAVG: hierarchical federated averaging algorithm; Fedsplit: federated split learning; LDP: local differential privacy; BR: Basic Rappor; DO: delay optimal; EO: energy optimal; CLAEKS: certificateless authenticated encryption with keyword search scheme; CL-SPKAE: certificateless searchable public key authenticated encryption scheme; Tc-PEDCK: time controlled public key encryption with delegated conjunctive keyword search; DQL: deep Q-learning; SMC: simple shared model scheme; SDC: simple shared data scheme; BAS: bilinear aggregate signature; ltAdap: adaptive federated learning in resource-constrained edge computing system; FLGDP: differentially private federated learning; FEEL: federated edge learning system; CAC: classic average consensus; AsyncFL: asynchronous federated learning; LPS: linear programming solver; SA-VNE: security-aware virtual network embedding; RL-VNE: reinforcement learning virtual network embedding; STEC: security tactic by virtualizing edge computing; GS-SNC: gold sequence-based secure network coding; DP-SNC: double prime numbers-based secure network coding; SPOC: secure practical network coding; LS: least square without considering security threats; LSST: least square considering security threats; SPA: sum-product algorithm.

also facilitating network navigation. The improvement is achieved by appropriately naming available devices. Firstly, they put forward a three-layer framework to analyze S-IoT. Furthermore, they defined different social relationships between IoT devices and presented a relationship model based on Trust, which is used to optimize resource management. Ultimately, an algorithm for optimal friendship selection based on Trust is proposed while introducing a loss function and minimizing it to select the optimal edge partners. Experimental results show that the presented S-IoT can achieve a minimum network distance, better degree distribution, and shorter average latency compared with other networks, such as random networks.

Chi et al. [68] presented an AIoT framework to tackle the challenges encountered in intelligent edge computing. They observed that deep learning models tend to be huge in scale, training and deployment of the models are disconnected, and the security of AIoT needs to be considered. Therefore, they proposed a four-layer construction that relies on MLP and blockchain. Initially, the framework utilizes lightweight models to conserve computing resources and simplify deployment at the edge. In addition, the solution of distributed identity and the decentralized MLP neural networks are used to optimize the global model. Ultimately, in response to the privacy and security problem, they presented integrating blockchain technology into the framework to guarantee the safeguarding of privacy and the sharing of data and parameters. The experimental findings demonstrate that the suggested framework is feasible and effective, as shown by the simulation experiments that apply carbon emissions data.

Guo et al. [132] investigated a method of distributing a finite number of edge servers among IoT devices in an Edge-assisted system using blockchain technology. They discovered that to deal with this problem, they will encounter three challenges: internet-based, truthful, and privacy-protected. Therefore, the MIDA-G mechanism utilizing differential privacy is presented to tackle the challenges. They first created a model of the problem and proposed an MIDA mechanism based on the double auction theory to attain resource allocation. Moreover, to prevent privacy breaches and protect privacy, they introduced differential privacy technology into the MIDA mechanism. Ultimately, they enhanced the MIDA mechanism to the MIDA-G mechanism for privacy preservation to accommodate increasingly intricate and practical scenarios. The experimental results have validated the correctness and efficiency of the designed mechanism in blockchain-based IoT systems and have also demonstrated the necessity of differential privacy.

Zhang et al. [492] explored the problem of the data security and computational complexity issues faced by cloud-edge-end cooperation-based computation offloading in the PIoT. They observed that many challenges need to be addressed when employing AI and blockchain technology in cloud-edge-end cooperation-based PIoT. Thus, a new PIoT framework emerged in response to the challenges. Firstly, they presented an architecture called BASE-PIoT using AI and blockchain technology to guarantee data protection, resource allocation, and intelligent calculation offloading. Furthermore, they investigated the compatibility between three representative blockchains and PIoT in order to select an appropriate blockchain for the proposed framework. Ultimately, a blockchain-enhanced task offloading algorithm based on federated actor-critic is proposed to improve security and latency. Experimental results show that, in contrast to DAC, the proposed BASE-PIoT achieves a reduction of 50.17% in overall queuing delay and a decrease of 18.99% in consensus delay.

Chen et al. [53] investigated the impact of different data-sharing mechanisms on security and computational efficiency. They found that conventional data-sharing mechanisms in the IoT environment present numerous problems and challenges, such as privacy protection, data security, etc. Consequently, they proposed an efficient and secure decentralized data-sharing scheme called Double Rainbows for IoT environments. Firstly, after introducing and analyzing some basic concepts and mathematical knowledge, they observed that the technology of IPFS and intelligent contracts are beneficial for improving the scheme. Secondly, based on the observation, they proposed a Double Rainbows scheme to enhance security, reliability, and efficiency. The scheme consists of seven stages, including initialization, model updating, and so forth. It's worth noting that it is built on cloud-edge-end construction and is based on Pairing-free SE. Performing security assessments and evaluating performance, experimental results show that Double Rainbows is safer, more efficient, and more suitable for real-world application scenarios.

Lai et al. [198] explored the precision, versatility, and security in smart diagnostic models for intelligent medical. They discovered that security is not guaranteed even when using federated learning, and privacy leakage persists. Furthermore, mobile healthcare devices have limited storage and computational resources, and model accuracy always needs improvement. Hence, a solution named EICPP is proposed to address existing problems and challenges. They first proposed a federated learning cooperative lightweight edge intelligent architecture called KubeFL to enable health tracking and assistive diagnosis. Besides, they put forward a cloud-edge-end federated learning model to ensure accuracy while mitigating the potential for privacy breaches. Finally, they utilized the differential privacy technology for model transmission between edge and cloud to guarantee the model security. Experimental results suggest that the proposed EICPP can not only achieve a recognition accuracy of 95% and efficiency but also ensure model security and protect user privacy.

Wu et al. [410] investigated a method of task offloading based on blockchain technology for an IoT-Edge-Cloud environment. They observed that existing research is too simplistic and always focuses solely on either MCC or MEC. Besides, security and privacy issues persist throughout the task offloading procedure. Therefore, they created an IoT-edge-cloud computational framework with blockchain support to address the problem. Firstly, the architecture simultaneously leverages both MEC and MCC to better utilize the advantages of low latency in MEC and the powerful computational capabilities of MCC. In addition, they conducted mathematical models of response time, energy consumption, and optimization problems. Lastly, they proposed the EEDTO algorithm, which utilizes the Lyapunov optimization method to dynamically select the optimal computational position for each task. Experimental results indicate that compared with IoT-Only, Edge-Only, Cloud-only, and LARCA, the proposed EEDTO can achieve minimal energy consumption and a relatively shorter response time.

Li et al. [204] explored a way to detect data injection attacks in a distributed network environment. They observed that existing methods do not adequately address user privacy and data island, and the past research has not considered the spatiotemporal correlation of data. Consequently, they designed a detection architecture that relies on collaboration between the end, edge, and cloud to identify FDIA in the distribution network. Firstly, they proposed an edge-cloud cooperative mechanism that uses federated learning technology to train the local models for FDIA detection, protecting user privacy and addressing the problem of data islands. Ultimately, a novel FDIA detection network called TSGCN, based on spatiotemporal information and a convolutional network, is proposed to enhance the performance of FDIA detection. Experimental results suggest that compared with the conventional method of FDIA detection, the presented method can fully leverage computational resources and improve accuracy.

Qin et al. [293] investigated the shortcomings of existing network traffic categorization techniques. They discovered that current schemes have the problem of efficiently consolidating distributed data, the scope of public dataset collection is limited, and the model's update rate is sluggish. Thus, a new architecture called Hier-SFL, based on split learning and federated learning, is proposed to address the problem in an end-edge cloud environment. Firstly, they combined end-edge-cloud systems to divide the neural network into three components: the representation layer, the hidden layer, and the task layer. Then, the clients train the representation layer parameters locally and periodically upload "smashed data". Furthermore, the edge servers train part of the hidden layers locally and aggregate client models periodically, achieving federated learning. Finally, the cloud center finishes global model training and distributes it to the edges and clients. Experimental results indicate that Hier-SFL can achieve an accuracy rate surpassing them by at least 5% and demonstrating greater stability, lower communication costs, and computational consumption compared with other methods.

Yang et al. [455] focused on a method of trajectory privacy protection for vehicles in the logistics sector. They observed that current privacy protection techniques overlook the spatiotemporal correlation in trajectory data, leading to significant errors in some aspects. Therefore, they proposed two new algorithms to protect user privacy and ensure data availability. At first, they created an intelligent logistics system that combines with the model of a semi-honest adversary and collaborates across the cloud, edge, and end. Additionally, they presented a location encoding method using spatial quadtree indexing that enables the adjustment to balance data availability and privacy. Finally, two new algorithms based on differential privacy are proposed: QLP and QJLP. QLP offers privacy protection at the location, while QJLP provides privacy protection at the trajectory. Experimental results show that QLP and QJLP are not only effective but also ensure user privacy and data availability.

Xiao et al. [419] explored the limitations of single-vehicle intelligence and the resources of edge servers. They found that single-vehicle intelligence lacks flexibility, is prone to failures, and edge server resources are expensive. Therefore, a new scheme utilizing platoon resources to establish a mobile edge computing platform is proposed. Firstly, they put forward a new cloud-edge-end concept called MEPC, which integrates with platoon resources to improve the system's flexibility and economy. Subsequently, based on the proposed MEPC, a cloud-edge-end orchestrated framework called M-CEE is proposed. In the end, they combined the M-CEE framework with various technologies, including artificial intelligence, blockchain, and others, to achieve security, effectiveness, and intelligence. They also used MEPC as a reference point to discuss the framework's applicability. Experimental results show that MEPC can manage the trade-off between latency and power consumption while ensuring performance compared with other strategies.

Fan et al. [105] investigated the shortcomings of searchable encryption schemes in intelligent grid scenarios. They discovered that the current searchable encryption scheme could not be suitable for the intelligent grid territory, and balancing security and efficiency when performing searches based on keywords is also a challenge. Consequently, they put forward a new searchable encryption scheme called MSIAP in the environment of cloud-edge-end collaboration. They first modified the original Apriori algorithm to mine the correlations among the vast intelligent grid data and construct a multi-layered indexing structure. Furthermore, encrypting keywords is achieved by utilizing both hash function and bilinear pairing. Finally, based on the multi-layered indexing structure and the correlations obtained through data mining, they realized high-performance subset retrieval and update of multi-keywords. Experimental results indicate that compared with prior studies, MSIAP can not only satisfy intelligent grid scenarios but also guarantee the safety and efficiency of retrieval and update.

Qiu et al. [294] designed to enhance the combination of the IoT and machine learning. They found that it is challenging to train models in the remote center, the IoT devices have resource limitations, collaborative training between IoT nodes is necessary, and the IoT nodes display heterogeneity and lack of trust. Hence, they presented an approach of collective Q-learning based on blockchain. Firstly, they utilized lightweight IoT nodes to train portions of neural networks and blockchain to share learning outcomes. Moreover, they enhanced conventional PoW to PoL in order to reduce the wastage of computational capacities and improve the learning ability of edge nodes. Ultimately, a resource allocation problem is proposed to test the performance of collective Q-learning in the IoT environment. The experimental findings suggest that the proposed approach can improve convergence performance, reduce training latency, and realize better resource utility compared with other methods.

Liu et al. [243] explored the application of blockchain technology and federated learning in data sharing. They observed that federated learning still faces security issues, blockchain technology struggles to handle massive data storage, and existing incentive mechanisms overlook the raw data providers. Therefore, a novel data-sharing framework that integrates blockchain technology and federated learning and achieves privacy preservation is put forward to address the challenges. Firstly, they proposed a cloud-edge-end cooperation-based layered framework and differential data sharing to tackle the lack of resources and heterogeneity. In addition, they presented an incentive mechanism utilizing a double-stage Stackelberg game to optimize the utility of data providers and requesters. Finally, a gradient-based algorithm is proposed to obtain the best solution. Experimental results indicate that the presented data-sharing framework can protect privacy, achieve better performance, and encourage more participants to share data.

Lin et al. [227] investigate the persisting security concerns within the domain of federated learning. They discovered that aggregation servers might provide inaccurate aggregated outcomes, and the current state of federated learning technology still has privacy leakage issues, such as information leakage through gradients. Consequently, they put forward an approach for federated learning named PPVerifier that ensures verifiable and privacy preservation in the environment of cloud-edge cooperation. Firstly, they combined Paillier encryption with technology for generating random numbers to accomplish gradient protection. Furthermore, a secret-sharing framework is proposed to defend against potential attacks. In the end, they presented a verification approach using a discrete logarithm-based hash function to detect lazy aggregators and validate the correctness of results. Experimental results show that the proposed PPVerifier can achieve gradient preservation and guarantee verifiability. It is also more efficient than the other method utilizing the MNIST dataset.

Guo et al. [138] developed the relationship between data privacy and efficiency in federated learning and the challenges it faces. They found that in federated learning, enhancing efficiency may introduce more security concerns, while focusing on privacy protection may decrease efficiency. Besides, combining privacy protection techniques with a hierarchical architecture is challenging. Thus, an adaptive federated learning, which is called AHFL based on hierarchical architecture, is proposed to address the problem. Firstly, they utilized federated learning in combination with a cloud-edge-end architecture to promote efficiency. Secondly, they implemented a dual-stage differential privacy protection mechanism to ensure privacy safeguards. Finally, they proposed a dynamic control algorithm to orchestrate the privacy protection measures and resource allocation to enhance overall system performance. Experimental results indicate that compared with other baselines, the proposed AHFL can reduce computation time by 8.58%, decrease communication time by 59.35%, lower memory consumption by 43.61%, and improve accuracy by 6.34%.

Fu et al. [116] explored a method to safeguard the privacy of device information from leakage for cloud-edge-based optimal energy management (OEM) in the smart grid. They observed that current privacy protection technologies still fall short in OEM tasks, such as analyzing data collected over multiple iterations, which may expose privacy. Therefore, in order to address the problem, they put forward a decentralized privacy-preserving OEM algorithm. Firstly, they presented an intelligent gird cloud-edge computing architecture where different layers accomplish distinct tasks. Subsequently, they designed a privacy-preserving average consensus algorithm to safeguard the initial state's privacy. Eventually, by applying the above algorithm, a decentralized OEM algorithm with privacy protection utilizing the generalized ADM algorithm is proposed to prevent information leakage. Experimental results show that the presented two algorithms in the smart grid can guarantee privacy security, realize the average consensus, and demonstrate that it is effective.

Liu et al. [241] investigated the shortcomings and issues of applying federated learning in the context of the AIoT. They discovered that federated learning in AIoT based on cloud-edge-end orchestration has low communication efficiency and slow model convergence in AIoT based on cloud-edge-end orchestration. For this reason, they presented a novel federated learning method named QuAsyncFL to deal with these challenges. Firstly, they employed asynchronous federated learning to better adapt to the cloud-edge-end architecture in their new approach and reduce network latency. Next, to protect privacy and enhance communication efficiency, they proposed integrating asynchronous FL with quantization techniques, which quantizes the local gradients based on a natural compression approach. In the end, an in-depth theoretical analysis of the convergence is offered to demonstrate the convergence of QuAsyncFL. Experimental results indicate that compared with the initial asynchronous FL approach, the proposed QuAsyncFL can achieve better communication efficiency and flexibility in cloud-edge-end orchestration-based AIoT.

Casola et al. [42] explored a method of optimizing the deployment of secure cloud-edge systems in an industrial IoT environment. They observed that the available research fails to consider data transfer costs and neglects latency limitations and potential security in industrial IoT environments. Therefore, to compensate for such deficiencies, they formalize, solve, and experimentally evaluate the existing issues. They first proposed a new formalization of the cloud–edge distribution problem that considered security, cost, and other factors. In addition, they defined three distinct optimization goals and focused on cost minimization. Finally, based on the above optimization goals, a greedy algorithm-based deterministic solver that is both suboptimal and efficient is proposed to find suboptimal solutions rapidly. Experimental results show that the presented solver can obtain the same solution in most cases while being significantly faster than a linear programming solver.

Liu et al. [242] presented the issues cross-domain data-sharing technologies face in a cloud-edge-end scenario with a zero-trust model. They found that existing research fulfills specific data-sharing properties but still ignores some properties, such as efficiency, fairness, security, and scalability. Consequently, they presented two data-sharing protocols and conducted theoretical analysis and experimental testing. Firstly, they devised a novel plaintext checkable encryption method based on ECC-Elgamal for verifying ciphertext validity in lightweight Internet of Things devices. Furthermore, a new sharding blockchains-based multi-domain cloud-edge-end framework and a protocol for sharing data across domains under partial-trust models are proposed to enhance scalability, performance, and security. At last, based on the above protocol, they put forward a protocol for sharing data across domains under zero-trust models to guarantee fairness. After conducting a theoretical and security analysis, experimental results show that the two proposed protocols can ensure superior performance, fairness, safety, and scalability.

Zhang et al. [497] investigated the feature of the cloud-edge-end orchestrations vehicular network and some virtual network embedding (VNE) algorithms related to network resource allocation. Within the research, it is uncovered that current VNE algorithms face difficulties in discovering the optimal solution and always neglect the distinct quality of users' service needs. Accordingly, they put forward a VNE algorithm utilizing deep reinforcement learning called DRLS-VNE to solve the resource allocation issue in CETCVN. First and foremost, they constructed a heterogeneous and multidimensional network model for further research. Moreover, they enhanced DRLS-VNE performance using a multi-layer DRL agent that dynamically extracts network feature properties. In the end, they proposed a mechanism for dynamically evaluating trustworthiness to perform real-time evaluations and establish embedding constraints. The experiments' findings demonstrate that the DRLS-VNE is effective and practical and can greatly improve the VNE solution's performance.

Fang et al. [109] explored an approach that can strike a balance between resource constraints and security requirements in cloud-edge-end collaboration-enabled artificial intelligence of things. They observed that the system is susceptible to attack because cloud-edge-end orchestration transmission media is open. Besides, limited end resources make some solutions infeasible. Therefore, they introduced a new gold sequence-based secure network coding strategy, which is lightweight and safe to address the above issues. They first utilized the gold sequence to produce the pseudo-random sequence, which was then used to scramble and descramble the original data. Secondly, they established a pre-coding matrix for the encoding and decoding the scrambled data. Lastly, random linear network coding is carried out by the intermediate nodes. According to the experimental results, the proposed GS-SNC performs better in computing complexity, security, and other aspects than SPOC and DP-SNC.

Wang et al. [398] introduced an intrusion detection approach for weak electromagnetic interference in the IIoT. They discovered that relevant research is scarce and that there is room for improvement in the design of detection models, detection performance, and resource constraints. Consequently, they suggested an approach based on deep learning for detecting WEMI intrusion in zero-touch networks. First, they proposed a fingerprint extraction approach based on moving averages and Kalman filters. To decrease the computational complexity, they extracted the features of the time and frequency domain from fingerprints. Next, they employed deep learning technology to achieve WEMI intrusion detection. At last, they presented a cloud-edge-end collaboration computing architecture and deployed the model to enhance detection efficiency. The experimental findings suggest that the detection accuracy of the proposed method based on the AdaBoost model can reach 92.4%, and the detection efficiency is faster compared with the centralized deployment approach.

Liu et al. [235] introduced the shortcomings and deficiencies of searchable encryption algorithms in a cloud-edge environment. Their exploration shows that single-keyword searchable encryption methods have low efficiency, while multi-keyword searchable encryption methods have a high computational cost. For this reason, a new cloud-edge orchestrations-based multi-keyword attribute-based searchable encryption (EMK-ABSE) approach is put forward to address the challenges. At first, in EMK-ABSE, the cloud server is responsible for storing encrypted data, while the edge nodes maintain the corresponding encrypted index in order to provide decryption assistance and carry out multi-keyword searches. In the end, they combined encryption with a hybrid online/offline approach to reduce computing costs. After conducting a theoretical and performance analysis, the findings from the experiments indicate that the proposed EMK-ABSE can ensure security, achieves encrypted multi-keyword retrieval, and exhibits superior efficiency and practicality compared to other methods.

Wang et al. [402] investigated the deficiencies and issues of tracking systems in the industrial Internet of Things. The research reveals that existing efforts still cannot address the issue of trustworthiness in the IIoT, which could potentially hinder the effective management of the systems. Thus, they put forward a federated control scheme based on expectation maximization (EM) and machine learning to solve existing problems. As a first step, they developed a hierarchical framework for tracking systems utilizing the theory of federated control to protect privacy. In addition, they integrated the above framework with machine learning-based localization and cloud-edge-end orchestration architecture in order to establish a federated scheme based on EM. Ultimately, based on the above steps, a model for trustworthy localization using the EM algorithm is proposed. The experimental results reveal that the proposed scheme's convergence and localization accuracy demonstrate superior performance compared to the conventional approach.

This section reviews and synthesizes work related to cloud edge-end collaboration systems from a security and privacy perspective. These works show that while cloud-edge-end collaboration systems offer many advantages, research on security and privacy still needs to be completed or even neglected. This underscores the continuing need to develop and optimize security and privacy measures in cloud-edge-end collaboration systems to defend against evolving threats and maintain user trust. Therefore, to optimize the security and privacy aspects of cloud-edge-end collaboration systems, many papers have proposed utilizing or optimizing techniques such as blockchain, federated learning, differential privacy, and others to ensure data security and user privacy with good results.

3.3.4 Optimization for Reliability of Cloud-Edge-End Systems

This section conducts a literature review on reliability optimization of cloud-edge-end systems. As listed in Table 3.4, we summarize these works and compare them in terms of optimization targets, technology, performance improvement, and benchmark schemes.

Yang et al. [445] explored the optimization problem of industrial robots in large-scale personalized product production in cloud manufacturing (CMfg) and proposed a CMfg cloud-edge-end collaboration framework. They discovered that industrial robots need more processing capabilities to produce large quantities of personalized products. Therefore, they need self-X (learning, awareness, and optimizing) ability. Deep learning can help them be smarter. However, the cost and data volume of common deep learning models are too high, and cloud-center computing cannot meet the demand. Based on this observation, they designed a CMfg framework for cloud-edge-end collaboration. They placed edge nodes as close to the data source as possible, enabling faster response times, greater broadband availability, and faster insights. Besides, they discussed vital technologies such as transfer learning and incremental learning, which provided excellent reference value for research in this field.

Huang et al. [169] discovered and explored a phenomenon: in the Industry 4.0 environment, multistate stochastic cloud/edge-based network (MECN) is increasingly important to businesses. As an important indicator of MCEN, MCEN reliability is crucial to evaluating MCEN's capabilities. A lot of methods have been used to predict MCEN reliability. However, no method has been found to predict MCEN reliability using a deep neural network architecture. Therefore, they built a reliability prediction model for an MCEN based on a DNN. For the DNN structure, they appropriately converted MCEN information into DNN format, determining DNN hyperparameters and related functions through Bayesian optimization. Last, through two instances (including Amazon Web Services instances), the prediction model's feasibility, applicability, and scalability are verified, providing managers with a new way to quickly understand MCEN's ability to predict MCEN's reliability instantly.

Ma et al. [256] discussed mobile edge computing of cloud-edge Internet of Things devices. They believed mobility is a significant difficulty in ensuring a reliable resource supply; the effectiveness of migration path selection is highly dependent on user mobility and stability among edge nodes. Therefore, they proposed a novel predictive and mobile trackaware approach to fault-tolerant service migration path selection in MEC. First, the study considers the impact of user movement on service migration by predicting whether a user will leave the connection range of the edge server with which they are contacted in time to premigrate and avoid service interruption. Furthermore, the study considers the time needed for startup and loading to establish a connection between edge users and edge services, as well as the impact of environmental and system factors on the processing power and

TABLE 3.4

Compare and Summarize Reliability Optimization from Multiple Aspects, Such as Optimization target, Technology, Performance Improvement, and Benchmark Solutions.

Ref.	Optimization Target	Technology	Performance	Benchmarks
[445]	Reliability	Deep learning, Incremental learning, and Transfer learning	*	*
[169]	Error	Deep Neural Network and Bayesian Optimization	2%↑	AWS
[256]	Reliability	Mobile Edge Computing		OTSM [286]
[433]	Network throughput	Software Defined Network	6.7%↑ and 33.2%↑	Greedy-RT and Greedy-NC
[20]	Delay, energy consumption, and reliability	Multi-objective optimization and Dynamic Bayesian Network	↑	JFP, 3OBJ, STG1 and FLP
[5]	Reliability and latency	Hybrid Cloud-Edge Computing	↑	CO, EO, CE, and ACE
[225]	Reliability and communication efficiency	Digital twin and Federated learning	29.41%↑ and 69.78%↑	WLFL and DNN-DTFL
[453]	Reliability	Deep learning and Deep reinforcement learning	↑	A3C
[420]	Energy consumption and reliability	Deep Neural Network	50%↑* and 30%↑ 20%↑	RSA and GDA
[7]	Availability	Cloud-Fog computing	8%↑ and 4%↑	CKM and RM
[276]	Reliability	Virtual network function placement	*	Real Data
[36]	Online performance	Long Short-term memory and DLS-MOEA	8.9%↓ and 43.5%↑	GEN and RAN
[38]	Latency	Monte-Carlo simulation, and Integer-linear-programming	18.3%↑	*
[498]	Cost and latency	Edge computing	↑	
[150]	Reliability	Cloud-edge-end collaboration and Deep mutual learning	*	*
[425]	Latency, load balancing, and reliability	Software-defined network	*	DDPG, DQN, and Louvain

Ref	Metric	Approach		Method
[281]	Reliability	Cloud-Edge continuum	*	*
[28]	Average miss prediction	Deep learning, Long Short-term memory, and Edge computing	8.9%↑	Kalman filter and Particle filter
[14]	Resource utilization	Cloud-Fog computing	↑	GA [2] SA [458] DRAM [434]
[305]	Reliability	Multi-modal communication and Cloud-Fog computing	↑	*
[246]	Total average data quality	Cloud-Edge-End Collaboration	25.48%↑ 27.10%↑	TAFR+TFGR [361] and Random method [153]

*: No relevant experiments and data are mentioned in the literature;

↑: Indicates data is superior compared to other benchmarks. If only the symbol is used, it indicates that the original text has a description but without specific values;

↓: Indicates data is inferior compared to other benchmarks. If only the symbol is used, it indicates that the original text has a description but without specific values;

AWS: Amazon Web Services; OTSM: This method predicts the movement according to their latest momentum and applies a best fit-based task allocation strategy; Greedy-RT: Greedy algorithm without data coding and packets are retransmitted when it is lost; Greedy-NC: Greedy algorithm encodes before data transmission; JFP: Joint Failure Probability; 3OBJ: Three-Objective Optimization; STG1: Sustainable and Resilient Edge proviSioning Stage-1 Only; FLP: Facility Location Problem; CO: Cloud-only; EO: Edge-only; CE: Cloud-edge; ACE: Adaptive cloud-edge; WLFL: wireless federated learning algorithm; DNN-DTFL: DNN-based DT-assisted user and resource scheduling algorithm for federated learning; A3C: Advantage Actor-Critic; RSA: Random-selection algorithm; GDA: Greed-decision algorithm; CKM: A checkpoint-based scheduling method; RM: A replication-based method scheduling; DLS-MOEA: Distributed local search multi-objective evolutionary algorithm; GEN: GEN adopts a well-known genetic algorithm to reconfigure the optimal mapping between all user groups and edge servers, as well as the task replication numbers of individual user groups at runtime; RAN: A simple online method that randomly remaps the optimal edge server for a user group when the difference between the offloading rates of offline and online tasks exceeds a specific threshold; Louvain: A community discovery algorithm based on modularity, which performs well in efficiency and effectiveness and does not require specifying the number of partitions [159]; GA: Genetic-based Approach; DRAM: Dynamic Resource Allocation Method; TAFR: Truth Finding by Attribute Reliability estimation; TFGR: An optimization framework for Truth Finding by Group Reliability estimation.

operational stability of edge servers. The system model adopts the structure model of the base station, edge server, and network switch. Select a cost-effective path to achieve service migration through moving vehicle trajectory prediction and service migration. Finally, a large number of simulations and data show that their proposed migration reliability and performance are better than those of the same type.

Xu et al. [433] believed that artificial intelligence and Space-Terrestrial Integrated Networks (STINs), essential infrastructure for 6G networks, have huge development potential in the military and communications fields. Because satellite in-orbit transmission has longer distances and faster speeds, data transmission is unreliable, data processing is not timely, and satellite resource utilization is low. In response to this situation, Tiansuan constellation introduces an edge computing platform to provide edge intelligence support for STINs. They suggested a cloud-edge aggregation AI architecture for STINs, consolidating centralized intelligent controllers on cloud servers and consolidating distributed intelligent agents on the satellite servers' edge. They presented an implementation approach for an intelligent network architecture by leveraging the Tiansuan constellation. The primary satellite of the Tiansuan constellation houses the controller, while the edge satellite accommodates the edge server. The controller and the edge server are integrated to accomplish tasks intelligently. Then, they discussed and experimented with three use cases: intelligent data transmission, intelligent remote sensing, and intelligent network slicing. Lastly, validating simulation outcomes demonstrates that the architecture can maximize the utilization of STINs resources and significantly enhance the overall network throughput.

Atakan Aral et al. [20] introduced a two-stage optimization algorithm, sustAinable and Resilient Edge proviSioning (ARES), to ensure the dependable placement of edge nodes in urban regions. Wireless sensor networks (WSN) find extensive applications in monitoring. Nevertheless, owing to the constraints posed by the sensor's inherent energy and processing capacities and restricted network bandwidth, WSN alone falls short of fulfilling the demands of contemporary IoT applications. Transferring sensor data to edge computing infrastructure presents a promising resolution. However, challenges like geographical distribution and temporary installation necessitate edge service providers to deploy a significant quantity of edge nodes within urban regions, which will exert a substantial impact. Therefore, they proposed the ARES algorithm. In ARES, multi-objective optimization and a reliability model based on dynamic Bayesian networks are used. Pareto-optimal solutions are derived during the initial phase for transmission time and energy efficiency. In the subsequent phase, the solution attained in the first stage is enhanced to achieve the desired level of reliability. Then, they also considered the impact of network failures, power outages, weather disturbances, and other issues on edge deployment that other researchers had not considered. Lastly, they performed simulation experiments utilizing data from the Vienna urban area. They substantiated that the ARES algorithm achieves superior equilibrium among transmission time, energy efficiency, and reliability within a shorter timeframe.

Washik Al Azad et al. [5] explored how to address the different latency requirements of Internet of Things applications. Recently, the rising popularity of IoT devices has led to a surge in the adoption of edge computing deployments. This approach leverages computing resources at the network edge to enable prompt processing of data generated by IoT devices and other user devices, thereby minimizing latency. These devices may need to be processed at the edge or close to the user. Therefore, they designed an information-centric hybrid cloud-edge computing framework called CLEDGE (CLoud + EDGE) based on Named Data Networking (NDN). Its purpose is to maximize the offloading of computing tasks for applications with different latency requirements. The design conceptualizes a computing landscape where computational resources are dispersed across the network edge and cloud infrastructure. Tasks can utilize the principles of Information-Centric Networking and NDN to discover suitable execution locations, either at the network edge or in the cloud, for executing computations that meet different latency requirements. The article also noted that CLEDGE's initial motivation originated from comprehending the requirements of the mixed reality community. In the design, they also recognized that CLEDGE is well-suited for mixed reality applications and applications requiring real-time or near-real-time processing. Consequently, CLEDGE can satisfy the requirements of diverse application scenarios, including IoT applications. Finally, their evaluation shows that CLEDGE can achieve over 90% task completion rate on time, far exceeding other benchmarks.

Liao et al. [225] explored the unreliability issues and low communication efficiency in digital twinning for managing low-carbon electrical equipment. DT is a cutting-edge technology for the intelligent optimization of electrical equipment. However, problems with DT's reliability and communication efficiency may still exist in practical applications. To address this, they proposed a solution called C3-FLOW for cloud-edge collaborative and reliable DT in managing electrical equipment with low carbon emissions. Jointly optimizing device scheduling, power control, channel allocation, and computing resource allocation can minimize the sum of the global loss function and time-averaged communication cost. Through the joint optimization of device scheduling, power control, channel allocation, and computing resource allocation, the combined global loss function and time-averaged communication cost can be minimized. In terms of the dynamic trade-off between reliability and communication efficiency, the dynamic balance between reliability and communication efficiency is accomplished by considering the weights assigned to the loss function, delay, and energy consumption. The deep actor-critic algorithm is also introduced to decrease communication costs further; in terms of collaborative resource allocation for cloud edge devices, externality problems are solved with lower complexity through power control and channel allocation algorithms based on packet switching matching. Finally, through simulation experiments, they verified C3-FLOW's advantages in terms of loss function, communication efficiency, and carbon emissions.

Yang et al. [453] investigated the matter of renewable energy management in smart grids. Due to renewable energy's intermittent nature, the renewable energy regulation problem, one of the important issues in achieving smart grid energy efficiency, is difficult to solve. A substantial volume of diverse data needs to be gathered to inform decision-making about the status of renewable energy sources. This data contains information about the status of renewable resources but also lacks values. To address these missing values and obtain the optimal strategy, they propose a fault-tolerant scheme for adapting renewable energy sources that rely on the cloud-edge orchestration computing approach. The first layer of the solution deploys deep learning-based algorithms and leverages GRU-based algorithms to restore missing data and forecast future values. The second tier employs a DRL algorithm to acquire the optimal strategy. They designed a deep reinforcement algorithm based on the PPO algorithm to solve the problem of maximizing renewable energy capacity. The agent can gradually improve its decision-making performance, achieving higher renewable energy capacity. They proved through experiments that the cloud-edge orchestration computing solution they proposed can achieve higher renewable energy regulation capabilities.

Xiao et al. [420] explored computing and deployment issues in Industrial Internet of Things applications. In scenarios characterized by a significant number of sensors, massive computing task requests contradict reasoning requirements, affecting operational efficiency and service reliability. In addition, due to the heterogeneity of cloud-edge system computing resources and the randomness of communication, calculating and deploying deep learning models under cloud-edge collaboration is also difficult. Therefore, they proposed a collaborative cloud edge service cognitive framework for deep neural network model service configuration to deliver adaptable computing services. Under this framework, DNN models are effectively transferred between cloud centers and edge nodes, offering dynamic and scalable storage and computing services. The framework establishes a revenue target to improve resource utilization efficiency by weighing metrics such as precision, response time, and power consumption. To maximize the revenue target, they converted the problem of optimizing revenue targets into a reinforcement learning problem involving partially observable DNN configurations. Introduce a self-adaptive DNN configuration algorithm based on the dueling DQN framework. The experimental results indicate that the mechanism exhibits effective external learning capabilities, adapts well to dynamic networks, and can handle latency and energy consumption while serving demand.

Abdulaziz Alarifi et al. [7] proposed a fault-tolerant aware scheduling under fog clouds. Centralized cloud computing systems provide cost-effective storage and computing. Still, they faced bandwidth depletion and service time problems caused by the significant geographical distance between data centers and customers. These difficulties can result in service delays and interruptions, particularly for time-sensitive and mission-critical applications like healthcare, manufacturing, and Internet of Things services. Fog computing was introduced as a decentralized computing model to address these challenges. Still, the dynamic and heterogeneous nature of IoT devices, as well as their interactions with the outside, may lead to various types of failures. Their proposed system takes into account the high dynamics and non-uniformity of the cloud and combines the computing core layer with the edge layer of the cloud. It embraces a three-tier architecture consisting of a cloud core, edge, and IoT device layer. Using a fault-tolerant model and appropriate scheduling strategies can improve business and reliability under fog clouds. The experimental findings demonstrate that the framework enhances cloud performance and enhances the availability of cloud resources.

HAN et al. [142] explored the reliability of multi-access edge computing (MAEC) in 5G communication networks. The deployment of 5G networks will incorporate MAEC, which facilitates the distribution of computing tasks and services from the central cloud to the edge cloud. This enables services to leverage low latency, power efficiency, context awareness, and improved privacy protection. The main goal of 5G is to provide flexible next-generation communication systems, so ensuring secure provisioning of all services at the network edge is imperative. Therefore, their proposal introduces a novel decentralized authentication architecture that enables flexible and cost-effective local authentication of network elements while incorporating context information awareness. Finally, they validated the effectiveness of this approach in achieving a flexible balance between network operating costs and MAEC reliability through simulations utilizing the backward link-based Markov model and the random walk model, considering mixed-class traffic scenarios.

Balázs Németh et al. [276] explored the optimization issues of Virtual Network Function (VNF) deployment in 5G infrastructure regarding stability and instability. Many IoT devices collaborate with different types of robots and humans, and these devices require reliable and available platforms to coordinate their collaboration or provide dynamic programmable control. These application scenarios impose strict latency and reliability requirements on the underlying network and cloud platform. Edge computing, fog computing, and multi-access edge computing can address these challenges by deploying computing resources close to end devices and offloading tasks between end devices and cloud servers, thereby reducing latency and network load. The underlying infrastructure (including public and private cloud/edge resources) provides execution for VNF for public 5G networks and private domain interconnections. In this regard, reasonable resource orchestration is indispensable for always finding the appropriate location of the software components that implement the service. In addition, decisions about VNF deployment must also consider factors such as safety, limited battery capacity, and VNF power consumption. Therefore, they proposed a VNF deployment optimization problem with the goal of cost minimization. They proposed a heuristic algorithm that considers the wireless coverage and battery limitations of fog computing devices, the first known algorithm to date. Algorithms that can achieve close to optimal results in practice. Finally, they confirmed the algorithm's benefits in terms of scalability, cost-effectiveness, and execution time through an extensive set of simulated experimental evaluations conducted in real-world scenarios.

Cao et al. [36] discussed the joint optimization of edge intelligence in large-scale network physical systems. In recent times, the swift advancement of information technology has facilitated the profound integration of physical and software components within systems. They are called network physical systems. Large-scale CPSs constitute a vital subfield within the broader domain of CPSs. Large-scale CPSs cover many fields, including production and daily life. At present, large-scale CPSs are challenging to deal with various types of data growth with low latency. Reliability management is also very important in large-scale CPSs. Large-scale CPS solutions' primary focus is optimizing service latency and reliability, but they ignore the negative impact on the system life cycle. They proposed an edge-intelligent solution that optimizes service latency and system lifecycle, taking into account reliability and energy consumption. The scheme is divided into two distinct phases: the offline phase and the online phase. During the offline phase, the LSTM technology is employed to forecast the task offload rate for each user group. Next, they developed an algorithm based on DLS-MOEA to determine the optimal static system configuration for calculating the offload mapping and task replication numbers. During the online phase, they designed an affinity-driven online solution to deal with the property of end users. Experiments show that their offline and online two-stage edge intelligence solutions are indeed superior to the most advanced benchmark testing solutions.

Cao et al. [38] discussed the service delay minimization of edge cloud computing coupled with cyber-physical systems under energy budget constraints and reliability requirements. They found that existing edge cloud computing based on delay awareness can not simultaneously consider energy budget and reliability requirements, which may greatly reduce the feasibility of CPS applications. Therefore, they proposed a two-stage reliable service delay optimization composed of static and dynamic. In the static phase, Monte-Carlo simulation and integer linear programming technology are used to determine the optimal calculation unload mapping and task backup number. During the dynamic phase, an adaptive backup mechanism is implemented to prevent unnecessary data transfer and execution, which results in additional energy savings and improved service latency. The experimental results demonstrate that compared with the typical baseline approach, this approach can delay the system service by 18.3%.

Zhang et al. [498] found that sensors usually have limited computing power in industrial production, so they cannot provide acceptable delays for computing-intensive detection tasks. Moreover, automatic detection in industrial settings typically requires timely responses, as even minor detection failures can result in significant production issues or accidents. To address this challenge, they proposed a risk-aware cloud-edge computing framework tailored explicitly for time-sensitive automatic manufacturing detection. To tackle the uncertainty stemming from channel access delay induced by 802.11ax, they employed conditional value at risk as a metric to assess inspection risk and ensure the dependable delivery of the uninstall service. Considering latency requirements and reliability thresholds, they have formulated a cost-minimization problem to attain the optimal deployment of the uninstall service. To address this issue, they introduced a branch-and-check (BNC) method to optimize and efficiently solve this mixed-integer nonlinear programming problem. Compared to the commonly used branch and bound method, the proposed BNC method significantly reduces the number of feasibility checks for worst-case conditional value at risk constraints, resulting in substantial computational time savings. Finally, they evaluated the proposed framework through simulation experiments and compared it with several other service location schemes, including edge-only execution, cloud-only execution, dedicated execution, and relaxation and recovery based on binary variables. The outcomes indicate that their proposed framework successfully fulfills the latency requirements of the Industrial Internet of Things.

He et al. [150] discussed the problem of the malfunction of the gearbox of the rotating machinery group. The gearbox is a critical component within the rotating machinery group, assuming a vital role in power transmission, speed modulation, torque transmission, and directional changes in large-scale industrial environments. The maintenance cost and downtime loss of large equipment groups will be significant due to equipment quantity, usage, and industrial technology upgrading. Consequently, a fault diagnosis scheme for parallel shaft gearboxes is employed, utilizing cloud-edge cooperation and deep mutual learning (DML). This scheme aims to mitigate property losses and enhance safety performance by effectively diagnosing gearbox faults. In their scheme, the local edge data is sent to the cloud to form a dataset, which is then normalized and transmitted into the image. DML is used to train two different layers of networks in the cloud, and then the tiny networks are distributed to all edge nodes. At the edge, the small network is retrained with local data to improve robustness and accuracy. Finally, experimental results have demonstrated the efficacy of the proposed method on the Drivetrain Prognostics Simulator platform, yielding favorable outcomes.

Xu et al. [425] discussed the deployment of dynamic controllers in mobile cloud-edge computing networks based on software-defined networks. The significance of applications in daily life is continually growing. However, due to the hardware conditions of equipment, it is difficult to adapt to the increasingly high demand in life. MECCNs can provide ultra-low latency service and sufficient computing resources. Nevertheless, the uneven distribution of users in mobile edge cloud computing networks results in an imbalanced network demand and load distribution. In large-scale networks, the control plane of edge layer multi-control architecture can handle network requests in time, but the fixed control plane can not dynamically demand network traffic. Therefore, they introduced a load-aware dynamic controller placement scheme based on SDNs-supported MECCNs deep reinforcement learning. They formulated the problem of dynamic controller deployment as a multi-objective combinatorial optimization problem. This model considers three key objectives: minimizing delay, balancing the network load, and ensuring control reliability. To ensure the control plane's high performance under the constantly changing network state and tackle the challenges above, they have developed a dynamic deployment algorithm based on the deep deterministic policy gradient algorithm. Experimental results show this algorithm is superior to others in delay load balancing and control reliability.

Adrián Orive et al. [281] have discovered that combining edge, fog, and cloud computing can deliver distinct features tailored to specific target applications, which can interact with and respond to the physical world promptly and store and process large amounts of data. However, this combination also brings some new challenges that need to be addressed, such as scalability, dynamics, and application QoS. To solve these problems, they proposed a distributed scheduling architecture called application-centric orchestration architecture (ACOA). The ACOA architecture includes several components, such as the system scheduler, the application scheduler, the node daemon, and the monitoring daemon. The system scheduler is responsible for deploying the application scheduler, which selects the best node for deployment according to the application's quality requirements and the system's state. The node daemon is responsible for executing and monitoring the containers on the node, and the monitoring daemon is responsible for monitoring the system and network status and updating the monitoring data to the state database. In the experimental part, they deployed two applications for verification: the smoke monitoring application and the speed configuration application. Each application has specific constraints and optimization strategies. The experimental results demonstrate that the ACOA-based deployment can better meet the application's QoS requirements, provide lower end-to-end response time, and increase reliability.

Mohammadreza Baharani et al. [28] found that as Internet of Things evolves, more devices and sensors are deployed on edge nodes, requiring reliability modeling capabilities. Traditional reliability modeling methods are subject to limitations, including the necessity of extensive data and computational resources and the inability to update the model in real-time. To solve this problem, they proposed a real-time reliability modeling framework based on deep learning. This framework utilizes a combination of edge and cloud computing to achieve device-specific reliability prediction and modeling on edge nodes and aggregation and analysis of advanced metadata through the cloud. They introduced Deep deep learning reliability awareness of converters at the edge (RACE) as an integrated online real-time reliability perception and modeling framework specifically designed for power electronics devices. Algorithm-wise, Deep RACE uses a recurrent neural network variant called LSTM. LSTM is used within the framework for aggregate training and device-specific inference. On the system side, Deep RACE proposes an IoT-based cloud-edge computing platform that pushes LSTM-based reliability inference to a single power converter. In the experiments, they used the Si-MOSFET power electron converter as the research object. They built a deep learning model by collecting and analyzing a large amount of device data. The results show that the Deep RACE system can accurately predict the device's health status in real-time, with a prediction error rate of 8.9% and a processing time of 26 milliseconds.

Fayez Alqahtani et al. [14] found that with the rapid development of the IoT, an increasing number of resources are being linked to the cloud edge, which leads to insufficient bandwidth of the cloud network and potential delay problems due to the long geographical distance between the edge resources and the cloud data center. These will affect the required quality of service for time-sensitive requests such as real-time requests. To solve this problem, they proposed a scheduling method based on load balancing, the load balanced service scheduling approach (LBSSA), which has better resource utilization, load balancing, and run-time performance in the cloud computing environment. By considering different types of requests and computing resources, LBSSA realizes the timely execution of real-time requests and the priority of important services. In addition, this method's request scheduling takes into account the failure rate of resources to provide high reliability for the requested service. The experimental outcomes demonstrate that, in comparison to alternative scheduling methods, their proposed approach requires fewer computing resources for request processing, enhances resource utilization, and outperforms other methods regarding load balancing disparities and execution time.

Ren et al. [305] studied the challenges faced by drones in carrying out missions; for instance, factors like restricted durability, computing and communication capacities, and external disruptions can potentially result in mission failure. To address these challenges, they proposed a task-oriented multimodal communication technology that employs the multimodal obfuscation reception model. This model enhances the demodulation capability of unmanned aerial vehicles for processing multimodal information. In addition, they have also introduced a collaborative computing model for UAVs that combines cloud and edge computing within the mission execution system of UAVs. The model mitigates the computational latency in demodulating multimodal information and the risk of task failure resulting from sequential task arrival or real-time task execution. When studying task-oriented multimode communication technology based on cloud-edge UAV cooperation, they introduced IQ modulation and multimode modulation methods. The experimental results show that in the same channel environment, IQ modulation increases the successful reception probability of a UAV single-mode receiver by 1.5 times. In various interference environments, IQ modulation increases the probability of successful reception by 14.99% under weak interference, 22.56% under normal interference, and 31.86% under strong interference. Through IQ modulation, the probability of malicious eavesdropping is significantly reduced from 98.11% to 10.02%.

Lu et al. [246] found the problem that unreliable workers submit malicious data in mobile crowd sensing systems. To ensure the development of high-quality applications, enlisting workers who adhere to honesty in the edge network is crucial. Therefore, they proposed a Mobile Crowd Sensing system based on artificial intelligence, a cloud-edge-end cooperative data collection scheme called MLM-WR, which is used in Artificial Intelligence of Things applications. MLM-WR addresses the following challenges for efficient and effective data collection for AIoT applications. (1) The findings of truthful workers: to discern workers' integrity, they modified their credibility based on the variance between the perceived data from workers and the actual ground truth data obtained through collaboration with unmanned aerial vehicles. (2) The discovery of sensing difference: by calculating attribute data errors, they acquire the reliability of workers' sensing attributes. Additionally, they determined the sensing quality of workers in various locations by combining absolute and relative sensing position preferences. (3) In MLM-WR, worker assignment is facilitated through a particle swarm optimization algorithm, which considers factors such as the reliability of perceived attributes, locations, and recruitment costs. This approach effectively resolves the tradeoff between recruitment costs and data quality, ensuring a balanced outcome. The experimental results show that MLM-WR saves 40.15% and 40.19% of the total cost compared with Random and TFAR+TFGR, respectively. In terms of total average data quality differences, MLM-WR is also better than these methods by 25.48% and 27.10%, respectively. Compared with TFAR+TFGR, MLM-WR improves by 36.12% in identifying honest workers.

These works show that deep learning, deep neural networks, and cloud-edge computing are ideal solutions for cloud-edge-end systems reliability. Some works focus on single-objective optimization, specifically for reliability-optimized cloud-edge computing. Research is also being conducted into combining deep learning and cloud-edge computing technologies to optimize reliability while maximizing performance.

3.3.5   Conclusions and Discussions

In the future, with the improvement of edge computing capability, the application of 5G technology, the optimization of artificial intelligence algorithms, and the progress of edge intelligent devices will bring higher performance and more expansive application fields to the cloud-edge-end system. These developments will drive cloud-edge-end systems to meet the growing latency, sustainability, security, and reliability demands, bringing more intelligent, reliable, and sustainable solutions to various industries.

4

Cloud-Edge-End Orchestrated Applications

In the context of rapidly advancing applications, cloud-edge-end orchestration garners increasing attention as a novel computing model. The cloud-edge-end orchestration is a distributed computing model that extends computing capabilities from the cloud to the network edge and end devices to meet application requirements for low latency, high bandwidth, and data privacy. In this context, the cloud refers to remote data centers, providing extensive computational and storage resources suitable for processing big data and complex computational tasks. The edge refers to network edge devices such as routers, switches, and base stations, which are closer to end devices and can provide low-latency services. The end refers to users' personal devices, such as smartphones, tablets, and IoT devices, which are capable of local computation and data processing. The primary characteristics of the cloud-edge-end orchestration are as follows:

- **Distributed:** The distributed characteristic implies that computational tasks are distributed across multiple devices and multiple layers, enabling parallel processing to enhance efficiency.

- **Heterogeneous:** The heterogeneous characteristic signifies that different devices possess varying computational capabilities and resources, necessitating appropriate resource allocation based on task requirements.

- **Dynamic:** The dynamic characteristic means that device states and network conditions may change over time, requiring dynamic adjustments to task allocation and resource management strategies.

Essentially, the fundamental concept of cloud-edge-end orchestration involves distributing computational tasks across cloud, edge, and end devices to achieve optimal performance and efficiency. Nevertheless, effectively designing and implementing applications under the cloud-edge-end orchestration remains a challenging issue involving the integration and optimization of various advanced technologies. It is our hope that this section provides a valuable reference for researchers and engineers to better understand and harness cloud-edge-end orchestration, thereby advancing the development of applications.

4.1 Cloud-Edge-End Computing for IoT Applications

IoT applications encompass the utilization of IoT devices, sensors, and technology to facilitate diverse practical use cases across various industries and domains. With the rapid evolution of digitalization and intelligent technologies, hundreds of billions, or even trillions, of end devices connected to the Internet are anticipated to generate an enormous volume of data at astonishing speeds in the future. The orchestration of cloud, edge, and end plays a pivotal role in providing essential computational architecture to drive IoT applications, fostering the prospect of establishing an entirely novel IoT innovation ecosystem.

Regarding cloud-edge-end-enabled IoT applications, each of the three layers: "cloud", "edge", and "end" plays a crucial role:

- **Cloud:** The cloud layer, often denoting a remote data center, offers extensive computational and storage resources, making it well-suited for tasks involving substantial data processing and complex computations. The cloud excels at managing heavy workloads, long-term data retention, and intricate analytical tasks that lack time sensitivity. Furthermore, it acts as a central control point for overseeing and orchestrating activities across edge devices and end devices.

- **Edge:** The edge layer refers to network edge devices such as routers, switches, and base stations that are closer to end devices. They can provide low-latency services by processing data closer to the source, reducing the amount of data that needs to be sent to the cloud, and thus saving bandwidth and reducing latency. Edge computing can handle real-time processing tasks, short-term data storage, and analytics that require low latency.

- **End:** The end layer refers to the user's personal devices, such as smartphones, tablets, and IoT devices. These devices can perform local computation and data processing, providing immediate feedback to user actions and enabling real-time applications. End devices are often the sources of data generation in IoT applications.

Each layer is essential in its own right and contributes to the overall performance of IoT applications. The cloud provides powerful computing resources for computation-intense tasks, the edge enables low-latency processing close to data sources, and the end devices generate and consume data in real-time. Together, they form a comprehensive computing architecture that can meet diverse application requirements in terms of latency, bandwidth, privacy, etc.

4.1.1 Cloud-Edge-End Computing for Intelligent IoT

The intelligence IoT application is an advanced technology that combines AI's data analysis capabilities with IoT's data collection and transmission capabilities. This combination allows for more efficient and intelligent processing of the vast amounts of data generated by IoT devices. Cloud-edge-end orchestration indicates the intelligent management and coordination of computing resources across the cloud, edge, and end devices, which is crucial for ensuring the efficient and effective operation of intelligent IoT applications.

Numerous practical scenarios for intelligent IoT applications that are either currently operational or in development exist. These applications take advantage of IoT techniques and various network architectures to enhance efficiency, convenience, and quality of life. In the subsequent discussion, several application scenarios will be explored.

- **Smart Manufacturing**

Smart manufacturing, also known as Industry 4.0 or the Fourth Industrial Revolution, is a manufacturing paradigm that harnesses advanced technologies to enhance productivity, quality, and operational efficiency within the manufacturing process. It entails the convergence of diverse digital technologies to establish a connected, data-driven production environment. The primary objective of smart manufacturing is to bolster efficiency, minimize waste, and facilitate agile responses to shifts in demand. Furthermore, it paves the way for adaptable, customized production processes. Embracing the tenets of smart manufacturing can yield industrial operations that are both more sustainable and competitive. Empowered by the cloud-edge-end architecture, it can efficiently meet the various functional requirements of smart manufacturing applications, such as control, scheduling, and prediction.

- **Control:** In centralized smart manufacturing systems, a multitude of monitor and control devices that generate a vast amount of industrial data are directly connected to the cloud. However, in such a scenario, massive data transmission puts a huge strain on the network, resulting in intolerable delays and privacy vulnerabilities. Incorporating intelligence at the network's edge is crucial for creating a decentralized system. Realizing this, Kulkarni et al. [197] proposed a decentralized asset monitoring and management platform in which intelligent edge devices possess the capabilities of sensing, local data processing, and executing control actions. Only data that requires further analysis is transmitted to the cloud, enabling autonomous operation at the edge.

- **Scheduling:** A major challenge is ensuring compliance with stringent low-latency and high-efficiency requirements in manufacturing systems. These challenges must be addressed by appropriately allocating and managing various computing resources in the cloud-edge-end architecture. The task scheduling problem in the field of smart manufacturing faces several challenges, including long scheduling times, high communication latency, and uneven node load distribution. Jian et al. [67] proposed a cloud-edge scheduling model and a bat algorithm with variable step size. Based on historical data, a long- and short-term memory network model is incorporated to predict the scheduling results quickly quickly.

- **Prediction:** Predictive maintenance enabled by artificial intelligence applications by monitoring, analyzing, and predicting the health of equipment, it is possible to increase efficiency, reduce maintenance costs, and extend equipment life in manufacturing organizations. This contributes to more sustainable and efficient smart manufacturing. Wang et al. [396] noticed several challenges in the cloud-based underwater acoustic sensor networks, including data transmission power and latency. They proposed a bidirectional prediction model based on end-edge-cloud orchestration, which harnesses edge elements for data analysis and prediction, thus diminishing the need for acoustic communication. Furthermore, a data collection protocol with edge elements relocates the centralized cloud to the distributed edge, leading to improved bandwidth utilization and reduced costs while preserving data accuracy.

- **Smart Transportation**

Intelligent IoT-enabled smart transportation refers to the use of artificial intelligence technologies and data analysis methods to intelligently enhance urban or transportation systems, aiming to improve the efficiency, safety, and sustainability of the transportation network. This concept encompasses various applications, including but not limited to autonomous driving, signal control, traffic monitoring and prediction, and emergency response. With the help of cloud-edge-end architecture, a smart transportation system enables the improved realization of advanced Internet of Vehicles (IoV), Advanced Driver Assistance Systems (ADAS), and ultra-low latency services.

- **Driving Assistance:** The driving assistance application is designed to aid drivers in various aspects of driving, such as improving safety, providing navigation assistance, optimizing fuel efficiency, and enhancing the overall driving experience. Driving assistance applications leverage sensors, cameras, radar, and other data sources to provide real-time information and assistance to drivers, helping them make informed decisions and avoid potential hazards. When considering the cloud-edge-end architecture in the research of driving assistance applications, several difficulties and challenges may arise. The driving environment can be unpredictable, with potential signal interferences, network dropouts, and adverse weather conditions. Ensuring the reliability of driving assistance systems under various circumstances, including edge and cloud computing failures, is vital. Lyu et al. [253] introduced a distributed and context-aware beaconing approach to improve broadcast reliability for safety applications. Within this approach, a vehicle initiates the process by employing machine learning algorithms to assess the link conditions with its neighboring vehicles. It then shares this link condition information with its neighbors and proceeds to identify the minimum number of helper vehicles required to retransmit its beacons to those neighbors experiencing poor link conditions.

- **Traffic Flow Forecasting:** By analyzing big data, accurate traffic flow prediction is crucial in providing recent traffic condition information, enabling smart vehicles to plan and adjust routes to prevent congestion. Consequently, numerous short-term traffic flow prediction models have been proposed to date. However, these models often focus on predicting the entire traffic network, leading to inefficient model training and computational stress on the central cloud. Lai et al. [199] introduced a spatial-temporal attention graph convolution network that operates on edge nodes, effectively enhancing short-term traffic flow prediction accuracy. Distributing the training of individual network components to specific roadside units removes the necessity of processing all data on the central cloud server.

- **Smart Home**

Smart homes have gained significant popularity in contemporary households, offering enhanced comfort and convenient services. Specifically, smart home applications utilize internet technology to remotely control home devices through devices such as cell phones, tablets, and computers. Furthermore, smart homes can also provide important services such as education, healthcare, and security.

- **Healthcare:** IoT technology offers an effective solution to tackle healthcare delivery challenges in home healthcare and remote patient monitoring. However, IoT generates vast amounts of data, typically processed in the cloud. In this manner, the latency introduced by data travels to the cloud and back to the home is deemed unacceptable. Verma et al. [377] proposed an edge-based method for remote health monitoring, utilizing embedded data mining, distributed storage, and notification services. This method processes real-time patient data and calculates a temporal health index. Li et al. [211] presented the ChainSDI framework, a solution for data interoperability and regulatory compliance challenges in home-edge-cloud healthcare applications. Utilizing blockchain and edge computing, it manages secure data sharing and computing on sensitive patient data, demonstrating the feasibility of software-defined infrastructures in healthcare.

- **Device Control:** Smart home-related intelligent IoT applications often rely on extensive data, which can be constrained due to privacy restrictions that limit data sharing. To enhance privacy in smart homes with sensitive data, Matsushita et al. [263] designed an edge-based lamp control system. This system can detect and recover from failures using camera image processing, with anomalies detected by monitoring cyclic outdoor brightness changes in window images. Li et al. [202] incorporated federated learning into smart home applications. The authors proposed a strategy for training models that leverages the combined computational

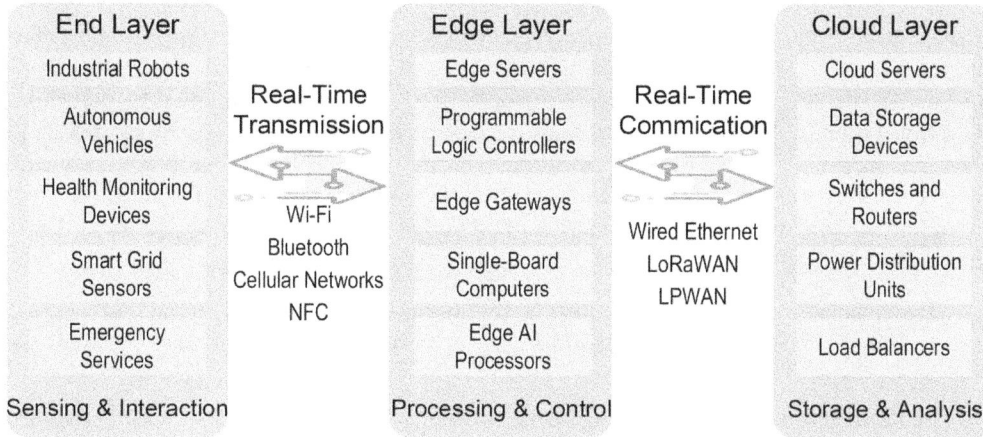

FIGURE 4.1

The cloud-edge-end system for time-sensitive IoT (Wi-Fi: wireless fidelity; NFC: near field communication; LoRaWAN: long range wide area network; LPWAN: low-power wide-area network).

power of edge nodes and the cloud platform. This approach, facilitated by split learning, ensures ample computational resources for the training of models. In brief, such a collaborative approach between cloud and edge computing offers a solution to address challenges commonly encountered in traditional cloud computing services, including high latency, traffic congestion, and privacy issues.

4.1.2 Cloud-Edge-End Computing for Time-Sensitive IoT

IoT technologies have revolutionized the way data is collected, analyzed, and utilized across various scenarios. While these technologies have opened the door to a world of possibilities, they have also exposed a critical challenge in handling time-sensitive requirements. As an essential category of IoT applications, time-sensitive IoT (TS-IoT) applications require real-time data processing, low-latency communication, and fast decision-making capabilities. Today, TS-IoT applications have covered a wide range of domains, including healthcare, transportation, industrial automation, smart cities, and more. Specifically, autonomous vehicles require real-time decision-making for self-driving cars based on sensor data, and healthcare applications require real-time patient monitoring and health tracking.

Traditional cloud-centric computing architectures, while effective in many contexts, fall short when it comes to TS-IoT applications that demand real-time data analysis and low-latency responsiveness. Edge computing positions data processing near the end device, favoring reduced latency and improved performance, but its limited computing resources are a shortcoming that cannot be ignored. By integrating cloud, edge, and end devices, cloud-edge-end computing focuses on enhancing the efficiency and responsiveness of TS-IoT applications to support stringent real-time requirements and enhance user experiences. Figure 4.1 provides an illustrative example of a cloud-edge-end system for TS-IoT, in which the cloud, edge, and end layers are responsible for IoT data storage and analysis, processing and control, and data acquisition from multiple sources, respectively.

- **Time-Sensitive Networking Design**

The foundational network of a time-sensitive cloud-edge-end system should offer exceptional reliability, guaranteeing predictable latency and minimal jitter. The well-known technique, IEEE 802.1 Time-Sensitive Networks (TSN), focuses on delivering deterministic and highly dependable packet forwarding for network services[112]. As surveyed in [273], the key features of TSN include deterministic communication, low latency, synchronization, quality of service, traffic shaping, seamless ethernet integration, standardization, and distributed clock synchronization. From the perspective of real-world applications, TSN can be applied to

- **Industrial Automation [365]:** TSN is widely used in industrial automation, enabling precise control and monitoring of manufacturing processes and machinery.

- **Automotive [87]:** TSN is used in automotive networks to support advanced driver-assistance systems (ADAS), autonomous vehicles, and in-car infotainment.

- **Professional Audio and Video [118]:** TSN ensures low latency and high-quality audio and video streaming in professional applications like live events and broadcasting.

- **Power Utility [349]:** TSN is employed in the power utility sector for the precise control of electrical grids and smart grid applications.

Although the beneficial effects of TSNs have been well-recognized globally and extensively tested for use in Local Area Networks (LANs), the use of TSNs in Metropolitan Area Networks (MANs) requires further attention, especially for cloud-edge-enabled latency-sensitive metropolitan industrial applications as well as smart city services. Tschöke et al. [365] analyzed utilizing Process Field Network (PROFINET) over TSN via dense wavelength division multiplex for remote machine park control through edge-supported virtual programmable logic controller. This enables the establishment of TSNs in metro networks, paving the way for future extensions to facilitate the dynamic configuration of TSN traffic and the implementation of multipath frame replication and elimination. Also realizing TSN's limitation in achieving deterministic transmission over extensive geographic areas, Tan et al. [349] presented a hierarchical network for end-to-end long-distance deterministic transmission. This approach utilized cyclic queuing forwarding and deterministic IP and developed joint scheduling by introducing traffic shaping at edges to maximize network throughput. In the future, TSN-enabled technologies would further enhance large-scale distributed machine control and accommodate ultra-reliable low-latency networks for large industrial areas and cities.

- **Real-Time Algorithms Design**

Real-time algorithms for cloud-edge-end systems play a crucial role in supporting time-sensitive IoT applications. These algorithms are designed to ensure that data processing, decision-making, and communication occur within strict time constraints. Various types of real-time algorithms are proposed to address the specific needs of time-sensitive applications in cloud-edge-end systems.

Real-time scheduling algorithms determine the order and timing of tasks to meet deadlines, aiming to schedule tasks efficiently. The scheduling algorithms are categorized as shown in Figure 4.2. Among these three algorithms, exact algorithms aim to find the optimal solution to scheduling problems, which explicitly considers all possible task assignments and exhaustively searches the solution space to determine the best schedule. While exact algorithms guarantee an optimal solution, their computational complexity often limits their

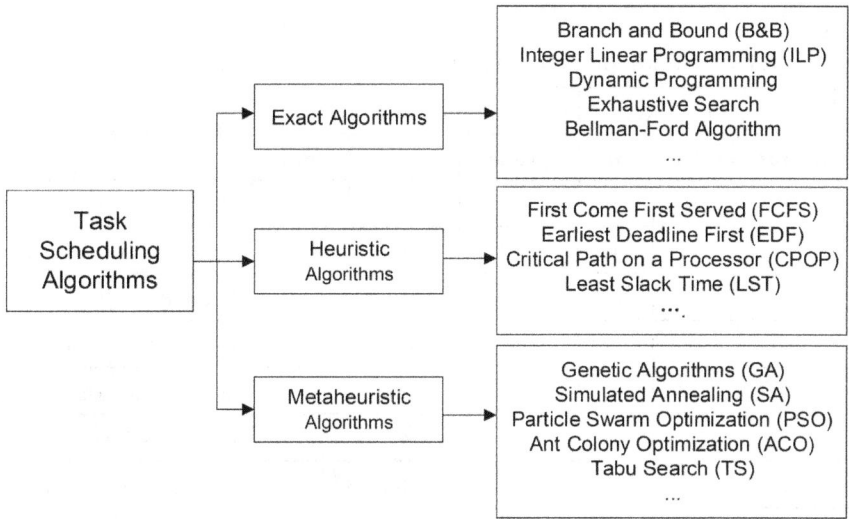

FIGURE 4.2
A categorization of task scheduling algorithms.

practical use to small-scale scheduling problems [41]. In such cases, heuristic or metaheuristic algorithms are often preferred to find near-optimal solutions more efficiently. Heuristic algorithms, which can find a good solution (not the optimal one) quickly, are approximate methods for solving task scheduling problems [205]. Metaheuristic algorithms are higher-level strategies for finding good solutions to complex optimization problems. They are particularly useful in task scheduling when the search space is vast and finding an optimal solution is challenging [463].

For effective scheduling performance, a variety of constraints might need to be considered for real-time scheduling in a cloud-edge-end system, depending on the particular use case and needs. In general, the metrics that are taken into account as constraints include the processing and memory capabilities of edge servers and end devices, communication bandwidth of the core and local network, and energy consumption of battery-operated end devices. Wei et al. [405] presented heuristics for handling data placement problems in light of the edge servers' storage limitations. Considering the latency requirements of tasks and the energy limitation, Yan et al. [442] proposed a deep reinforcement learning-based cloud-edge-end load balancing method for choosing the appropriate edge or cloud server for offloading.

Table 4.1 summarizes the primary characteristics of existing cloud-edge-end-driven TS-IoT scheduling algorithms and highlights their distinguishing properties. In brief, cloud-edge-end-driven TS-IoT represents a shift in IoT architecture to meet the demands of real-time applications by leveraging the strengths of cloud, edge, and end computing tiers. Scholars and practitioners are working on innovative solutions to enable the deployment of TS-IoT applications. These applications aim to provide low latency, high responsiveness, and efficient processing, offering significant advantages across various domains.

4.1.3 Cloud-Edge-End Computing for Internet of Vehicles

The Internet of Vehicles (IoV) is an application area based on IoT technologies. It aims to connect vehicles and road traffic systems to the Internet to achieve an intelligent, interconnected, and automated transportation system. Cloud computing is a computing model

TABLE 4.1
Summary of existing cloud-edge-end-driven TS-IoT scheduling algorithms.

Ref.	System Model			Considerations			Methodology
	Cloud	Edge	End	Latency	Energy	Radio	
[504]	+	+	+	+	−	−	Convex Programming
[527]	−	+	+	−	+	−	Approximate Algorithm
[167]	−	+	+	−	+	−	Iterated Local Search
[463]	+	+	+	−	+	−	Firefly Algorithm
[442]	+	+	+	+	+	−	Deep Reinforcement Learning
[472]	−	+	+	+	−	+	Distributed Algorithm
[152]	+	−	+	+	+	−	Dynamic Programming
[274]	−	+	+	+	+	+	Deep Reinforcement Learning
[215]	−	+	+	+	+	+	Convex Programming
[6]	+	+	+	+	−	−	Decomposition Algorithm
[317]	+	+	+	−	−	−	Composite Optimization
[224]	−	+	+	+	+	+	Decomposition Algorithm

based on cloud data centers, providing extensive data storage and processing capabilities suitable for a wide range of applications. However, in the context of IoV, cloud computing faces several challenges.

- **Real-Time Response:** Vehicles require real-time data processing and decision-making, such as traffic flow monitoring, intelligent navigation, and autonomous driving decisions. Timeliness is crucial in IoV applications, whereas cloud computing typically involves data transmission to the cloud for processing, potentially falling short of the low-latency requirements of vehicles.

- **Network Communication:** The volume of data generated by vehicles is substantial, demanding significant bandwidth for data transmission, which can lead to network congestion and resource wastage.

- **Privacy & Security:** Vehicle data involves personal privacy and security issues, requiring cloud computing to ensure data security, adding to the system's complexity.

To address the challenges mentioned above, MEC has emerged as a promising solution. MEC involves pushing computing resources closer to the data source to reduce latency and alleviate network burdens. It emphasizes deploying computing resources near data sources to achieve faster responses to real-time demands and reduce data transmission. Such characteristics of MEC are in line with the goals of IoV applications. However, MEC-enabled IoV also faces several challenges.

- **Limited Computing Resources:** Edge nodes are typically deployed with small-scale servers with limited computing resources. This limitation may restrict

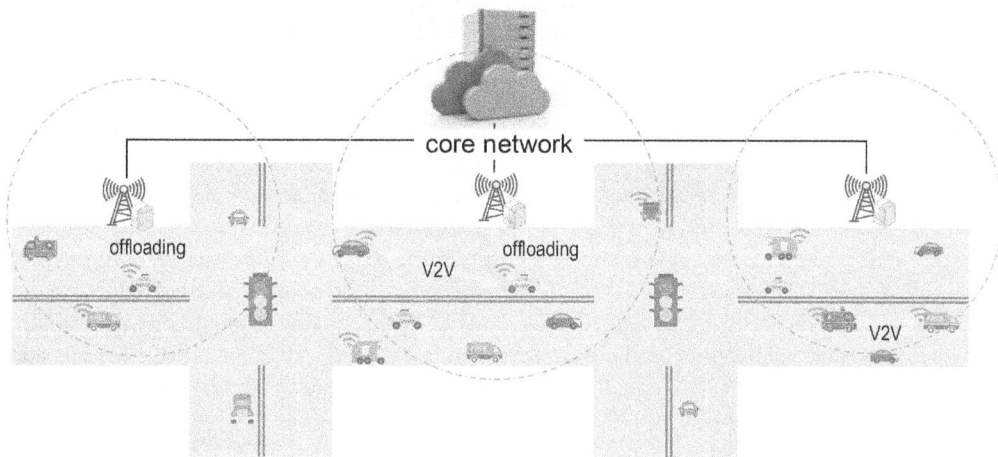

FIGURE 4.3
Cloud-edge-end computing for Internet of Vehicles.

executing complex computational tasks by IoV applications at the edge, especially those demanding large-scale data processing and deep learning models.

- **Limited Storage Resources:** Vehicles generate substantial data during their operation, including sensor data, camera data, and location information. Edge devices may struggle to handle such extensive datasets, leading to performance degradation or necessitating efficient data compression algorithms.

- **Limited Collaborative Capability:** Edge nodes are typically deployed in a distributed manner, and their collaborative capabilities among one another are limited. Some IoV applications may require collaboration among vehicles to share information, such as traffic flow data or hazard alerts, and the collaborative potential in an edge computing environment may be constrained.

As shown in Figure 4.3, for IoV applications, cloud-edge-end collaborative computing becomes an innovative solution. It combines cloud computing, edge computing, and the vehicles' resources to achieve more efficient data processing and analysis. This collaborative approach enables vehicles to perform initial data processing on their local and edge devices and transmit crucial data to cloud resources for in-depth analysis. The following four primary categories of IoV applications may be distinguished based on their use scenarios.

- **Vehicle Safety:** These applications focus on enhancing the safety of vehicles and drivers. They include driver assistance systems such as automatic braking, blind-spot monitoring, and others to reduce traffic accidents and improve driver safety. The automotive industry is advancing integrated safety systems to improve driver and passenger protection. Considering the lack of accuracy of existing bounding box methods for edge collision detection, some novel vehicle profile estimation algorithms have emerged.Mothershed et al. [269] developed an evaluation framework and found that the convex hull method and the three-arc method performed the best, each having advantages in different situations. In order to achieve safe driving, Xun et al. [440] established a vehicle-edge-cloud system for driving behavior assessment. In this system, the vehicle, the edge, and the cloud play different roles. The vehicle generates data and transmits it to the edge, the edge generates behavior rankings based on the driving behavior assessment model and returns to the car, and the cloud continuously trains the model and periodically transmits the updated model to the edge.

- **Vehicle Entertainment:** These applications provide entertainment and information services to drivers and passengers, including features such as music, video, internet connectivity, and application integration, with the goal of enhancing the driving and riding experience. By combining the resources of mobile-end devices (vehicles) and fixed-edge servers (road infrastructures), Hou et al. [157] proposed a reliable edge-aided system for IoV applications, involving partial computation offloading and task allocation with reprocessing. The proposed algorithm is based on the search strategy of the particle swarm optimization algorithm, which maximizes system reliability while guaranteeing the task delay constraint. As hardware technology evolves, the computational power of the vehicles themselves needs to be taken into account in order to efficiently utilize spare resources for processing computational tasks. To this end, Sharma et al. [319] explored the efficient deployment of data collection and processing tasks on these moving vehicles. A task-based mathematical model for distributed service placement is developed that scales with resource requirements and mobility to efficiently optimize processing and communication costs.

- **Autonomous Driving:** These applications enable vehicles to autonomously drive and operate without a driver. The vehicle uses a combination of sensors, cameras, radar, lidar, and sophisticated algorithms to perceive its surroundings and make moving decisions. In recent years, many researchers have been working on optimizing automatic driving systems. However, there are limitations in the deployment and use of some methods considering the diversity of driving modes and the limitations of the computational and storage capabilities of the vehicles themselves. Regarding real-time detection and logging of vehicle events, Ke et al. [189] explored a real-time method based on edge AI. They proposed a linear complexity algorithm for detecting and tracking objects, which is not only efficient but also works well with various camera settings and can be used on different vehicle cameras. Additionally, they developed a mechanism for logging events across multiple onboard systems, making it easier to apply to different systems.

- **Traffic Management:** These applications are designed to support navigation, reduce traffic, and maximize traffic flow. To improve traffic efficiency and road conditions, they include intelligent traffic signal control, route planning, navigation apps, and real-time traffic flow monitoring. Generally, traffic management systems collect extensive video data for incident detection, straining network paths to the traffic management center (TMC). Such a centralized, cloud-only approach will inevitably exacerbate the core network burden, leading to low reliability and high latency. Liu et al. [230] designed a cloud-edge framework for supporting two typical traffic monitoring applications: speed detection and congestion detection. The framework promises to achieve better trade-offs that account for TMCs' variable network conditions and edge servers' limited processing power.

4.1.4 Conclusions and Discussions

IoT applications have grown significantly in recent years, with an increasing number of connected devices generating massive amounts of data. To efficiently handle this data and meet real-time processing demands, Cloud-edge-end computing has emerged as a promising paradigm by leveraging computing resources at different layers of the network hierarchy.

The cloud layer offers extensive computational and storage resources, suitable for tasks involving substantial data processing and complex computations. The edge layer delivers low-latency services by processing data closer to end users, reducing the amount of data sent to the cloud. The end layer includes user devices that can perform local computation and data processing or offload computation tasks onto edge or cloud servers for execution.

This chapter aims to provide insights into the integration of cloud-edge-end computing into a variety of practical applications, emphasizing the importance of cloud-edge-end computing in enabling efficient and effective IoT applications. We highlight three application scenarios in intelligent IoT: smart manufacturing, smart transportation, and smart home. For each scenario, we discuss the challenges and solutions related to cloud-edge-end computing. We also discuss the significance of time-sensitive IoT applications and the development of time-sensitive scheduling approaches, in which various optimization goals, performance constraints, and problem-solving techniques are taken into account to achieve reliable and low-latency communication. In addition, we analyze the limitations of cloud- and edge-assisted IoVs and reveal that the collaboration of clouds, edges, and end devices is the most promising solution to tackle the data storage and processing challenges in IoV applications.

In summary, the integration of cloud-edge-end computing into IoT applications offers numerous benefits, revolutionizing the way that IoT devices collect, process, and utilize data. By bringing computational resources closer to IoT devices, edge computing can reduce latency and enable real-time data processing. This is very important to time-sensitive applications such as autonomous driving and healthcare monitoring systems. Cloud computing offers virtually unlimited resources through virtualization technology, allowing IoT systems to scale dynamically on demand. The close collaboration among end devices, edge nodes, and the cloud data center, through effective task scheduling and resource management techniques, not only offers scalable resources for storing and analyzing large volumes of IoT sensor data but also reduces the requirement for centralized processing. Also, by distributing computational tasks across multiple layers, resource utilization and network throughput can be improved, ensuring high-quality services for IoT applications.

Despite the advantages of maximizing the efficiency, scalability, and intelligence of connected IoT systems, the full utilization of cloud-edge-end computing in IoT scenarios is still facing many technique challenges. For instance, distributing computation may raise security concerns, such as data breaches and unauthorized access, necessitating robust encryption and access control mechanisms tailored for the distributed processing of IoT sensor data. Also, integrating heterogeneous devices and platforms at different layers requires standardized protocols and frameworks to ensure seamless cross-layer communications. Moreover, Managing and processing large volumes of IoT data distributed across multiple layers demands efficient and intelligent data storage, retrieval, and analysis techniques.

4.2 Cloud-Edge-End Computing for CPS Applications

Characterized by embedding computing, communication, and control capabilities (3C) into physical assets, cyber-physical systems (CPSs) are rapidly spreading in various fields, such as smart cities, service recommendation, video/audio surveillance, cancer diagnosis, medicine discovery, and industrial automation. By combining sensing, actuation, and control mechanisms with advanced computing technologies, CPSs represent a transformative integration of physical processes with computing infrastructures, computational algorithms, and communication networks, creating interconnected systems that bridge the gap between the physical and digital domains. As illustrated in Figure 4.4, the CPS mechanism involves

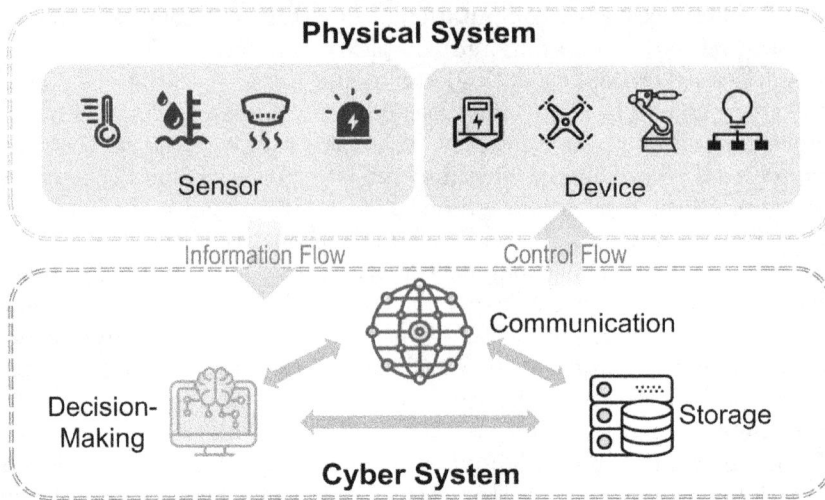

FIGURE 4.4
The interaction between physical processes and cyber components in CPSs.

complex interactions among physical components, computational elements, and communication infrastructure to achieve real-time monitoring, analysis, and decision-making of physical processes.

Nonetheless, the ongoing increase in the number of interconnected end devices presents a critical challenge because the massive amount of data generated by these devices may need to be promptly transmitted and scrutinized by CPS applications to offer fast and precise feedback. In addition to the demand for low latency, CPSs also require mobility support and location awareness. As a growing communication trend, cloud-edge-end computing can satisfy these demands well by integrating mobile edge computing (MEC) and cloud computing into CPSs. As illustrated in Figure 4.5, an advanced cloud-edge-end-assisted CPS can be constructed with three layers, each of which takes into account various design and optimization concerns. The sensory level concerns manipulating end devices, e.g., energy generation, and electrical control. The communication level, built upon wired and/or wireless links, concentrates on latency reduction, bandwidth allocation, congestion control, etc., to support reliable communications among interconnected end devices and cloud/edge computing resources. As the application level, deploying cloud and edge servers leverages the strengths of both centralized and distributed computing paradigms, contributing to the overall efficiency, scalability, and reliability of the system. On the one hand, the clouds not only enable CPS applications to benefit from the high-performance computing capability and massive storage infrastructure but also empower CPS users to build applications that can use and manage smart CPS devices. On the other hand, the edge servers facilitate the process of CPS applications at the network edge close to the IoT devices, relieving the latency and security issues in the centralized cloud infrastructure models. The fusion of cloud, edges, and devices can offer the highly desired balance and flexibility to distribute the computation load of CPS applications jointly at cloud and edge servers in an intelligent and efficient fashion. The substantive issues of interest at the application level include storage management, task scheduling, service migration, privacy protection, etc., for the purpose of catering to the diverse requirements of modern CPS applications.

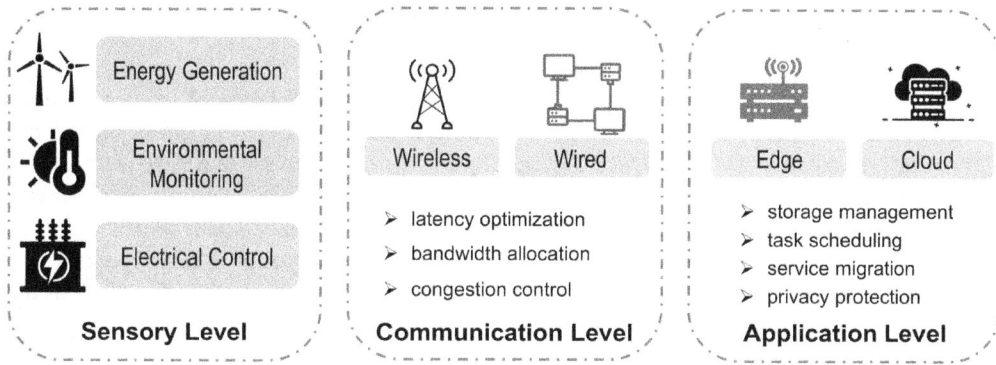

FIGURE 4.5
An example of a cloud-edge-end-assisted CPS composed of three layers.

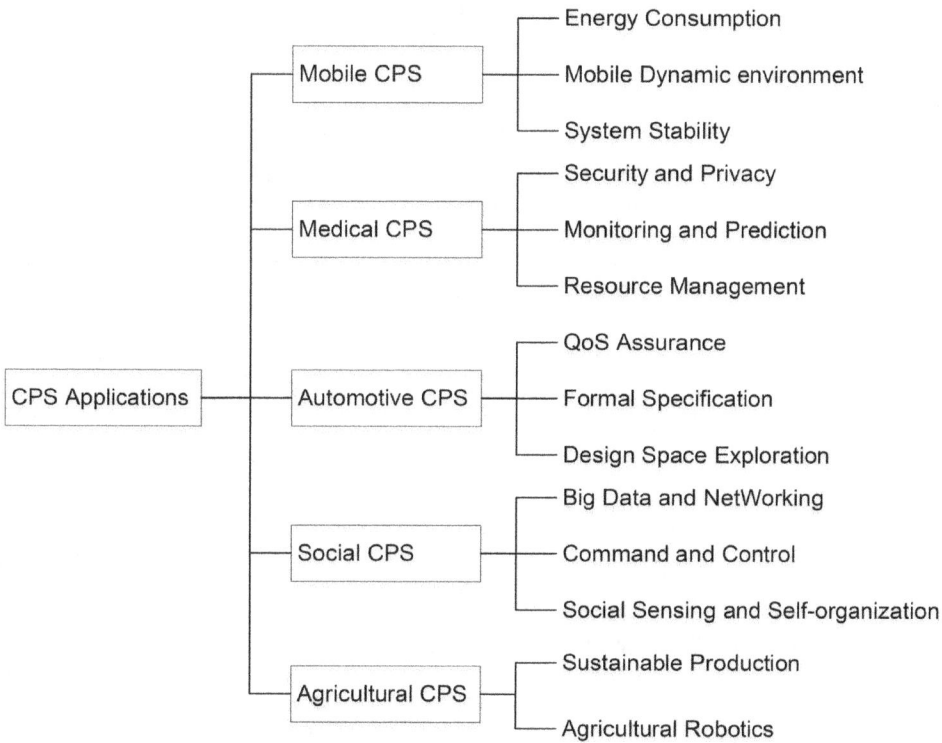

FIGURE 4.6
A categorization of CPS applications in various fields.

This section aims to provide a thorough investigation of recent cloud-edge-end computing advancements achieved in various advanced CPS application fields, including mobile CPS, medical CPS, automotive CPS, social CPS, and agricultural CPS. Figure 4.6 summarizes representative CPS applications and significant challenges related to these fields, covering security, stability, manageability, sustainability, energy efficiency, etc.

4.2.1   Cloud-Edge-End Computing for Mobile CPS

A variety of CPS applications have emerged for enhancing production efficiency and enabling real-time responses to mobile users in industry and academia. The key benefits of mobility technologies in CPS applications lie in their ability to easily deploy, adapt, and evaluate real-world functions with mobility, dynamism, agility, and adaptability in a flexible and timely fashion. Owing to the characteristics mentioned above, mobile CPSs are expected to be vital to developing advanced CPSs. Many research efforts have been made to conduct in-depth investigations to improve the applicability and intelligence of mobile CPS applications. We discuss below in detail the three most critical challenges, i.e., energy efficiency, dynamic behaviors in mobile environments, and system stability of mobile CPSs.

- **Energy Efficiency**

    Since most physical components can be charged with power supplies when running, energy is usually not a common concern in traditional CPS. However, energy consumption has become a critical concern in mobile CPS. Due to their mobility, mobile devices often operate without a constant power supply. Under this circumstance, CPS cannot accomplish sensing and communicating when task scheduling requires a significant amount of energy [137]. For those energy-eager cyber components, several studies have been developed to manage energy consumption in mobile CPS efficiently. In [370], the authors provided a comprehensive survey on energy-efficient solutions for mobile devices from 1999 to May 2011 from multiple different perspectives, including energy and power modeling, interactions among mobile users, communication mechanisms and protocols, resource management, and computation offloading.

    Although MEC and MCC have been proven to save the battery life of energy-limited mobile devices and improve computational efficiency, effectively transferring data from mobile devices to edge servers and cloud data centers is a crucial challenge. Cuervo et al. [77] designed an energy-efficient system called MAUI that achieves substantial energy savings by means of a fine-grained offloading strategy while minimizing the changes required by mobile applications. Xia et al. [413] attempted to tackle an online location-aware offloading problem in a two-tiered mobile cloud computing environment. They devised an efficient online algorithm for location-aware task offloading to minimize each mobile device's energy consumption while satisfying each task's service-level agreement (SLA) requirement.

- **Mobile Dynamic Environment**

    Mobile CPS faces the unique challenge of coping with a dynamic environment caused by its inherent mobility. The highly dynamic characteristics of mobile cyber components (e.g., smartphones and laptops) pose many challenges to mobility models, routing protocols of networks, self-adaptive software systems, data transmission, and communications. For the mobile dynamic CPS, we have, for instance, the work of Hu et al. The work in [165] suggested an agent-based multilayer approach integrating context-aware semantic services (CSS) to create and implement context-aware applications for vehicular networks comprised of mobile devices. The proposed approach is autonomous and intelligent for self-adapting to rapidly changing network and dynamic contexts of VSN users. Conti et al. [75] discussed the human mobility characterization for mobility models and provided a replication strategy to cope with opportunistic routing in the mobile CPS environment. Besides, they investigated middle-ware and applications using utility-based forwarding techniques in mobile ad hoc networks. Vasconcelos et al. [372] proposed a CPS building method to manage distributed dynamic adaptation and develop adaptation policies. In the proposed method, a scalable data distribution layer provides compositional dynamic software adaption, and dynamic software adaptation allows an application to respond to new application requirements and/or context changes. A dynamic path-planning strategy for unmanned aerial vehicle (UAV) swarms was developed by Neishaboori et al. [275]. They decomposed the dynamic path planning problem into a series of static combinatorial optimization problems and introduced a dynamic branch and price algorithm for solving the decomposed sub-problems.

- **System Stability**

Equipment unavailability and unreliability caused by an operating system crash and battery exhaustion will negatively impact the system and hinder the widespread use of mobile CPSs. Thus, system stability is also a critical concern in mobile CPSs. Here, we briefly discuss several related studies that seek solutions to maintaining the stability of mobile CPSs. Peng et al. [285] proposed an autonomous flight control policy for a mobile CPS consisting of small-scale UAVs. This three-stage policy can be divided into kernel control, command generation, and flight scheduling stages. The first stage is to guarantee the asymptotic stability of the UAV motion under the impact of surrounding air. At the command generation stage, the proposed policy employs dynamic inversion to cope with the nonlinearity in affine systems when generating flight commands. The flight scheduling module finally makes a decision on the flight arrangements under certain flight conditions. Hu et al. [164] introduced an efficient task scheduling manner for crowdsensing applications in mobile CPS. They constructed a flexible and universal architecture across mobile devices and cloud computing platforms to efficiently deploy and manage multiple mobile crowdsensing tasks for both application developers and end users. Aiming to minimize execution time and energy consumption of processing component services, Deng et al. [82] provided a novel offloading mechanism that takes into account dependency relations and fault tolerance for mobile CPS services. Based on the offloading model, they devised an offloading algorithm based on the genetic algorithm. Bertran et al. [30] conducted a study on achieving the tradeoff between energy efficiency and flight performance for fixed-wing UAVs. They introduced a reference UAV model that jointly considers system stability and energy consumption as a function of UAV size and power.

Deriving from traditional CPS, mobile CPSs have now drawn much attention and have advanced dramatically in the past few years. Owing to the mobility characteristic of end devices, mobile CPSs are advantageous over traditional CPSs by leveraging the capabilities to extend and enhance the interaction between the cyber and physical worlds. It is believed that the mobile CPS will be deployed to more innovative application fields by handling the issues mentioned above well in the future.

4.2.2 Cloud-Edge-End Computing for Medical CPS

Medical cyber-physical systems (MCPSs) have successfully incorporated the design of medical implementations in health care with CPS, which have been used in hospitals and clinics to automate the devices [272]. MCPS is a networked infrastructure that integrates medical devices, smart sensors, and software systems to monitor, diagnose, and treat patients remotely or in real-time, offering personalized and customized quality healthcare services.

As illustrated in Figure 4.7, with the assistance of cloud-edge-end computing, MCPS integrates medical devices, including monitoring-based devices (e.g., heart-rate monitors), statistics-based devices (e.g., drug outdated reports), knowledge-based devices (e.g., chronic disease diagnosis), and prediction-based devices (e.g., cancer prediction) [499]. Despite the considerable effort directed toward CPS-centric application scenarios, there are still challenges and open issues in the medical scenario that need to be addressed. In this part, we discuss several predominant challenges in cloud-edge-end-assisted MCPSs, including security and privacy, monitoring and prediction, and resource management.

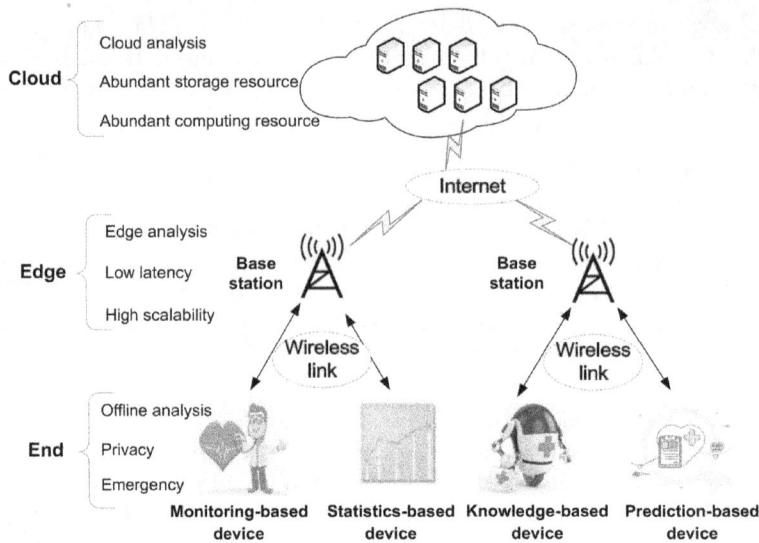

FIGURE 4.7
The cloud-edge-end framework for MCPSs.

- **Security and Privacy**

   Procedures for tamper detection and input validation generated by the integrated sensors' end users (i.e., doctors and patients) are vital to prevent adversaries from recruiting in the control inputs to the MCPS. Consequently, high security and privacy should be guaranteed while transferring patient files/reports.

   Regarding data privacy, Kocabas et al. [194] first reviewed conventional and emerging schemes for designing MCPS over the last two decades and suggested a new framework for MPCS. This framework has four layers with different hardware and communication capabilities: data collection layer, data pre-processing layer, cloud analytics layer, and decision support layer. In each layer, this framework selects an appropriate encryption mechanism to protect user data privacy. Sharma et al. [318] developed an energy-energy MCPS to ensure high security while satisfying the SLA constraint. This solution employed an ad hoc on-demand distance vector (AODV) protocol to monitor the communication process and remove any identified malicious devices. Privacy-preserving integrity verification for MCPS has been extensively researched. In particular, Xu et al. [428] proposed a data integrity verification model for MCPS that takes data security and privacy-preserving into account. This model is established upon lightweight streaming authenticated data structures. The detailed implementation, e.g., the design principle, architecture design, formal and security definitions, and communication protocols, can be referred to in [428]. Tailored for cloud-assisted MCPS, a lightweight certificate privacy-preserving integrity verification method was presented in [505]. This method incorporates conditional anonymity to achieve conditional identity privacy of patients and trace the real identities of misbehaving patients by a secure key exchange mechanism.

- **Monitoring and Prediction**

   As a sensing, processing, and communication-embedded healthcare platform, MCPSs provide real-time monitoring with feedback control services. To alleviate healthcare problems, the development of MCPS plays a vital role in improving infrastructures using monitoring and control technologies.

Devarajan et al. [84] presented an intelligent fog-supported-MCPS for Parkinson's disease prediction by means of voice sample analysis to suggest appropriate medication. In this MCPS, fog computing nodes are deployed between the end devices and the cloud server to achieve data privacy as well as to reduce communication overheads. Ramasamy et al. [303] considered an artificial intelligence (AI)-enabled MCPS that can be utilized by doctors to monitor and detect patient diseases. Machine learning techniques, especially deep learning models, are used for AI to cope efficiently with the large amount of data collected for fall events in the elderly. Simeone et al. [324] developed a dynamic health monitoring system that can be personalized to gather and analyze physiological and environmental data for assessing health risks. They also introduced a real-time assessment approach that combined data processing and fuzzy-based data fusion methods to facilitate intelligent health risk prediction. Akter et al. [4] constructed a CPS-based heart disease prediction model. They suggested a methodology using different ML algorithms to process the real-time data by analyzing the dataset and adopting the decision tree classifier, in which the heart disease dataset is collected from Kaggle resources. Alrowais et al. [15] conducted a research project to detect cyber attacks while generating healthcare data in MCPS. The advantage of this work is the capability of minimizing the detection error rate with high computational efficiency. For this purpose, they devised a fuzzy C-Means approach combined with a swarm intelligence algorithm to monitor and collect healthcare data through a network of remotely interconnected medical sensor devices.

- **Resource Management**

The following network generation will exploit various resources with substantial sensing abilities to be extended beyond the physically connected computers to comprise multimodal information from cognitive, biological, and social networks. Hence, healthcare applications necessitate computing resources to make intelligent decisions based on patient data.

Regarding resource management, Rolim et al. [307] introduced a cloud computing solution established upon a network of wireless healthcare sensors to automate clinical diagnostics and monitoring capabilities. Lin et al. [129] devised a resource management approach for minimizing the total monetary cost in an MCPS supported by mobile edge-cloud computing. This approach jointly considers task offloading, task migration, and resource allocation to construct a complex mixed-integer nonlinear optimization problem. They decomposed the formulation problem into a set of linear problems and designed a two-phase heuristic to find a near-optimal solution to the original resource management problem. To address the healthcare resource allocation problem, Soraia et al. [329] established a healthcare reward model to characterize the reaction to the satisfaction factors of MCPS and accordingly provided a reward maximization approach to enhance the delivery and utilization of healthcare resources. Hossain et al. [155] implemented an MCPS relying on machine learning techniques to predict the health risks of pregnant women. In this MCPS, essential healthcare data are gathered through medical sensors measuring temperature, blood pressure, heart rate, etc., with resource-constrained configurations to analyze the key risk factors. Verma et al. [376] presented a scalable AI-driven MCPS built upon a fog-cloud platform to support early detection, prevention, and control of the COVID-19 outbreak. The users' real-time analysis results and information, as well as geographic information, are uploaded to the cloud server for intelligent and efficient decision-making, in order to arrange the demanded resources in a timely fashion.

The various MCPS applications discussed above, with the integration of cloud-edge-end computing, reveal that MCPS will gradually and independently replace traditional medical equipment scientifically and efficiently. Precisely, how to assure the security and privacy of healthcare CPS, monitor and predict diseases by patients' data, and utilize computation resources are fundamental problems to be solved.

4.2.3 Cloud-Edge-End Computing for Automotive CPS

As a new-generation in-vehicle network, automotive CPS (ACPS) embeds various sensors and distributed computing devices into complex communication networks. ACPSs are complex systems that combine continuous and discrete dynamics, exhibit heterogeneous characteristics due to component variations, and operate across multiple domains. They are composed of diverse embedded systems and physical processes interconnected through communication networks. The functionality of automotive systems is shifting toward electronic implementations rather than traditional hydraulic or mechanical systems. In a contemporary vehicle, as many as 70 electronic control units can communicate by exchanging over 2500 signals across up to 5 different communication systems [151]. Particular focus has been put on identifying and creating characteristics of ACPS in current research trends.

- **Sustainable Transportation**

    Sustainable transportation in the research field of ACPS integrates seamlessly with the objectives of a sustainable future, encompassing zero emissions, using renewable energy sources, crash prevention, secure social networking capabilities while driving, and the option for autonomous driving. The future vehicles powered by ACPS will offer diverse designs, resulting in shorter and more predictable travel durations. These vehicles optimize space and time for parking, enhance roadway efficiency, contribute to quieter cities, ensure pedestrian and cyclist safety, promote equitable access, and offer cost-effective transportation solutions. Underwood [369] developed a roadmap for automated vehicles that takes a CPS-based strategy to investigate sustainable transportation to achieve efficient door-to-door mobility in the United States. This research employs the Delphi expert survey technique to predict the vision and chart the course of action with input from twenty experts specializing in automotive robotics.

- **CAN with Flexible Data-Rate**

    As the next-generation in-vehicle network standard, adopting controller area network with flexible data-rate (CAN FD) in the research field of ACPS enhances the capabilities of automotive cyber-physical systems. Its support for higher data phase bit rates and larger message payloads enables faster, more robust, and versatile communication, ultimately contributing to improved performance, safety, and functionality within the automotive industry. Xie et al. [422] developed a CAN FD-optimized design method for ACPSs to reduce bandwidth consumption while adhering to signal timing restrictions. They defined two slack assessment metrics for quantitatively evaluating possible packing options, devised a clustering-based signal packing algorithm, and validated the utilization of signals and packed messages.

- **Formal Specification**

    Currently, formal specification technology is crucial in ensuring the accurate design and development of reliable and secure vehicle systems. It serves as a crucial tool in verifying and validating their correctness. Formal specification refers to using mathematical models and languages to precisely describe a system's behavior, structure, and properties. In the context of ACPSs, formal specification technology allows engineers to define and analyze the functionalities, interactions, and constraints of various components within the system. Zhang et al. [484] modeled ACPS using the new formal specification formwork called hybrid relation calculus, which can deal with not only continuous systems with differential equations but also discrete systems without differential equations. The proposed hybrid relation calculus can be used to represent physical components and their interactions with intelligent units through communication modeling.

- **Safety Assurance**

Safety assurance in ACPS refers to the process of ensuring that these systems are designed, developed, and operated in a manner that guarantees safe and reliable operation. Given the increasing complexity and interconnection of modern vehicles, safety assurance is of paramount importance to mitigate risks and prevent accidents or failures. Tang et al. [354] solved the cost-aware reliability task scheduling problem by developing a cost-criticality and reliability-assurance scheduling algorithm based on fault-tolerant techniques for mixed multiple-function ACPS. The proposed algorithm aims to minimize the metric of deadline missing while ensuring reliability demands and adhering to hardware cost limitations. Izosimov et al. [180] presented a method for systematically considering security attributes in developing cyber-physical systems illustrated with an automotive case study. Besides demonstrating our experiments on the actual truck, they discussed ways to include security into a vehicle manufacturer's design flow efficiently, utilize a requirement elicitation process for security requirements, and perform analysis using attack trees. Ali et al. [9] implemented a dependable safety-critical CPS-on-chip (CPSoC) for automotive applications. They accomplished the goal of low downtime and high dependability by applying online monitoring of digital processor cores and intellectual properties (IPs). This ACPS can promptly implement counteractions using embedded instruments based on the IEEE IJTAG standards with appropriate software.

- **Quality Assurance**

Colledani et al. [74] presented an innovative automotive electric vehicle production approach to assemble in-line quality-driven rotors. This approach relies on an ACPS that determines the assembly process with magnetized stack quality, evaluated through data collected by in-line inspection. Dominka et al. [92] proposed a comprehensive method comprising four distinct quality assurance strategies: the general quality design, the central quality coordinator, the coupling metric analysis approach, and the dynamic quality testing method, to manage undesired behaviors and enhance desired interactions in software-intensive ACPSs. All these strategies were evaluated in a natural automotive environment.

- **Design Space Exploration**

In the ACPS research field, design space exploration (DSE) refers to the process of systematically exploring and evaluating different design alternatives to identify the optimal configuration that meets specified system requirements. It involves analyzing various design parameters, architectural choices, and trade-offs to achieve desired system performance, efficiency, safety, and cost-effectiveness. DSE aims to find the most suitable combination of hardware and software components, communication protocols, control algorithms, and system architectures. This exploration process helps engineers make informed decisions and optimize the overall design of the automotive system. Zhang et al. [502] established a coordinated simulation model for time-triggered ACPS. This model focuses on a comprehensive network/platform layer formwork and is incorporated with a design tool to facilitate quick prototyping. To reduce the cost and time-to-market, they allow DSE to receive practical control performance insights during the initial design phases. Wan et al. [379] proposed a co-design methodology for an ACPS involving multiple interacting domains, capable of simultaneously representing the physics and control aspects in ACPS architectures. Originating from the ACPS functional model, this co-design framework combines system design principles with high-level synthesis techniques to guide the exploration and validation of design spaces for multi-domain ACPS applications. A multi-functional codesign strategy that originates from the functional-level CPS, which can effectively represent multiple aspects such as physics and control concurrently, particularly in automotive applications. This strategy combines system engineering theories with high-level synthesis technologies from the design automation field, leading to a novel co-design approach that supports DSE and the evaluation of complex automotive architectures.

- **Adaptive Development**

Regarding an iterative and flexible approach to system design and development, adaptive development enables the ACPS to adapt and evolve over time. It involves continuously monitoring and responding to changes in requirements, technologies, and environmental conditions to ensure that the ACPS remains effective, safe, and up-to-date throughout its lifecycle. In the context of automotive CPS, adaptive development recognizes that the requirements and challenges faced by these systems can evolve rapidly due to factors such as changing regulations, advancements in technology, emerging cybersecurity threats, or evolving customer expectations. Therefore, more than a rigid and fixed development approach may be required to address these dynamic aspects effectively. Guo et al. [421] introduced an adaptive development scenario for ACPS based on integrated digital twinning techniques. They developed a clone flow to ensure each physical component has a corresponding cloned digital twin. Additionally, they proposed an intelligent digital twinning framework to facilitate close interaction between the digital twin and its corresponding physical component. Baek et al. [27] created an intelligent lightweight message authentication mechanism to identify cyber-attacks adaptively in modern ACPSs. The proposed mechanism is an integral component of a set of software decisions, which utilizes several safety measures to enhance security within ACPS environments.

The ACPSs integrate the conventional state-based discrete control model with the continuous physical model. Differential equations can typically express the continuous models in ACPSs, as they are effective tools for characterizing physical entities. Moreover, ACPSs involve intricate interactions between physical systems and cyber components. The development of operational methodology and control mechanisms in the cyber domain plays a crucial role in determining the performance of the physical systems. As a result, ACPSs can be classified as hybrid systems that incorporate sophisticated algorithms [484].

4.2.4 Cloud-Edge-End Computing for Social CPS

As a rapidly evolving research field, cyber-physical-social systems (CPSS) paves the way for incorporating the social dimension into CPS, positioning humans as its fundamental components. The CPSS consists of nature, cyber-space, and society, establishing a sustainable framework for exchanging data and insights regarding materials, products, manufacturing methods, societal engagements, and human expertise. The CPSS enables self-synchronization, parallel operations, and management across physical, informational, cognitive, and social fields. It is regarded as a perfect platform for facilitating the design and creation of intelligent environments with management organizations. The evolution of integrating social and human aspects into CPS research is leading to several research challenges. Notably, we identified three major categories of works: big data and networking, command and control, social sensing, and self-organization, which can summarize the applications of the CPSS in literature.

- **Big Data and Networking**

CPSS expands upon CPS or IoT by incorporating human social behaviors, thus promoting a synergistic interplay between computational processes and human experiences [461]. Consequently, Big Data Collectors (BDCs), Service Organizers (SOs), and users are integrated with CPSS to establish a data-driven computing infrastructure.

Aiming to combine human agents with big data and edge computing approaches, Dautov et al. [79] studied the data processing pattern in CPSSs through clustered edge computing using stream processing technology. A mechanism framework is provided to manage social involvements and contributions in the context of CPSSs to minimize the volume of data transmitted across the network. Wang et al. [390] presented an edge-cloud-assisted CPSS framework for smart-city applications, which can migrate data processing from the cloud layer to the edge layer for real-time processing and service delivery. Zhang et al. [493] proposed a distributed k-nearest neighbor (D-kNN) algorithm in a cloud-edge system to process Large-scale datasets in CPSS. The kNN algorithm is limited by a single computer's storage capacity and computation power, so the proposed D-kNN performs distributed calculations of k neighbors on each storage stack. Xu et al. [430] developed a CNN to train speckle images in cloud-edge environments for CPSS. The proposed CNN architecture can enhance the resolution of reconstructed images while maintaining high computing efficiency, which is beneficial to solving the problem of a random speckle correlation imaging system. Shen et al. [320] proposed a statistical behavior-based block allocation (SBOA) scheme for processing CPSS data, the key idea of which is to estimate the cache block characteristics according to the statistic behavior of data re-referencing for reads/writes.

- **Command and Control**

CPSS comprises the computer system, the controlled entity, and social components, enabling the management of physical objects to realize social processes via online computation platforms.

For traffic control, Cuo et al. [136] introduced a hierarchical architecture of CPSS-based intelligent transportation system (ITS) in terms of five layers: perception, communication, computing, control, and application. Society is considered an active entity in the ITS, which is involved in sensing, computing, communicating, controlling, and managing processes. Sánchez et al. [309] investigate and propose a methodology for the design of CPSS co-simulation tools, which enables users to implement the co-simulator most appropriate for all applications. Candra et al. [33] proposed performance metric models and brought into effect the concept of quality of data (QoD) to monitor the performance metrics of CPSS flexibly. The proposed monitoring system is evaluated in a real-world smart city scenario to justify its efficacy and applicability. Xin et al. [423] constructed a stochastic game theory-based model for CPSS with blockchain and cloud services. They have studied the stochastic stability of the Cournot duopoly game using the Lyapunov method and singular boundary theory. Suhardi et al. [339] studied the CPSS technology for sustainable fuel supply services. They designed an inventory control system for public gas stations to fulfill the Service expectations and improve customer satisfaction.

- **Social Sensing and Self-Organization**

As an integrated framework, CPSS blends elements from digital and physical space, human intellect, and societal-cultural aspects. It facilitates a dynamic interaction between digital information and reality's tangible and cognitive dimensions, along with a virtual representation that mirrors various aspects of the actual world. Moreover, CPSS seamlessly merges the spaces of the physical, digital, and social, aiming to deliver anticipatory and customized services for humans.

Zhou et al. [521] worked on designing practicable algorithms and scalable platforms for high-performance parallel computing across CPSSs. Semeraro et al. [315] evaluated the case studies conducted by Zhou et al., and aimed to address the state-of-the-art research issue and explore the impacts of a CPSS on sustainability performance. Fazel et al. [18] developed a holistic and modular framework that holds an ontological approach for identifying the optimal collaboration of humans and CPSS in a problem-solving ontology. Su et al. [337] proposed an incentive computing solution that relied on the behaviors and reputations of social users to deliver high-quality crowdsourcing services for CPSS. The proposed solution includes a comprehensive crowdsourcing computing model, an incentive scheme to encourage contributing user data, and an auction game model for determining the most appropriate social user to acquire the required data. Chakraborty et al. [44] explored a centralized framework that incorporated sensing, communication, and computing approaches to implement on-demand management of the electric grid within the CPSS. Flexible power loads with operational constraints and utility functions are modeled as a centralized optimization problem to minimize the overall power consumption of legible loads. Alternatively, the power loads can be controlled using a decentralized management approach established from a non-cooperative game theory model.

In general, research on CPSS has been conducted to encourage and support innovative directions, alleviate CPSS's social dynamic problems, and ultimately ensure human-machine synergy. In the future, we believe that CPSS research will gradually apply big data and networking, command and control, social sensing, and self-organization with social aspects and open opportunities for multidisciplinary efforts.

4.2.5   Cloud-Edge-End Computing for Agricultural CPS

Advanced agriculture enables sensors, actuators, and digital platforms to coordinate and collaborate with cyber principles and tools. With the advancement of computing paradigms and cybernetics, emerging agricultural CPSs are making great efforts to organize physical devices to offer computation, control, and communication functionalities in the agricultural field. According to recent studies, agricultural CPS can be affected by multiple important factors, e.g., humans, sensors, robots, and agricultural plants, as well as the data generated by the interaction among them [134]. Below, we summarize the challenges of sustainable production and agricultural robotics in the context of agricultural CPS.

- **Sustainable Production**

  With respect to promoting productivity, sustainable production without affecting natural resources is an environmentally protected way to generate more revenue for enterprises and enhance the production of higher-quality products. Research efforts relying on advanced computing paradigms have emerged to achieve this goal.

  Sarkar et al. [312] discussed a cyber-agricultural system (CAS) to facilitate the fusion of CPS, digital twins, with traditional agriculture for sustainable production. This work introduced the development status of CAS and the future innovative trends. Delgado et al. [80] conducted a big data analysis of sustainable agriculture based on the geospatial cloud. The authors believe the future of precision agriculture is expected to be propelled by sustainable precision agriculture and environment (SPAE), by combining existing technologies with big data analytics. Guo et al. [134] designed an agricultural CPS framework for greenhouse stress management incorporated with MDR (monitoring, detecting, responding). The proposed MDR-CPS adopted collaborative control theory to facilitate requirements planning and address conflicts, enabling a collaborative system to improve interactions in CPS. Metta et al. [265] designed a comprehensive framework that merges social, cybernetic, and physical elements to evaluate participatory settings that address interconnected sustainability challenges across forestry, agriculture, and rural development. They first operationalize the agricultural CPS concept within a flexible assessment model that can adapt to various analytical depths, situations, and objectives. Then, they implemented this framework in practical settings involving multiple stakeholders to enhance their insights into the effects of digital transformation.

- **Agricultural Robotics**

To provide farmers with efficient and effective avant-garde-driven options, technologies and advances in crop production, monitoring, and phenotype have generated interest in using robots to achieve automated operations.

Chen et al. [56] built a service-oriented infrastructure for agricultural CPS with integrated geospatial services. They apply sensor networks and automated event-driven approaches to respond to field emergencies quickly and appropriately. Dusadeerungsikul et al. [100] stated a collaborative control protocol for agricultural CPS that facilitates the integration of robotics within CPS for intelligent and precise farming. This approach focuses on identifying and diagnosing plant stress using hyperspectral imaging techniques. Zhu et al. [524] coped with the joint problem of multi-agent computation offloading for robotic applications in a cloud-edge-end CPS system. In particular, they adopted the Nash equilibrium strategy to offload robotic tasks to cloud or edge servers. Wakchaure et al. [378] concluded with various AI techniques and robotics approaches in agriculture. They pointed out that the application of robots and autonomous systems is raising the standard of farming and becoming more popular. Huang et al. [168] proposed a robot-based approach for intelligent management in CPS agriculture. In this method, a hardware component employing a binary neural network enables the precise identification of crops while the encryption hardware safeguards sensor data communication.

With the help of literature on sustainability and automation, agricultural CPSs can effectively act on a crop's growth cycle, which goes through three primary stages: cultivation, monitoring, and harvesting. The development of agricultural CPS in the future can push the boundary further by providing greater flexibility of scale, higher functionality, increased resilience, greater profitability, and enhanced autonomy [312].

4.2.6 Conclusions and Discussions

The deployment of CPS applications at scale depends heavily on communication and computation infrastructures. The fusion of cloud-edge-end computing with CPSs holds immense promise for revolutionizing how we design, deploy, and manage distributed computing environments. By leveraging the strengths of each computing paradigm, including the vast resources of the cloud, the low latency of edge computing, and the direct interaction with physical processes at the end devices, this integration offers unprecedented opportunities to enhance the efficiency, reliability, and intelligence of CPS applications.

This chapter investigates the advancements in cloud-edge-end computing in various CPS application fields, including mobile CPS, medical CPS, automotive CPS, social CPS, and agricultural CPS. Through this investigation, we have explored the advantages of combining clouds, edges, and ends in CPS, such as real-time decision-making, scalability, resilience, and enhanced intelligence. However, we have also highlighted the challenges and considerations, including security and privacy concerns, communication and interoperability issues, and resource management challenges. Addressing these challenges will be crucial to unlocking the full potential of cloud-edge-end orchestrated CPS applications.

In addition to robust security and privacy mechanisms as well as standardized protocols and frameworks catering to the distributed nature of cloud-edge-end computing, which have been discussed in the context of IoT applications, advancements in resource optimization techniques are crucial for maximizing the efficiency and performance of CPS applications deployed across cloud, edge, and end layers. This includes developing intelligent task scheduling algorithms, workload balancing strategies, and adaptive resource provisioning resource management mechanisms. Also, further advancements in edge intelligence will enable CPS to perform complex analytics and decision-making at the network edge, reducing the reliance on centralized cloud resources and enhancing responsiveness. By pursuing these future research directions, we can potentially pave the way for the development of next-generation CPS solutions that are efficient, reliable, and intelligent.

4.3   Cloud-Edge-End Computing for Smart Cities

In an era marked by burgeoning urbanization and an ever-increasing demand for more efficient and sustainable urban environments, the concept of "Smart Cities" has emerged as a beacon of hope. As populations continue to concentrate in urban areas, innovative solutions to enhance the quality of life, sustainability, and overall urban management have become imperative. By seamlessly integrating cloud computing, edge computing, and end devices, cloud-edge-end computing is poised to revolutionize urban living, making cities smarter and more livable, efficient, and responsive to residents' needs. This section delves into the intricacies of cloud-edge-end computing in smart cities by exploring its development for smart buildings, AI applications, intelligent transportation systems, smart grids, and empowering data analytics.

4.3.1   Cloud-Edge-End Computing for Smart Buildings

The convergence of technology and architecture has created a new concept of "Smart Buildings", with its worldwide market size expected to expand from USD 96.96 billion in 2023 to USD 408.21 billion by 2030[1]. Equipped with many sensors, networks, and automation systems, smart buildings are redefining how we interact with our built environment, as shown in Figure 4.8. More significantly, the true potential of smart buildings is unlocked through the strategic integration of cloud-edge-end computing, which offers a comprehensive solution for optimizing buildings' performance, efficiency, and sustainability.

Several research directions and motivations exist in the field of cloud-edge-end computing-enabled smart buildings, such as building management systems (BMSs), real-time control, security privacy, energy efficiency, and user-centric approaches. Among these, the development of BMSs aims to monitor building operations, enabling predictive maintenance and improving occupant comfort. With the growing demand for fast response in smart buildings, developing real-time control systems has become essential, specifically for applications like heating, ventilation, air conditioning (HVAC), and catastrophe early warning. Considering smart buildings collect and process sensitive data, security privacy has emerged as a significant concern, raising issues like data encryption, access control, and protection against cyber threats. Aiming to improve user experience and occupant satisfaction, user-centric approaches focus on the development of personalization of building settings, user feedback systems, and smart environments adapted to individual preferences. From technology integration to sustainable design, researchers are working to create more effective and efficient built environments. The above research themes are not independent, and researchers must focus on as many aspects as possible to achieve a well-working smart building.

- **Building Management Systems**

    Regarding BMSs, Seitz et al. [313] developed a cloud-edge-end architecture for location-aware conflict negotiation and decision support for building occupants in smart buildings.

---

[1]https://www.fortunebusinessinsights.com/industry-reports/smart-building-market-101198

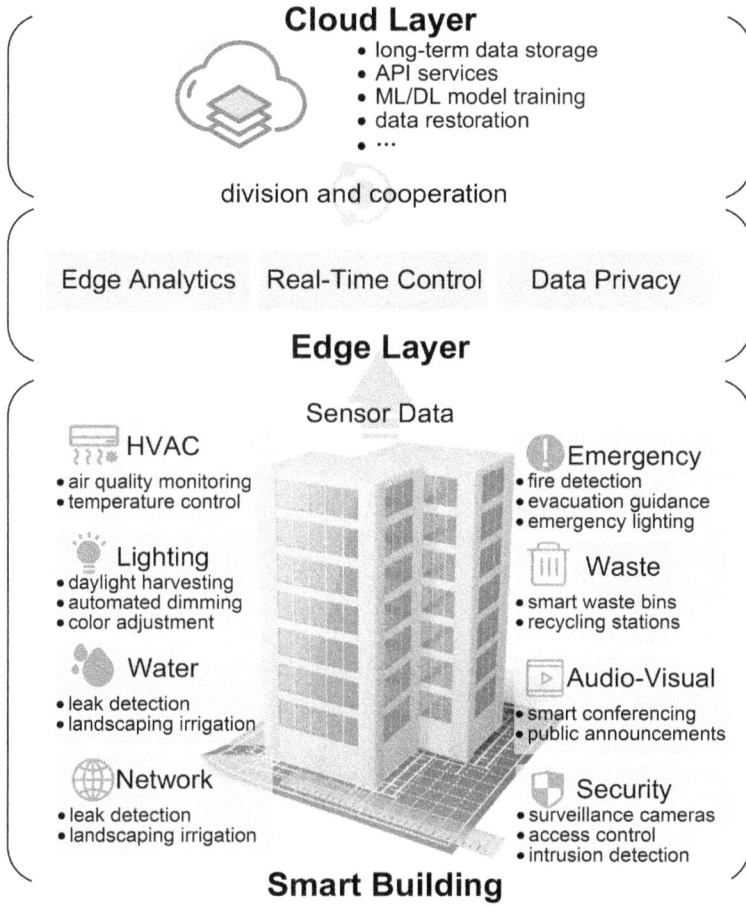

FIGURE 4.8
The cloud-edge-end framework for smart buildings.

Through the implementation of a decentralized cyber-physical system (CPS) with building occupants, actuators, and sensors in place of the centralized software architecture, this approach enhanced security and resilience while lowering latency and improving the quality of service. Santana et al. [311] presented an easily replicable system architecture for crowd recognition in smart buildings, which is a privacy-aware platform that enables artificial intelligence algorithms to analyze crowd behavior inside buildings based on sensed Wi-Fi traces. As a crowd management system, the strict privacy and lightweight design have been proven in two real-world buildings. Taking the office as an application scenario, Wang et al. [399] established a smart building IoT management system based on the edge architecture and the actual parameters of various power-consuming devices. An energy-saving model based on nonlinear optimization and particle swarm optimization algorithm is integrated and shown to be of great significance for office energy monitoring and management.

- **Real-Time Control**

In terms of real-time control, Dutta et al. [101] suggested an intelligent, green, and connected building that would allow users to operate all of the appliances in real-time. It offers a promising smart building solution by using the cloud service and fog gateway server to manage all the devices and immediately notify the user in real-time. Ding et al. [88]

presented a customized building control system that manages four building subsystems (i.e., HVAC, blinds, windows, lights) to satisfy three needs (i.e., indoor air quality, thermal comfort, visual comfort) for human. It uses the benefits of reinforcement learning-based control, such as quick adaptation to new structures, real-time actuation, and the capacity to manage a sizable state space.

- **Energy Efficiency**

About 30% of the world's total energy consumption and carbon emissions come from buildings, which has major adverse effects on the environment and the energy supply. Therefore, it is crucial and urgent to create effective smart building energy management technologies in order to increase energy efficiency. Plageras et al. [288] constructed a topology-architecture system for a smart building in order to provide an energy-efficient solution using the data gathered and managed by the sensors. With the help of software-defined networking, edge computing, and D2D-enabled communications, Ibrar et al. [177] proposed an improved resource allocation and energy management method for smart buildings. This method optimizes resource allocation and offloading decisions to reduce energy usage and delay in a smart building. Energy efficiency while meeting the delay restriction is the main issue when using edge computing to carry out tasks generated by end devices. In light of this, Sahoo et al. [308] investigated an energy and delay minimization problem in an IoT-enabled smart city and presented a three-layer network architecture. Based on this architecture, an auction-based edge resource allocation strategy is designed to provide low processing latency and energy-efficient service for delay-sensitive applications.

- **Security**

Security has been a significant concern in smart buildings. This concern is now much more crucial regarding data sharing and communication. Zhang et al. [480] studied the security computing resource allocation problem in serverless multi-cloud edge computing systems. A security protection method is developed to protect data integrity and privacy, taking into account the heterogeneous characteristics of computing resource nodes, such as geographic locations, computing capabilities, and unit energy consumption costs. Li et al. [214] explored the security of data sensing in smart buildings and proposed a fog-based data aggregation approach. The proposed approach combines fog computing for energy sustainability with AI-based blockchain to improve edge device security. However, it should be emphasized that many smart building technologies overlook concerns about security and privacy. Future studies need to focus more on this area to reduce network attacks and provide methods for quickly identifying and reporting privacy violations.

4.3.2 Cloud-Edge-End Computing for AI Applications in Smart Cities

With burgeoning technological advancements, the concept of "Smart Cities" has emerged as a transformative force, redefining the way we interact with our urban environments. The heart of this revolution lies in the fusion of artificial intelligence (AI) and cloud-edge-end computing paradigms. By leveraging these technologies, smart cities are devoted to enhancing infrastructure, public services, and quality of life for their inhabitants. In the context of artificial intelligence applications in smart cities, exploring the intricate technological advantages and synergistic potentials between cloud, edge, and end computing is crucial to driving cities toward efficient and sustainable development.

Smart Cities encompass a wide range of aspects that aim to use technology and data-driven solutions to enhance the quality of urban living. From the point of view of actual functionality, some of the key aspects involved in smart cities include healthcare [95], industry [395, 341, 368, 299], agriculture [8], environmental sustainability [13, 454], healthcare [192], economic development [300, 355], etc.

- **Industry**

  The convergence of cutting-edge technologies has given rise to the concept of Smart Industry, revolutionizing the landscape of manufacturing and production. Smart Industry leverages the innovative integration of AI and cloud-edge-end computing to enhance efficiency, productivity, and responsiveness within the manufacturing sector. Using cloud servers and edge devices, Sun et al. [341] proposed a data stream cleaning system. The system is used in two scenarios: monitoring base stations and injection molding machines. It successfully maintains processing speed and cleaning effectiveness even when the number of accessed edge devices rises. Ullah et al. [368] proposed a framework for an IIoT-based violence detection network using artificial intelligence to automate the surveillance system in terms of recognizing human activities. Only the frames containing objects are sent for detailed analysis in the cloud, and features are extracted using convolutional long-short-term memory. Rahman et al. [299] merged several AI algorithms to automatically classify the captured events and objects and provide analytics, reports, and warnings from the IIoT data in real-time on the distributed edge and cloud nodes.

- **Economy**

  Driven by innovation and efficiency, cities are now at the forefront of integrating AI into their economic fabric. AI-enabled economy services in a smart city is a burgeoning field with immense promise for revolutionizing how cities operate and provide services to inhabitants. Rahman et al. [300] used cognitive fog nodes to host and process transactions from mobile edge and IoT nodes as well as offloaded geo-tagged multimedia content. Additionally, to support sharing economy services, AI is used to analyze and extract information about critical events, create semantic digital analytics, and preserve the findings in blockchain and decentralized cloud repositories. With the aim to meet the demands of various users in edge computing, Tang et al. [355] developed a pricing model for dynamic resource overbooking by creating on-demand, daily, auction, and new spot billing techniques. It has been demonstrated that an auction strategy with price rules and winner selection rules ensures individual rationality, computational effectiveness, and veracity. A dynamic resource overbooking system, which incorporates a cancellation policy and a resource forecast technique, is proposed in order to leverage better the auction strategy to utilize idle resources.

- **Environmental Sustainability**

  AI-enabled environmental sustainability in a smart city leverages the power of data-driven insights to promote responsible urban development, reduce resource consumption, minimize environmental impact, and create more resilient, eco-friendly urban environments. It involves using AI to collect, analyze, and act upon data to optimize resource management, reduce environmental impact, and improve the overall ecological well-being of urban areas. Alqahtani et al. [13] developed an urban trash management system that utilized a cuckoo search-based long short-term recurrent neural network to gather and interpret data about a city. The waste management center is alerted by analyzing the type of waste, the size of the truck, and the source of the waste so that appropriate action can be taken. Yang et al. [454] proposed an AI-driven visual end-edge-cloud architecture that improves upon the traditional design in terms of fusing humans and machines and reducing carbon emissions. In this architecture, systematic analysis and intelligent computing techniques are created for carbon emission, facilitating the delivery of AI intelligence for 6G networks using hybrid hierarchical optimization techniques.

- **Video Surveillance**

AI-enabled video surveillance in smart cities based on cloud-edge-end computing aims to provide real-time situational awareness, enhance security, and improve the efficiency of city operations. It optimizes the use of resources and bandwidth, reducing response times and enabling intelligent decision-making based on analyzed video data. However, addressing privacy concerns, data management, and regulatory compliance is crucial while implementing and operating such systems in smart cities. Chiu et al. [69] proposed a federated and semisupervised learning-based edge learning system. The system trains AI models at edge devices using a semisupervised learning method and regularly uploads the training results to the cloud server to create a single model based on a federated learning method. Taking into account that it is challenging to migrate computation and data-intensive tasks from the cloud to the edge due to the high computational need, Nikouei et al. [278] presented a hybrid lightweight tracking strategy to achieve intelligent surveillance as an edge service. This strategy uses a lightweight convolutional neural network in conjunction with a decision tree-based hybrid kernelized correlation filter technique for high performance. Chen et al. [64] proposed a real-time smart video surveillance system based on edge-cloud computing for object detection. Edge computing integrates media data from networked distributed edge devices for AI in the cloud. Once the cloud-based AI discovers global knowledge, it shares it with the edge nodes, enabling low-latency surveillance.

As a dynamic and evolving field, several exciting future directions regarding the integration of cloud-edge-end computing and AI technologies for smart cities can be anticipated in the coming years. First, deploying 6G and future-generation wireless networks will significantly impact the capabilities of cloud-edge-end computing in smart cities. Research will focus on harnessing the full potential of high-speed, low-latency networks for AI applications. Second, developing specialized hardware for edge AI, such as AI accelerators and neuromorphic chips, will continue to be a research priority to improve the efficiency and performance of edge devices. Third, developing AI-driven solutions for enhancing the reliability and resilience of networks to ensure consistent performance, especially in situations where there are natural disasters or other disruptions. To address more complex challenges and explore better solutions, the research and practice of AI application-enabled smart cities will continue to evolve with the aim of creating more efficient, sustainable, and citizen-friendly urban environments.

4.3.3 Cloud-Edge-End Computing for Intelligent Transportation Systems

With the rapid development of urbanization, cities filled with dynamic flows of people and vehicles made efficient and intelligent transportation systems a top priority. The integration of cloud, edge, and end computing into intelligent transportation systems is a promising technology, which is expected to enable seamless, efficient, and safe transportation and will also have a profound impact on how cities operate and citizens commute. In the literature, some recent studies emphasize the need for extremely low latency in real-time traffic management and security-critical applications. Some studies emphasize effective data storage and management, which is essential to provide a comprehensive view of transportation systems. Some studies emphasize optimizing resource allocation between edge and cloud computing, focusing on dynamic resource allocation algorithms that adapt to changing traffic conditions and demands. Considering cloud-edge-end computing for intelligent transportation systems, several challenges and complexities exist that need to be addressed for successful implementation.

- **Traffic Management and Prediction**

  Balancing the trade-off between processing data at the edge for low latency and utilizing cloud resources for more extensive analysis is a complex challenge in intelligent transportation systems. Real-time decision-making and prediction require minimal latency, especially in safety-critical scenarios. Considering that increased traffic monitoring has brought significant obstacles to storing, communicating, and processing traditional cloud-based transportation systems, Chen et al. [50] proposed an edge-based traffic flow detection strategy. This strategy incorporates a vehicle detection algorithm based on the YOLOv3 model and a multiobject vehicle tracking algorithm based on deep, simple online and real-time tracking. Integrating edge and cloud together, Liu et al. [230] proposed a two-tier traffic monitoring scheme that takes into account the trade-off between the limited computing capability of edge nodes and the unstable network condition of the cloud-based traffic management center. Lai et al. [199] propose a spatial-temporal attention graph convolution network on edge nodes, achieving accurate short-term traffic flow prediction. Assigning each network component to a specific roadside unit for training eliminates the need to handle all data on the central cloud server. Jiang et al. [183] proposed an end-edge-cloud computing framework for vehicular cooperative control, aiming to facilitate vertical and horizontal cooperation. This framework employs a two-stage reinforcement learning process to find the optimal vehicle control solution.

- **Network Communication**

  Achieving low latency in real-time traffic communication is essential for delay-sensitive applications. Traffic network communication should adapt to changing conditions, including the movement of vehicles and the addition of new network nodes. Developing systems that can handle dynamic network topologies is complex and challenging. Yang et al. [443] proposed a deep reinforcement learning-based channel allocation and task offloading method in the temporary unmanned aerial vehicle (UAV)-assisted vehicular edge computing networks (VECNs), considering the situations where traffic roads become congested or roadside units (RSUs) are inaccessible outside of communication range. Zhao et al. [141] concentrated on the problem of overloaded edge nodes and proposed a multipath transmission workload optimization strategy. Multipath transmission supports communication between vehicles and edge nodes and the real-time virtual machines (VMs) migration between edge nodes. Garg et al. [120] presented a composite architecture for intelligent transportation systems, incorporating a distributed software-defined networking and mobility management method. Compared to the conventional mobility management method, it has also been enhanced by adding the optimal routing decision module.

- **Storage Architecture**

  The high volume and velocity of data generated by various traffic sensors, cameras, and connected vehicles can overwhelm storage systems. Data lifecycle management, data access, data Retrieval, data security, reliable, and scalability are all key challenges in the development of an intelligent transportation system. Lu et al. [248] presented a storage-elastic blockchain method based on various edge storage capacities that facilitates re-write operations to replace an old block with a new one while maintaining the hash connections. In this method, the hot data can be efficiently accessed from edge servers without significantly increasing communication costs or the chain's overall size. Qiao et al. [292] proposed a distributed trustworthy storage architecture based on edge computing for intelligent transportation. The storage method uses reinforcement learning to store data dynamically, improving resource scheduling and storage space allocation. An identity authentication mechanism based on trapdoor hashing is also incorporated to provide security for the transportation network. Ling et al. [228] detailed the fundamental elements of a distributed multilevel storage infrastructure to acquire, store, and analyze the expanding volumes of data from intelligent transportation systems securely and reliably.

- **Computation Offloading**

  There are many aspects to consider when effectively offloading traffic-related computation tasks to the cloud, edge, or end devices, such as response time, dynamic workload distribution, edge server and end device constraints, and resource heterogeneity. All of these are crucial for optimizing traffic analysis and decision-making processes. Ma et al. [254] presented a vehicular-cloud-assisted MEC offloading method to utilize resources and maximize task throughput fully. This method uses deep learning to predict vehicle trajectories and generates a vehicle cloud to consolidate computational resources. Deep reinforcement learning and the Lyapunov optimization-based framework are developed to deal with the coupling between variables and time slots. Xia et al. [415] focused on a distributed dynamic computation offloading problem for an intelligent transportation system that is enabled by heterogeneous edge servers. A Stackelberg game-based computation offloading algorithm is developed to find the optimal power allocation and computation offloading solutions through centralized training and decentralized execution. Ko et al. [193] proposed a belief-based task offloading technique for determining the computing edge server and communications subchannel for each vehicle, considering the changing characteristics and limited resources of vehicular edge computing systems. In order to fend off attacks from privacy attackers with prior information, Gao et al. [119] proposed a privacy task offloading approach where the local computing model, channel model, and privacy loss model are created, aiming to quantify assessment indicators in terms of privacy, time, and energy.

- **Resource Management**

  Amidst the swift progress in the Internet of Vehicles and artificial intelligence technologies, the cooperative intelligent transportation system (C-ITS) has garnered considerable focus lately. For delivering a highly reliable and minimal latency computational performance in C-ITS, leveraging computation offloading to edge-cloud servers is regarded as critical. In light of the unbalanced distribution of access vehicles as well as the enormous amount of data, Duan et al. [99] presented a hierarchical architecture with quality of service-aware and power-aware resource management for the edge-enabled vehicular system. Based on a minimal latency with a migration loads strategy, the load-unbalancing issue can be resolved by choosing the proper response time threshold and migrating loads from overloaded edge servers to idle ones. Wei et al. [404] focused on resource allocation in vehicular cloud computing by optimizing resource assignment from both the provider's and users' viewpoints, formulated as a constrained multi-objective optimization problem.An improved non-dominated sorting genetic algorithm II is designed to solve this multi-objective problem, which attempts to increase the acceptance rate and reduce the cost of the provider cloud. Tang et al. [350] proposed a three-layer vehicular network made up of the vehicular cloud, the roadside unit-based cloudlet, and the remote cloud to cope with the rapidly growing number of computationally expensive and time-sensitive vehicle-related applications. From the perspective of a cloudlet, an approximate approach is developed to optimize the utility value more efficiently than meta-heuristics. Within the 6G-enabled transport system context, Cao et al. [34] investigated the autonomous-aware software resource management and allocation framework. The proposed architecture based on reinforcement learning would achieve intelligent resource management and allocation after training. In short, improving the safety, effectiveness, sustainability, and overall performance of transportation systems is the main objective of resource management in intelligent transportation systems.

These challenges and complexities in Cloud-Edge-End Computing for Intelligent Transportation Systems represent ongoing research areas and active development efforts. Addressing these issues is essential for successfully deploying efficient, safe, and sustainable transportation systems in modern cities. Researchers and practitioners continue to work toward innovative solutions to overcome these challenges and improve the performance and reliability of intelligent transportation systems.

4.3.4 Cloud-Edge-End Computing for Smart Grids

As a digitally enabled system, the smart grid collects and utilizes data about consumer and supplier behaviors to enhance the performance of electricity services. In the context of the smart grid, edge-cloud-end computing can process and analyze data from various sources, such as smart meters, sensors, and IoT devices, in real-time. This can help optimize grid operations, predict and manage energy demand, and improve the reliability, efficiency, and energy sustainability of the power system. The state-of-the-art in this field involves using advanced machine learning and data analytics techniques to process and analyze smart grid data. This includes methods for anomaly detection to identify faults or attacks on the grid, predictive modeling for demand response and energy management, and optimization algorithms for resource allocation in the grid.

- **Integration Framework Development of Computing, Communication, Perception for Smart Grid**

  The development of an integrated framework that combines computing, communication, and perception capabilities at the cloud-edge-end is important for processing and analyzing data in a smart grid. Wu et al. [411] presented a cloud-edge-end-based six-layer architecture for the smart meters system, incorporated with a re-configurable mixed-signal controller. This architecture can speed up processing and improve the accuracy of arc fault detection under various loading circumstances. Bachoumis et al. [26] designed a cloud-edge framework to enhance computations for distributed energy resources load forecasting and power flow optimization. The proposed framework allows local energy market operators to respond to service procurement signals sent by the balancing authority, even when sub-second service delays are required. Tzanis et al. [367] introduced a cloud-edge computing-driven smart grid framework to facilitate calculations for transient state estimate, which is a computation-intensive monitoring tool for failure data analytics and obtaining trustworthy knowledge. This framework orchestrates virtual machines running on virtualized and non-virtualized server nodes. Realizing the limitations in conventional cloud-based power systems, such as bandwidth and latency, Chen et al. [60] proposed an edge-enabled smart grid system, considering both hardware and software design. With the help of edge computing, the smart grid will achieve the connection and management of physical ends, providing real-time analysis and data process and promoting the intelligentization of the smart grid.

- **Terminal Key Management and Access Authentication Technology for Smart Grid**

  The approaches for managing cryptographic keys and authenticating devices are critical for ensuring the security and integrity of communications within a cloud-edge-end-based smart grid network. Based on the edge-cloud architecture and federated-learning technology, Su et al. [338] proposed a smart grid system for security-aware energy data sharing. A two-layer deep reinforcement learning algorithm was developed to provide security trust and facilitate owner engagement, enabling energy data owners to collaboratively train shared AI models without disclosing their personal data. Liu et al. [232] proposed a privacy-preserving smart grid system for aggregate communication and function query based on edge computing structure. The proposed system, which employs globally distributed edge nodes and a centrally located cloud server to accomplish low-latency communication and electricity data storage, is encrypted using a double trapdoor cryptosystem. Considering the security issues arising from the information exchange between devices involved in smart grids, Fu et al. [116] proposed a privacy-preserving energy management approach based on the cloud-edge computing paradigm, which makes use of the multipliers alternating direction techniques. Energy management is formulated as a social welfare maximization problem, including supply and demand, while maintaining the supply-demand balance and satisfying operating constraints. To enable low-latency and real-time services, Wang et al. [385] introduced a blockchain-based security assurance for the smart grid. This protocol leverages the capabilities of edge computing to support efficient conditional anonymity and flexible key management and further mitigate challenges in improving the quality of service as the system scales.

- **Real-Time Monitoring, Control, and Management of Electrical Power**

  Real-time monitoring, control, and management of electrical power refer to the continuous and immediate supervision, adjustment, and operation of the electrical grid and its associated components to ensure the efficient delivery of electrical power. Based on the cloud-edge-end architecture, designing a power dispatching strategy, energy consumption optimization strategy, and fault-tolerant strategy becomes essential to making real-time decisions and responding to changing conditions within the electrical grid. Wang et al. [387] proposed a cloud and edge computing framework to achieve complex energy operation control and massive information processing for power grids with unpredictability and widely distributed resources. This framework leverages the capabilities of cloud computing for large-scale data processing and storage, whereas it leverages edge computing for real-time decision-making. Siddiqui et al. [323] presented a knowledge-based strategy for smart meters in cloud-based smart grids to optimize energy utilization in a large-scale distributed computing environment, which is a critical factor in green smart grids. Given the large amounts of data in heterogeneous forms collected by massive end devices, Yang et al. [453] proposed a cloud-edge-end orchestrated computing scheme for fault-tolerant renewable energy accommodation in a smart grid. Cloud servers use large-scale data computation in deep learning algorithms to sense features and fix missing values. Edge servers employ deep reinforcement learning algorithms to obtain optimal policies and enable low-latency and real-time services. To improve the performance of the grid system, Hu et al. [161] proposed an artificial intelligence-assisted grid event classification method that integrates an edge-cloud sharing policy to balance load as well as reduce processing time.

- **Resource Allocation and Task Scheduling of Smart Grid**

  The development of effective and efficient algorithms for resource allocation and task scheduling in the cloud-edge-enabled smart grid system is crucial for optimizing grid operations and improving the grid system's service quality. Resource allocation and task scheduling in the smart grid aim to optimize resource utilization, improve grid efficiency, enhance reliability, and ensure that electricity supply meets demand while considering economic and environmental factors. Zhou et al. [514] investigated the joint optimization of computational offloading and service caching in edge computing-based smart grids to handle data generated by exponentially growing smart devices. This work involves determining how to offload computational tasks from devices to edge servers and deciding which services to cache on edge servers for efficient retrieval. Aiming at system cost minimization, a mixed-integer non-linear program optimization model is formulated and further solved by proposing a gradient descent allocation algorithm and a game theory-based approach. Considering end-users, edge

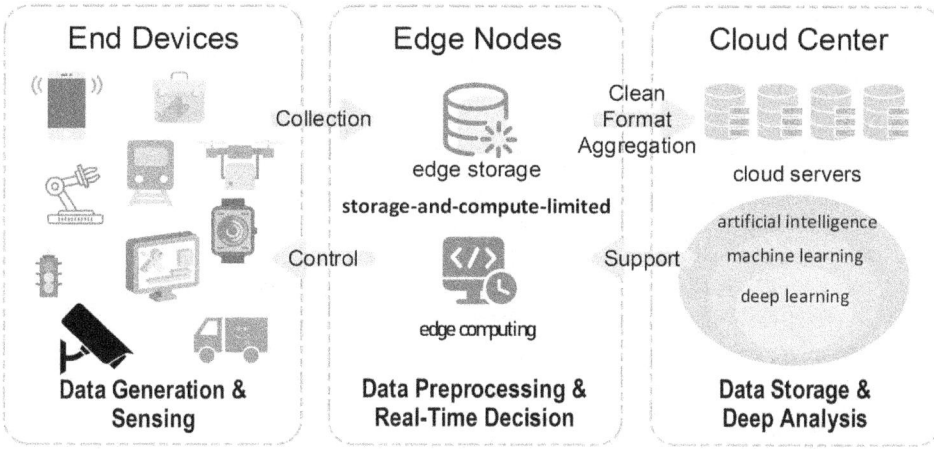

FIGURE 4.9
The cloud-edge-end framework for data analysis in smart cities.

nodes, cloud centers, and multi-edge alliances, Sun et al. [346] proposed a cooperative scheme to solve the problem of collaborative task allocation among multiple edges, with the aim of maximizing the social welfare of the multi-edge system without compromising the interests of each edge node. Gunaratne et al. [131] proposed an edge intelligence-driven monitoring paradigm that utilizes end data to monitor large variance drifts in control variables. The proposed method enables the grid to delegate decision-making to the edge layer, minimizing latency and data integrity issues while enabling quicker monitoring capabilities. Since an efficient edge computing system requires multiple resource-limited edge nodes to cooperate with each other to optimize the workload distribution, Niu et al. [279] introduced a balanced initialization, resource allocation, and task allocation strategy for edge computing-enabled power IoT systems, to minimize the service latency. Both task allocation among edge nodes and resource allocation within edge nodes are essential in this regard.

Smart grids increasingly rely on advanced technologies, including data analytics and machine learning, to make informed decisions and adapt to the dynamic nature of the modern energy landscape. To fully use the large volumes of data generated by smart grids and the advantages of edge-cloud-end computing architecture, many challenges remain to be addressed. These include issues related to data privacy and security, the need for standardized protocols and interfaces for data exchange between different grid components, and the development of scalable and robust systems.

4.3.5 Cloud-Edge-End Computing Empowering Data Analytics for Smart Cities

Data analytics for smart cities is an interdisciplinary academic field that focuses on harnessing data to drive positive urban transformations, enhance services, and make cities more livable and efficient. It plays a crucial role in addressing the complex challenges faced by modern urban environments. Cloud computing, on the one hand, helps manage the enormous quantity of data generated in cities and is used to create livable, resource-effective cities. For instance, it can reduce costs by identifying more effective business practices and better business judgments. Edge computing, on the other hand, brings services near the edge of a network, extending the possibilities of cloud computing. As shown in Figure 4.9, by transferring processing, data, and services from the centralized cloud to edge servers, edge computing performs well in communication latency and traffic reduction. It also enables decision-making and real-time processing at the network's edge, nearer to where data is created. There are four categories regarding the types of data analytics applications as follows. Each type of analytics serves a different purpose and can provide valuable insights depending on the organization's specific needs or questions.

- **Descriptive Analytics**

    Descriptive analytics in the research field of smart cities involves the use of data analytic techniques to provide insights into historical and current urban conditions, patterns, and trends. It focuses on summarizing and visualizing data to answer the "what happened" question, allowing researchers, city planners, and policymakers to better understand urban phenomena and make informed decisions [347].

- **Diagnostic Analytics**

    Diagnostic analytics in the research field of smart cities involves using data analytic techniques to investigate and understand the causes of urban issues, anomalies, and problems. It focuses on answering the "why did it happen" question by identifying the root causes of specific urban challenges or unusual events. Diagnostic analytics is a critical component of the broader analytics process in smart cities and is used to support urban planning, resource allocation, etc[226].

- **Predictive Analytics**

    Predictive analytics in the research field of smart cities involves the use of data analytic techniques to anticipate future events, trends, and outcomes in urban environments. It focuses on answering the question of "what is likely to happen" by leveraging historical and real-time data to make informed predictions. Predictive analytics is a valuable tool for optimizing urban services, resource allocation, and decision-making in smart cities[282].

- **Prescriptive Analytics**

    Prescriptive analytics in the research field of smart cities involves using advanced data analytic techniques to recommend actions and strategies for optimizing urban processes. It goes beyond predictive analytics, which forecasts future outcomes, to provide actionable insights on "what should be done" to achieve specific urban objectives. Prescriptive analytics is a powerful tool for urban planning, resource allocation, and policy development in smart cities[373].

    Considering cloud-edge-end-enabled smart cities, the physical entities for data analytics can be the cloud layer, edge layer, end device layer, and combinations of these three layers.Regarding this, a comprehensive categorization and discussion of the recent approaches to cloud-edge-end computing empowering data analytics for smart cities is provided below.

- **Cloud-Enabled Data Analytics for Smart Cities**

    In the context of smart cities, large volumes of data from multiple sources are gathered, processed, integrated, and shared based on cloud computing technologies. The role of cloud computing is critical, providing the infrastructure needed to transform city data into actionable insights and inform the decision-making process. Mohammadi et al. [267] proposed a deep reinforcement learning approach that is suitable for smart city applications, using both labeled and unlabeled data to enhance the learning agent's performance. This work marks a pioneering exploration into the extension of deep reinforcement learning into the realm of semisupervised paradigm. Popa et al. [289] introduced a modular platform designed to harness the capabilities of cloud services for the collection, aggregation, and storage of data originating from smart environments. By constructing sophisticated neural network models, this platform can offer recommendations for optimizing daily habits to reduce energy consumption, resulting in cost savings. Zhao et al. [507] proposed admission control and profit optimization algorithms in a cloud-based Analytics-as-a-Service system, which can admit data analytics requests, optimize resource allocation, and ensure QoS. Skourletopoulos et al. [328] introduced elasticity debt analytics, a concept designed to manage resource provisioning in mobile cloud environments while maintaining service quality. Furthermore, this work proposed a green-centric, game-theoretic strategy to reduce elasticity debt in mobile cloud offloading situations, taking into account variables such as processing time, energy consumption, and overhead. In essence, cloud-enabled data analytics for smart cities harnesses the cloud's computational power, scalability, and data management capabilities to facilitate efficient data analysis and support the development of data-driven, efficient, and sustainable urban environments.

- **Edge-Enabled Data Analytics for Smart Cities**

    Although cloud-enabled data analytics allows for deep analysis and can handle large data sets, however, it may not be suitable for real-time applications due to potential latency issues. Edge-enabled data analytics, in which data is processed closer to where it is generated, can reduce latency and allow for real-time insights. He et al. [149] presented a multitier fog computing framework designed to efficiently analyze IoT sensor data in smart city applications, which incorporates both ad-hoc and dedicated fog computing layers. Based on the simulation results, this work revealed the effectiveness of fog-based analytics services, highlighting the advantages over cloud-only models in terms of job-blocking probability and service utility. Muhammad et al. [271] introduced a resource-efficient convolutional neural network system that leverages edge computing for fire detection in challenging surveillance scenarios. This system utilizes lightweight neural networks devoid of dense layers, making it a practical solution for mobile and embedded vision applications in uncertain IoT environments. Guim et al. [130] emphasized autonomous life cycle management of converged edge platforms, streamlining workload orchestration to promote green computing initiatives. An edge-based intelligent resource configuration methodology is proposed to support multi-tenant services, ensuring that service-level objectives and supporting sustainability goals. Realizing the preprocessing challenge associated with raw data from multiple edge sites, Jin et al. [185] delved into the joint optimization of data placement and task assignment to accelerate collaborative analytics and mitigate network traffic. A hybrid approach involving convex relaxation and two-stage optimization is proposed, addressing issues like non-convexity and the uncertainty of query characteristics. Unlike cloud-enabled data analytics, edge-enabled data analytics for smart cities is particularly beneficial in scenarios where quick decision-making is required or bandwidth is limited.

- **Cloud-Edge-Enabled Data Analytics for Smart Cities**

    The cloud-edge-enabled data analytics approaches combine the strengths of both cloud and edge computing. Specifically, the cloud layer is well suited for resource-intensive and long-term data analytics, acting as a central repository for historical and real-time data that can be used to perform complex data analysis and machine learning tasks. The edge layer is a distributed network of edge devices, such as IoT sensors, gateways, and localized computing resources deployed closer to the data sources in a smart city. Its primary role is to process data at or near the source, providing real-time analytics and reducing the need to transmit all data to the central cloud. The edge layer enhances responsiveness and minimizes latency, making it ideal for time-critical applications like smart traffic management, public safety, and autonomous vehicles. Babar et al. [25] introduced an architecture for IoT-enabled big data analytics, seamlessly integrating edge and cloud computing. This architecture consists of two layers: IoT-edge and cloud processing. It also incorporates an optimized MapReduce parallel algorithm for data injection and storage and a resource negotiator for cluster management. Ali et al. [10] presented an edge-enhanced cloud system for stream analytics, comprising filtration and identification phases that reduce data amount and leverage deep learning inference. Compared to centralized cloud-only approaches, this system significantly economizes time and bandwidth for analyzing large-scale data streams. Ghosh et al. [123] investigated a hybrid approach that combines edge and cloud computing for IoT data analytics. This approach partitioned the trained autoencoder in the feature learning phase, placing the encoder part at the edge and the decoder part in the cloud. Yu et al. [466] introduced a big data ecosystem featuring a three-layer architecture for predictive maintenance – comprising edge, cloud, and application layers. This ecosystem allocates tasks between the edge and cloud layers, improving reliability and scalability. Additionally, a distributed edge computing-assisted autoencoder is proposed to enhance system performance and efficiency. In essence, these approaches process data at the edge for real-time applications and send select data to the cloud for in-depth analysis and long-term storage, effectively supporting a wide range of smart city applications.

The research directions in data analytics for smart cities are diverse and continually evolving to meet the evolving needs of urban environments. As smart city projects continue to expand, interdisciplinary collaboration among data scientists, urban planners, policymakers, and domain experts is essential to address complex challenges and unlock the potential of data-driven urban transformation.

4.3.6   Conclusions and Discussions

Smart cities are rapidly evolving to leverage technology to enhance the quality of life, sustainability, and efficiency of urban environments in the face of urbanization and increasing demands for efficiency. Central to this transformation is the integration of cloud-edge-end computing, a distributed computing paradigm that brings powerful computing capabilities to the network's edge, closer to the end devices and users that need them. This can provide significant benefits for smart cities, which increasingly depend on real-time data processing and analysis to improve efficiency and safety.

This chapter explores the concept of smart cities and the role of cloud-edge-end computing in revolutionizing urban living. The collaboration of clouds, edges, and ends is crucial for optimizing smart buildings' performance, efficiency, and sustainability through building management systems, real-time control, energy efficiency, and user-centric approaches. AI applications in smart cities benefit from the integration of cloud-edge-end computing, enhancing infrastructure, public services, and quality of life in areas such as healthcare, industry, agriculture, environmental sustainability, and economic development. Intelligent transportation systems can be improved through the integration of cloud-edge-end computing, enabling seamless, efficient, and safe transportation with low latency and effective data storage and management.

In the field of smart buildings, future work directions involve improving building management systems for predictive maintenance and occupant comfort, developing real-time control systems for applications, addressing security and privacy concerns, and proposing approaches for personalized building settings and occupant satisfaction. For intelligent transportation systems, future work directions include balancing the trade-off between edge and cloud computing for low latency real-time traffic management, developing communication systems that adapt to changing network topologies, and optimizing computation offloading for traffic analysis and decision-making processes. Furthermore, future work directions in resource management for intelligent transportation systems involve resolving load-unbalancing issues, optimizing resource allocation from both the provider's and users' viewpoints, and achieving intelligent resource management and allocation through reinforcement learning. Overall, addressing these challenges and complexities is crucial for the successful deployment of efficient, safe, and sustainable smart cities and intelligent transportation systems.

5

Summary and Future Research

5.1 Summary

This book focuses on the modeling, optimization, and applications of cloud-edge-end systems. Regarding modeling, it covers the architecture and performance modeling of cloud computing, edge computing, end devices, and cloud-edge-end orchestrated systems. Performance optimization covers the optimization of latency, energy, security, privacy, and reliability in cloud-edge-end computing systems. Regarding applications, it covers the applications of cloud-edge-end computing in IoT, CPS, and smart cities. The summary of each chapter is below.

In Chapter 1, we first introduce classifications and concepts of cloud computing, categorized into IaaS, PaaS, and SaaS according to architecture and public cloud, private cloud, and hybrid cloud based on deployment. Next, we present the concept and key technologies of edge computing, including edge offloading, edge caching, edge inference, and edge training. Finally, we introduce the hardware and performance metrics of IoT end devices.

In Chapter 2, we first introduce the evolution from mobile cloud computing to mobile edge computing, cloud-edge computing, and cloud-edge-end computing. Next, we discuss blockchain, network function virtualization, energy harvesting technologies, and hierarchical and horizontal architectures of cloud-edge-end computing. Finally, we study server placement, data training, and resource management in cloud-edge-end computing.

In Chapter 3, we first review the challenges faced by edge-to-cloud technologies in terms of latency, security, reliability, and energy consumption. Next, we describe the mathematical models for latency, energy, security, and reliability in cloud-edge-end systems. Finally, we explore the performance optimization of cloud-edge-end systems in depth by examining the optimization of cloud-edge-end systems in terms of latency, energy, security, and reliability.

In Chapter 4, we first introduce the applications of cloud-edge-end computing in IoT, including intelligent IoT, time-sensitive IoT, and the Internet of Vehicles. Next, we discuss the applications of cloud-edge-end computing in CPS, including mobile CPS, medical CPS, automotive CPS, social CPS, and agricultural CPS. Finally, we present the applications of cloud-edge-end computing in smart cities, including smart buildings, AI applications, intelligent transportation, smart grids, and data analytics.

5.2 Future Research

Cloud-edge-end computing represents a novel computing paradigm and is a broad research field. It integrates cloud computing, edge computing, and device computing, leveraging the strengths of each layer to provide efficient, scalable, and reliable services. Future research can be carried out from the following aspects.

DOI: 10.1201/9781003540281-5

- The Fusion of Cloud-Edge-End Computing and Artificial Intelligence. The development of artificial intelligence (AI) algorithms brings revolutionary changes to the intelligent processing of big data. At the same time, AI tasks impose higher requirements on real-time performance, energy consumption, and accuracy of computing devices. Through the network, the cloud-edge-end computing orchestrated system leverages each level's advantages to provide better solutions for AI applications. Considering the characteristics of AI models and data, further research is needed to establish mathematical models for AI tasks in the cloud-edge-end system.

- Security and Privacy Protection. The cloud-edge-computing system involves processing and transmitting large amounts of user data, including personal and sensitive data. Therefore, ensuring the data security and user privacy of dispersed nodes in the cloud edge architecture, maintaining data integrity during processing and transmission, and designing robust authentication and authorization mechanisms for dynamic and heterogeneous environments are crucial for the system's reliability, stability, and long-term development.

- Energy Efficiency and Sustainability. With the construction of many computational infrastructures in the cloud-edge-end system, energy efficiency and sustainability have become important challenges for the system. Therefore, there is an urgent need for research in energy-efficient hardware, intelligent energy management algorithms, and sustainable energy utilization to enhance the energy efficiency of devices, particularly at the edge and end levels, reduce carbon footprints, and promote green and sustainable computing.

Bibliography

[1] Cloud white: Detecting and estimating qos degradation of latency-critical workloads in the public cloud. *Future Generation Computer Systems*, 138:13–25, 2023.

[2] Raafat O. Aburukba, Mazin AliKarrar, Taha Landolsi, and Khaled El-Fakih. Scheduling internet of things requests to minimize latency in hybrid fog-cloud computing. *Future Generation Computer Systems*, 111:539–551, 2020.

[3] Mike Adams, Shannon Bearly, David Bills, Sean Foy, Margaret Li, Tim Rains, Micheal Ray, Dan Rogers, Frank Simorjay, Sian Suthers, et al. An introduction to designing reliable cloud services. *Microsoft Corporation*, pages 1–14, 2014.

[4] Fahmida Akter, Mohammod Kashem, Md Islam, Mohammad Chowdhury, Md Rokunojjaman, and Jia Uddin. Cyber-physical system (CPS) based heart disease's prediction model for community clinic using machine learning classifiers. *Journal of Hunan University Natural Sciences*, 48(12):86–93, 2021.

[5] Md Washik Al Azad, Susmit Shannigrahi, Nicholas Stergiou, Francisco R Ortega, and Spyridon Mastorakis. Cledge: A hybrid cloud-edge computing framework over information centric networking. In *2021 IEEE 46th Conference on Local Computer Networks (LCN)*, pages 589–596. IEEE, 2021.

[6] Hyame Assem Alameddine, Sanaa Sharafeddine, Samir Sebbah, Sara Ayoubi, and Chadi Assi. Dynamic task offloading and scheduling for low-latency IoT services in multi-access edge computing. *IEEE Journal on Selected Areas in Communications*, 37(3):668–682, 2019.

[7] Abdulaziz Alarifi, Fathi Abdelsamie, and Mohammed Amoon. A fault-tolerant aware scheduling method for fog-cloud environments. *PloS One*, 14(10):e0223902, 2019.

[8] Hatem A. Alharbi and Mohammad Aldossary. Energy-efficient edge-fog-cloud architecture for IoT-based smart agriculture environment. *IEEE Access*, 9:110480–110492, 2021.

[9] Ghazanfar Ali, Hassan Ebrahimi, Jerrin Pathrose, and Hans G. Kerkhoff. Design and implementation of a dependable cpsoc for automotive applications. In *IEEE Industrial Cyber-Physical Systems*, pages 246–251, 2018.

[10] Muhammad Ali, Ashiq Anjum, Omer Rana, Ali Reza Zamani, Daniel Balouek-Thomert, and Manish Parashar. RES: Real-time video stream analytics using edge enhanced clouds. *IEEE Transactions on Cloud Computing*, 10(2):792–804, 2022.

[11] Ali Alnoman and Alagan Anpalagan. Computing-aware base station sleeping mechanism in h-cran-cloud-edge networks. *IEEE Transactions on Cloud Computing*, 9(3):958–967, 2019.

[12] Abdulrahman Saad Alqahtani. Performance computation and implementation of distributed controllers for reliable software-defined networks. *The Journal of Supercomputing*, pages 1–11, 2021.

[13] Fayez Alqahtani, Zafer Al-Makhadmeh, Amr Tolba, and Wael Said. Internet of things-based urban waste management system for smart cities using a cuckoo search algorithm. *Cluster Computing*, 23:1769–1780, 2020.

[14] Fayez Alqahtani, Mohammed Amoon, and Aida A. Nasr. Reliable scheduling and load balancing for requests in cloud-fog computing. *Peer-to-Peer Networking and Applications*, 14:1905–1916, 2021.

[15] Fadwa Alrowais, Heba G. Mohamed, Fahd N. Al-Wesabi, Mesfer Al Duhayyim, Anwer Mustafa Hilal, and Abdelwahed Motwakel. Cyber attack detection in healthcare data using cyber-physical system with optimized algorithm. *Computers and Electrical Engineering*, 108:108636, 2023.

[16] Amal Alzahrani and Maolin Tang. A microservice-based saas deployment in a data center considering computational server and network energy consumption. pages 505–515, 2023.

[17] Amazon. Project Website: `http://aws.amazon.com/`.

[18] Fazel Ansari, Marjan Khobreh, Ulrich Seidenberg, and Wilfried Sihn. A problem-solving ontology for human-centered cyber physical production systems. *CIRP Journal of Manufacturing Science and Technology*, 22:91–106, 2018.

[19] AppEngine. Project Website: `http://appengine.google.com`.

[20] Atakan Aral, Vincenzo De Maio, and Ivona Brandic. Ares: Reliable and sustainable edge provisioning for wireless sensor networks. *IEEE Transactions on Sustainable Computing*, 7(4):761–773, 2021.

[21] Atakan Aral, Rafael Brundo Uriarte, Anthony Simonet-Boulogne, and Ivona Brandic. Reliability management for blockchain-based decentralized multi-cloud. In *2020 20th IEEE/ACM International Symposium on Cluster, Cloud and Internet Computing (CCGRID)*, pages 21–30, 2020.

[22] Onur Ascigil, Truong Khoa Phan, Argyrios G. Tasiopoulos, Vasilis Sourlas, Ioannis Psaras, and George Pavlou. On uncoordinated service placement in edge-clouds. In *2017 IEEE International Conference on Cloud Computing Technology and Science (CloudCom)*, pages 41–48, 2017.

[23] Gagangeet Singh Aujla and Anish Jindal. A decoupled blockchain approach for edge-envisioned iot-based healthcare monitoring. *IEEE Journal on Selected Areas in Communications*, 39(2):491–499, 2021.

[24] Azure. Project Website: `http://www.microsoft.com/azure/`.

[25] Muhammad Babar, Mian Ahmad Jan, Xiangjian He, Muhammad Usman Tariq, Spyridon Mastorakis, and Ryan Alturki. An optimized IoT-enabled big data analytics architecture for edge–cloud computing. *IEEE Internet of Things Journal*, 10(5):3995–4005, 2023.

[26] Athanasios Bachoumis, Nikos Andriopoulos, Konstantinos Plakas, Aristeidis Magklaras, Panayiotis Alefragis, Georgios Goulas, Alexios Birbas, and Alex Papalexopoulos. Cloud-edge interoperability for demand response-enabled fast frequency response service provision. *IEEE Transactions on Cloud Computing*, 10(1):123–133, 2022.

[27] Youngmi Baek and Seo-Hee Park. Adaptive and lightweight cyber-attack detection in modern automotive cyber-physical systems. In *IEEE Consumer Communications & Networking Conference*, pages 949–950, 2023.

[28] Mohammadreza Baharani, Mehrdad Biglarbegian, Babak Parkhideh, and Hamed Tabkhi. Real-time deep learning at the edge for scalable reliability modeling of si-mosfet power electronics converters. *IEEE Internet of Things Journal*, 6(5):7375–7385, 2019.

[29] Tayebeh Bahreini and Daniel Grosu. Efficient algorithms for multi-component application placement in mobile edge computing. *IEEE Transactions on Cloud Computing*, 10(4):2550–2563, 2022.

[30] Eduard Bertran and Alex Sànchez-Cerdà. On the tradeoff between electrical power consumption and flight performance in fixed-wing uav autopilots. *IEEE Transactions on Vehicular Technology*, 65(11):8832–8840, 2016.

[31] Neda Bugshan, Ibrahim Khalil, Mohammad Saidur Rahman, Mohammed Atiquzzaman, XunYi, and Shahriar Badsha. Toward trustworthy and privacy-preserving federated deep learning service framework for industrial internet of things. *IEEE Transactions on Industrial Informatics*, 19(2):1535–1547, 2023.

[32] Rajkumar Buyya, Chee Shin Yeo, and Srikumar Venugopal. Market-oriented cloud computing: Vision, hype, and reality for delivering it services as computing utilities. In *2008 10th IEEE International Conference on High Performance Computing and Communications*, pages 5–13. IEEE, 2008.

[33] Z. C. Muhammad Candra, Hong-Linh Truong, and Schahram Dustdar. On monitoring cyber-physical-social systems. In *IEEE World Congress on Services*, pages 56–63, 2016.

[34] Haotong Cao, Sahil Garg, Georges Kaddoum, Satinder Singh, and M. Shamim Hossain. Softwarized resource management and allocation with autonomous awareness for 6G-enabled cooperative intelligent transportation systems. *IEEE Transactions on Intelligent Transportation Systems*, 23(12):24662–24671, 2022.

[35] Jie Cao, Lanyu Xu, Raef Abdallah, and Weisong Shi. EdgeOS_H: A home operating system for internet of everything. In *2017 IEEE 37th International Conference on Distributed Computing Systems (ICDCS)*, pages 1756–1764, 2017.

[36] Kun Cao, Yangguang Cui, Zhiquan Liu, Wuzheng Tan, and Jian Weng. Edge intelligent joint optimization for lifetime and latency in large-scale cyber–physical systems. *IEEE Internet of Things Journal*, 9(22):22267–22279, 2021.

[37] Kun Cao, Liying Li, Yangguang Cui, Tongquan Wei, and Shiyan Hu. Exploring placement of heterogeneous edge servers for response time minimization in mobile edge-cloud computing. *IEEE Transactions on Industrial Informatics*, 17(1):494–503, 2020.

[38] Kun Cao, Tongquan Wei, Mingsong Chen, Keqin Li, Jian Weng, and Wuzheng Tan. Exploring reliable edge-cloud computing for service latency optimization in sustainable cyber-physical systems. *Software: Practice and Experience*, 51(11):2225–2237, 2021.

[39] Kun Cao, Junlong Zhou, Peijin Cong, Liying Li, Tongquan Wei, Mingsong Chen, Shiyan Hu, and Xiaobo Sharon Hu. Affinity-driven modeling and scheduling for makespan optimization in heterogeneous multiprocessor systems. *IEEE Transactions on Computer-Aided Design of Integrated Circuits and Systems*, 38(7):1189–1202, 2018.

[40] Kun Cao, Junlong Zhou, Tongquan Wei, Mingsong Chen, Shiyan Hu, and Keqin Li. A survey of optimization techniques for thermal-aware 3d processors. *Journal of Systems Architecture*, 97:397–415, 2019.

[41] Antonio Caruso, Stefano Chessa, Soledad Escolar, Xavier del Toro, and Juan Carlos López. A dynamic programming algorithm for high-level task scheduling in energy harvesting IoT. *IEEE Internet of Things Journal*, 5(3):2234–2248, 2018.

[42] Valentina Casola, Alessandra De Benedictis, Sergio Di Martino, Nicola Mazzocca, and Luigi Libero Lucio Starace. Security-aware deployment optimization of cloud–edge systems in industrial iot. *IEEE Internet of Things Journal*, 8(16):12724–12733, 2020.

[43] Alberto Ceselli, Marco Premoli, and Stefano Secci. Mobile edge cloud network design optimization. *IEEE/ACM Transactions on Networking*, 25(3):1818–1831, 2017.

[44] Pratyush Chakraborty and Pramod P. Khargonekar. A demand response game and its robust price of anarchy. In *2014 IEEE International Conference on Smart Grid Communications (SmartGridComm)*, pages 644–649, 2014.

[45] Younès Chandarli, Nathan Fisher, and Damien Masson. Response time analysis for thermal-aware real-time systems under fixed-priority scheduling. In *2015 IEEE 18th International Symposium on Real-Time Distributed Computing*, pages 84–93, 2015.

[46] Xiaolin Chang, Bin Wang, Jogesh K. Muppala, and Jiqiang Liu. Modeling active virtual machines on iaas clouds using an m/g/m/m+k queue. *IEEE Transactions on Services Computing*, page 408–420, 2016.

[47] Xiaolin Chang, Ruofan Xia, Jogesh K. Muppala, Kishor S. Trivedi, and Jiqiang Liu. Effective modeling approach for iaas data center performance analysis under heterogeneous workload. *IEEE Transactions on Cloud Computing*, 6(4):991–1003, 2018.

[48] Hernani D. Chantre and Nelson L. S. da Fonseca. Multi-objective optimization for edge device placement and reliable broadcasting in 5g nfv-based small cell networks. *IEEE Journal on Selected Areas in Communications*, 36(10):2304–2317, 2018.

[49] Faouzi Ben Charrada and Samir Tata. An efficient algorithm for the bursting of service-based applications in hybrid clouds. *IEEE Transactions on Services Computing*, 9(3):357–367, 2016.

[50] Chen Chen, Bin Liu, Shaohua Wan, Peng Qiao, and Qingqi Pei. An edge traffic flow detection scheme based on deep learning in an intelligent transportation system. *IEEE Transactions on Intelligent Transportation Systems*, 22(3):1840–1852, 2021.

[51] Chien-An Chen, Myounggyu Won, Radu Stoleru, and Geoffrey G. Xie. Energy-efficient fault-tolerant data storage and processing in mobile cloud. *IEEE Transactions on Cloud Computing*, 3(1):28–41, 2015.

[52] Lixing Chen, Cong Shen, Pan Zhou, and Jie Xu. Collaborative service placement for edge computing in dense small cell networks. *IEEE Transactions on Mobile Computing*, 20(2):377–390, 2021.

[53] Long Chen, Lingyan Xue, Haiping Huang, Wenming Wang, Mengxun Cao, and Fu Xiao. Double rainbows: A promising distributed data sharing in augmented intelligence of things. *IEEE Transactions on Industrial Informatics*, 19(1):653–661, 2022.

[54] Meng Chen, Jiaxin Hou, Yongpan Sheng, Yingbo Wu, Sen Wang, Jianyuan Lu, and Qilin Fan. Ha-d3qn: Embedding virtual private cloud in cloud data centers with heuristic assisted deep reinforcement learning. *Future Generation Computer Systems*, 148:1–14, 2023.

[55] Min Chen, Yongfeng Qian, Yixue Hao, Yong Li, and Jeungeun Song. Data-driven computing and caching in 5g networks: Architecture and delay analysis. *IEEE Wireless Communications*, 25(1):70–75, 2018.

[56] Nengcheng Chen, Xiang Zhang, and Chao Wang. Integrated open geospatial web service enabled cyber-physical information infrastructure for precision agriculture monitoring. *Computers and Electronics in Agriculture*, 111:78–91, 2015.

[57] Qinglin Chen, Zhufang Kuang, and Lian Zhao. Multiuser computation offloading and resource allocation for cloud–edge heterogeneous network. *IEEE Internet of Things Journal*, 9(5):3799–3811, 2021.

[58] Sheng Chen, Qihang Zhang, Xiaodong Dong, Xiaoyi Tao, Keqiu Li, Tie Qiu, and Ivan Lee. Deep reinforcement learning based cooperative partial task offloading and resource allocation for iiot applications. *IEEE Transactions on Industrial Informatics*, 19(9):9855–9866, 2023.

[59] Siguang Chen, Jiamin Chen, Yifeng Miao, Qian Wang, and Chuanxin Zhao. Deep reinforcement learning-based cloud-edge collaborative mobile computation offloading in industrial networks. *IEEE Transactions on Signal and Information Processing over Networks*, 8:364–375, 2022.

[60] Songlin Chen, Hong Wen, Jinsong Wu, Wenxin Lei, Wenjing Hou, Wenjie Liu, Aidong Xu, and Yixin Jiang. Internet of Things based smart grids supported by intelligent edge computing. *IEEE Access*, 7:74089–74102, 2019.

[61] Xing Chen, Jianshan Zhang, Bing Lin, Zheyi Chen, Katinka Wolter, and Geyong Min. Energy-efficient offloading for dnn-based smart iot systems in cloud-edge environments. *IEEE Transactions on Parallel and Distributed Systems*, 33(3):683–697, 2021.

[62] Xu Chen, Lei Jiao, Wenzhong Li, and Xiaoming Fu. Efficient multi-user computation offloading for mobile-edge cloud computing. *IEEE/ACM Transactions on Networking*, 24(5):2795–2808, 2015.

[63] Ying Chen, Yongchao Zhang, Yuan Wu, Lianyong Qi, Xin Chen, and Xuemin Shen. Joint task scheduling and energy management for heterogeneous mobile edge computing with hybrid energy supply. *IEEE Internet of Things Journal*, 7(9):8419–8429, 2020.

[64] Yung-Yao Chen, Yu-Hsiu Lin, Yu-Chen Hu, Chih-Hsien Hsia, Yi-An Lian, and Sin-Ye Jhong. Distributed real-time object detection based on edge-cloud collaboration for smart video surveillance applications. *IEEE Access*, 10:93745–93759, 2022.

[65] Guoli Cheng, Shi Ying, and Bingming Wang. Tuning configuration of apache spark on public clouds by combining multi-objective optimization and performance prediction model. *Journal of Systems and Software*, 180:111028, 2021.

[66] Nan Cheng, Feng Lyu, Wei Quan, Conghao Zhou, Hongli He, Weisen Shi, and Xuemin Shen. Space/aerial-assisted computing offloading for iot applications: A learning-based approach. *IEEE Journal on Selected Areas in Communications*, 37(5):1117–1129, 2019.

[67] Jing Ping Chengfeng Jian and Meiyu Zhang. A cloud edge-based two-level hybrid scheduling learning model in cloud manufacturing. *International Journal of Production Research*, 59(16):4836–4850, 2021.

[68] Cheng Chi, Zihang Yin, Yang Liu, and Senchun Chai. A trusted cloud-edge decision architecture based on blockchain and mlp for aiot. *IEEE Internet of Things Journal*, 2023.

[69] Te-Chuan Chiu, Yuan-Yao Shih, Ai-Chun Pang, Chieh-Sheng Wang, Wei Weng, and Chun-Ting Chou. Semisupervised distributed learning with non-IID data for AIoT service platform. *IEEE Internet of Things Journal*, 7(10):9266–9277, 2020.

[70] Tianshu Chu, Jie Wang, Lara Codecà, and Zhaojian Li. Multi-agent deep reinforcement learning for large-scale traffic signal control. *IEEE Transactions on Intelligent Transportation Systems*, 21(3):1086–1095, 2020.

[71] M. Ciappa, F. Carbognani, P. Cova, and W. Fichtner. Lifetime prediction and design of reliability tests for high-power devices in automotive applications. In *2003 IEEE International Reliability Physics Symposium Proceedings, 2003. 41st Annual*, Dec 2003.

[72] Cisco. Cisco Global Cloud Index: Forecast and Methodology (2016–2021) White Paper, 2018. `https://virtualization.network/Resources/Whitepapers/0b75cf2e-0c53-4891-918e-b542a5d364c5_white-paper-c11-738085.pdf`.

[73] Gabriella Colajanni and Patrizia Daniele. A mathematical network model and a solution algorithm for iaas cloud computing. *Networks and Spatial Economics*, 22(2):267–287, 2022.

[74] M. Colledani, D. Coupek, A. Verl, J. Aichele, and A. Yemane. A cyber-physical system for quality-oriented assembly of automotive electric motors. *CIRP Journal of Manufacturing Science and Technology*, 20:12–22, 2018.

[75] Marco Conti and Silvia Giordano. Mobile ad hoc networking: milestones, challenges, and new research directions. *IEEE Communications Magazine*, 52(1):85–96, 2014.

[76] Stefania Costache, Djawida Dib, Nikos Parlavantzas, and Christine Morin. Resource management in cloud platform as a service systems: Analysis and opportunities. *Journal of Systems and Software*, 132:98–118, 2017.

[77] Eduardo Cuervo, Aruna Balasubramanian, Dae-ki Cho, Alec Wolman, Stefan Saroiu, Ranveer Chandra, and Paramvir Bahl. MAUI: Making smartphones last longer with code offload. In *International Conference on Mobile Systems, Applications, and Services*, pages 49–62, 2010.

[78] Jie Cui, Bei Li, Hong Zhong, Geyong Min, Yan Xu, and Lu Liu. A practical and efficient bidirectional access control scheme for cloud-edge data sharing. *IEEE Transactions on Parallel and Distributed Systems*, 33(2):476–488, 2021.

[79] Rustem Dautov, Salvatore Distefano, Dario Bruneo, Francesco Longo, Giovanni Merlino, and Antonio Puliafito. Data processing in cyber-physical-social systems through edge computing. *IEEE Access*, 6:29822–29835, 2018.

[80] Jorge A. Delgado, Nicholas M. Short, Daniel P. Roberts, and Bruce Vandenberg. Big data analysis for sustainable agriculture on a geospatial cloud framework. *Frontiers in Sustainable Food Systems*, 3:54, 2019.

[81] Shijun Deng, Zhiwen Chen, Fuhua Kuang, Chunhua Yang, and Weihua Gui. Optimal control of chilled water system with ensemble learning and cloud edge terminal implementation. *IEEE Transactions on Industrial Informatics*, 17(11):7839–7848, 2021.

[82] Shuiguang Deng, Longtao Huang, Javid Taheri, and Albert Y. Zomaya. Computation offloading for service workflow in mobile cloud computing. *IEEE Transactions on Parallel and Distributed Systems*, 26(12):3317–3329, 2015.

[83] Prasad Ramesh Desai, S. Mini, and Deepak K. Tosh. Deep reinforcement learning based cooperative partial task offloading and resource allocation for iiot applications. *IEEE Internet of Things Journal*, 9(19):18898–18907, 2022.

[84] Malathi Devarajan and Logesh Ravi. Intelligent cyber-physical system for an efficient detection of parkinson disease using fog computing. *Multimedia Tools and Applications*, 78:32695–32719, 2019.

[85] R. Dhaya and R. Kanthavel. Ioe based private multi-data center cloud architecture framework. *Computers and Electrical Engineering*, 100:107933, 2022.

[86] Chuntao Ding, Ao Zhou, Yunxin Liu, Rong N. Chang, Ching-Hsien Hsu, and Shangguang Wang. A cloud-edge collaboration framework for cognitive service. *IEEE Transactions on Cloud Computing*, 10(3):1489–1499, 2022.

[87] Peng Ding, Dan Liu, Yun Shen, Huibin Duan, and Qiuhong Zheng. Edge-to-cloud intelligent vehicle-infrastructure based on 5G time-sensitive network integration. In *IEEE International Symposium on Broadband Multimedia Systems and Broadcasting*, pages 1–5, 2022.

[88] Xianzhong Ding, Wan Du, and Alberto Cerpa. OCTOPUS: Deep reinforcement learning for holistic smart building control. In *ACM International Conference on Systems for Energy-Efficient Buildings, Cities, and Transportation*, pages 326–335, 2019.

[89] Yan Ding, Kenli Li, Chubo Liu, and Keqin Li. A potential game theoretic approach to computation offloading strategy optimization in end-edge-cloud computing. *IEEE Transactions on Parallel and Distributed Systems*, 33(6):1503–1519, 2021.

[90] Yan Ding, Kenli Li, Chubo Liu, Zhuo Tang, and Keqin Li. Budget-constrained service allocation optimization for mobile edge computing. *IEEE Transactions on Services Computing*, 16(1):147–161, 2021.

[91] Yan Ding, Chubo Liu, Xu Zhou, Zhao Liu, and Zhuo Tang. A code-oriented partitioning computation offloading strategy for multiple users and multiple mobile edge computing servers. *IEEE Transactions on Industrial Informatics*, 16(7):4800–4810, 2020.

[92] Sven Dominka, Dominik Ertl, Michael Dübner, Romana Wiesinger, and Hermann Kaindl. Taming and optimizing feature interaction in software-intensive automotive systems. In *IEEE Industrial Cyber-Physical Systems*, pages 324–329, 2018.

[93] Luobing Dong, Weili Wu, Qiumin Guo, Meghana N. Satpute, Taieb Znati, and Ding Zhu Du. Reliability-aware offloading and allocation in multilevel edge computing system. *IEEE Transactions on Reliability*, 70(1):200–211, 2019.

[94] Wei Dong, Qiang Yang, Wei Li, and Albert Y. Zomaya. Machine-learning-based real-time economic dispatch in islanding microgrids in a cloud-edge computing environment. *IEEE Internet of Things Journal*, 8(17):13703–13711, 2021.

[95] Yueyu Dong, Fei Dai, and Mingming Qin. A privacy-preserving deep learning scheme for edge-enhanced smart homes. In *IEEE International Conference on Dependable, Autonomic and Secure Computing, International Conference on Pervasive Intelligence and Computing, International Conference on Cloud and Big Data Computing, International Conference on Cyber Science and Technology Congress*, pages 1–6, 2022.

[96] Sandeep M. D'souza and Ragunathan (Raj) Rajkumar. Thermal Implications of Energy-Saving Schedulers. In Marko Bertogna, editor, *29th Euromicro Conference on Real-Time Systems (ECRTS 2017)*, volume 76 of *Leibniz International Proceedings in Informatics (LIPIcs)*, pages 21:1–21:23. Schloss Dagstuhl–Leibniz-Zentrum fuer Informatik, 2017.

[97] RuiZhong Du, Cui Liu, Yan Gao, PengNan Hao, and ZiYuan Wang. Collaborative cloud-edge-end task offloading in noma-enabled mobile edge computing using deep learning. *Journal of Grid Computing*, 20(2):14, 2022.

[98] Rubing Duan, Radu Prodan, and Xiaorong Li. Multi-objective game theoretic schedulingof bag-of-tasks workflows on hybrid clouds. *IEEE Transactions on Cloud Computing*, 2(1):29–42, 2014.

[99] Wei Duan, Xiaohui Gu, Miaowen Wen, Yancheng Ji, Jianhua Ge, and Guoan Zhang. Resource management for intelligent vehicular edge computing networks. *IEEE Transactions on Intelligent Transportation Systems*, 23(7):9797–9808, 2022.

[100] Puwadol Oak Dusadeerungsikul, Shimon Y. Nof, Avital Bechar, and Yang Tao. Collaborative control protocol for agricultural cyber-physical system. *Procedia Manufacturing*, 39:235–242, 2019.

[101] Joy Dutta and Sarbani Roy. IoT-fog-cloud based architecture for smart city: Prototype of a smart building. In *International Conference on Cloud Computing, Data Science & Engineering - Confluence*, pages 237–242, 2017.

[102] Ibrahim A. Elgendy, Wei-Zhe Zhang, Chuan-Yi Liu, and Ching-Hsien Hsu. An efficient and secured framework for mobile cloud computing. *IEEE Transactions on Cloud Computing*, 9(2):844–844, 2021.

[103] Marwa Elsayed and Mohammad Zulkernine. Offering security diagnosis as a service for cloud saas applications. *Journal of Information Security and Applications*, 44:32–48, 2019.

[104] Haolong Fan, FarookhKhadeer Hussain, Muhammad Younas, and OmarKhadeer Hussain. An integrated personalization framework for saas-based cloud services. *Future Generation Computer Systems*, 53:157–173, 2015.

[105] Kai Fan, Qi Chen, Ruidan Su, Kuan Zhang, Haoyang Wang, Hui Li, and Yintang Yang. Msiap: A dynamic searchable encryption for privacy-protection on smart grid with cloud-edge-end. *IEEE Transactions on Cloud Computing*, 2021.

[106] Wenhao Fan, Mingyu Hua, Yaoyin Zhang, Yi Su, Xuewei Li, Bihua Tang, Fan Wu, and Yuan'an Liu. Game-based task offloading and resource allocation for vehicular edge computing with edge-edge cooperation. *IEEE Transactions on Vehicular Technology*, 72(6):7857–7870, 2023.

[107] Chao Fang, Zhaoming Hu, Xiangheng Meng, Shanshan Tu, Zhuwei Wang, Deze Zeng, Wei Ni, Song Guo, and Zhu Han. Drl-driven joint task offloading and resource allocation for energy-efficient content delivery in cloud-edge cooperation networks. *IEEE Transactions on Vehicular Technology*, 2023.

[108] Chao Fang, Xiangheng Meng, Zhaoming Hu, Fangmin Xu, Deze Zeng, Mianxiong Dong, and Wei Ni. Ai-driven energy-efficient content task offloading in cloud-edge-end cooperation networks. *IEEE Open Journal of the Computer Society*, 3:162–171, 2022.

[109] Weidong Fang, Chunsheng Zhu, and Wuxiong Zhang. Toward secure and lightweight data transmission for cloud-edge-terminal collaboration in artificial intelligence of things. *IEEE Internet of Things Journal*, 2023.

[110] Chuan Feng, Pengchao Han, Xu Zhang, Bowen Yang, Yejun Liu, and Lei Guo. Computation offloading in mobile edge computing networks: A survey. *Journal of Network and Computer Applications*, 202(103366), 2022.

[111] Massimo Ficco and Francesco Palmieri. Leaf: An open-source cybersecurity training platform for realistic edge-iot scenarios. *Journal of Systems Architecture*, 97:107–129, 2019.

[112] Norman Finn. Introduction to time-sensitive networking. *IEEE Communications Standards Magazine*, 6(4):8–13, 2022.

[113] Kaihua Fu, Wei Zhang, Quan Chen, Deze Zeng, and Minyi Guo. Adaptive resource efficient microservice deployment in cloud-edge continuum. *IEEE Transactions on Parallel and Distributed Systems*, page 1825–1840, 2022.

[114] Kaihua Fu, Wei Zhang, Quan Chen, Deze Zeng, Xin Peng, Wenli Zheng, and Minyi Guo. Qos-aware and resource efficient microservice deployment in cloud-edge continuum. In *2021 IEEE International Parallel and Distributed Processing Symposium (IPDPS)*, pages 932–941, 2021.

[115] Rao Fu, Yuanguo Bi, Guangjie Han, Chuan Lin, and HaiZhao. A continuous object tracking scheme based on two-stage prediction in industrial internet of things. *IEEE Internet of Things Journal*, 10(19):17022–17034, 2023.

[116] Weiming Fu, Yanni Wan, Jiahu Qin, Yu Kang, and Li Li. Privacy-preserving optimal energy management for smart grid with cloud-edge computing. *IEEE Transactions on Industrial Informatics*, 18(6):4029–4038, 2022.

[117] Steve W. Fuhrmann and Robert B. Cooper. Stochastic decompositions in the m/g/1 queue with generalized vacations. *Operations Research*, 33(5):1117–1129, 1985.

[118] Tatsuya Fukui, Yuki Sakaue, Katsuya Minami, and Tomohiro Taniguchi. Delay-based shaper with dynamic token bucket algorithm for deterministic networks. *IEEE Access*, 10:114424–114433, 2022.

[119] Honghao Gao, Wanqiu Huang, Tong Liu, Yuyu Yin, and Youhuizi Li. PPO2: Location privacy-oriented task offloading to edge computing using reinforcement learning for intelligent autonomous transport systems. *IEEE Transactions on Intelligent Transportation Systems*, 24(7):7599–7612, 2023.

[120] Sahil Garg, Kuljeet Kaur, Syed Hassan Ahmed, Abbas Bradai, Georges Kaddoum, and Mohammed Atiquzzaman. MobQoS: Mobility-aware and QoS-driven SDN framework for autonomous vehicles. *IEEE Wireless Communications*, 26(4):12–20, 2019.

[121] Jeremy Geelan et al. Twenty-one experts define cloud computing. *Cloud Computing Journal*, 4:1–5, 2009.

[122] Hamoun Ghanbari, Bradley Simmons, Marin Litoiu, and Gabriel Iszlai. Feedback-based optimization of a private cloud. *Future Generation Computer Systems*, 28(1):104–111, 2012.

[123] Ananda Mohon Ghosh and Katarina Grolinger. Edge-cloud computing for internet of things data analytics: Embedding intelligence in the edge with deep learning. *IEEE Transactions on Industrial Informatics*, 17(3):2191–2200, 2021.

[124] Saibal Ghosh and Dharma P. Agrawal. A high performance hierarchical caching framework for mobile edge computing environments. In *2021 IEEE Wireless Communications and Networking Conference (WCNC)*, pages 1–6, 2021.

[125] Yanqi Gong, Kun Bian, Fei Hao, Yifei Sun, and Yulei Wu. Dependent tasks offloading in mobile edge computing: A multi-objective evolutionary optimization strategy. *Future Generation Computer Systems*, 2023.

[126] Dan Gonzales, Jeremy M. Kaplan, Evan Saltzman, Zev Winkelman, and Dulani Woods. Cloud-trust—a security assessment model for infrastructure as a service (iaas) clouds. *IEEE Transactions on Cloud Computing*, 5(3):523–536, 2017.

[127] Mohammad Goudarzi, Huaming Wu, Marimuthu Palaniswami, and Rajkumar Buyya. An application placement technique for concurrent iot applications in edge and fog computing environments. *IEEE Transactions on Mobile Computing*, 20(4):1298–1311, 2021.

[128] Edge Computing Task Group. Introduction to edge computing in iiot, 2018. `https://www.iiconsortium.org/pdf/Introduction_to_Edge_Computing_in_IIoT_2018-06-18.pdf`.

[129] Lin Gu, Deze Zeng, Song Guo, Ahmed Barnawi, and Yong Xiang. Cost efficient resource management in fog computing supported medical cyber-physical system. *IEEE Transactions on Emerging Topics in Computing*, 5(1):108–119, 2017.

[130] Francesc Guim, Thijs Metsch, Hassnaa Moustafa, Timothy Verrall, David Carrera, Nicola Cadenelli, Jiang Chen, David Doria, Chadie Ghadie, and Raül González Prats. Autonomous lifecycle management for resource-efficient workload orchestration for green edge computing. *IEEE Transactions on Green Communications and Networking*, 6(1):571–582, 2022.

[131] Nadeera Gnan Tilshan Gunaratne, Mali Abdollahian, Shamsul Huda, Marwan Ali, and Giancarlo Fortino. An edge tier task offloading to identify sources of variance shifts in smart grid using a hybrid of wrapper and filter approaches. *IEEE Transactions on Green Communications and Networking*, 6(1):329–340, 2022.

[132] Jianxiong Guo, Weili Wu, Tian Wang, and Weijia Jia. An online multi-item auction with differential privacy in edge-assisted blockchains. *IEEE Internet of Things Journal*, 2023.

[133] Li Guo, Jia Yu, Ming Yang, and Fanyu Kong. Privacy-preserving convolution neural network inference with edge-assistance. *Computers & Security*, 123:102910, 2022.

[134] Ping Guo, Puwadol Oak Dusadeerungsikul, and Shimon Y. Nof. Agricultural cyber physical system collaboration for greenhouse stress management. *Computers and Electronics in Agriculture*, 150:439–454, 2018.

[135] Songtao Guo, Qiucen Jiang, Yifan Dong, and Quyuan Wang. Taskalloc: Online tasks allocation for offloading in energy harvesting mobile edge computing. In *2019 IEEE Intl Conf on Parallel & Distributed Processing with Applications, Big Data & Cloud Computing, Sustainable Computing & Communications, Social Computing & Networking (ISPA/BDCloud/SocialCom/SustainCom)*, pages 116–123, 2019.

[136] Wei Guo, Yi Zhang, and Li Li. The integration of cps, cpss, and its: A focus on data. *Tsinghua Science and Technology*, 20(4):327–335, 2015.

[137] Yanxiang Guo, Xiping Hu, Bin Hu, Jun Cheng, Mengchu Zhou, and Ricky Y. K. Kwok. Mobile cyber physical systems: Current challenges and future networking applications. *IEEE Access*, 6:12360–12368, 2018.

[138] Yeting Guo, Fang Liu, Tongqing Zhou, Zhiping Cai, and Nong Xiao. Privacy vs. efficiency: Achieving both through adaptive hierarchical federated learning. *IEEE Transactions on Parallel and Distributed Systems*, 34(4):1331–1342, 2023.

[139] Ramyad Hadidi, Jiashen Cao, Michael S. Ryoo, and Hyesoon Kim. Toward collaborative inferencing of deep neural networks on internet-of-things devices. *IEEE Internet of Things Journal*, 7(6):4950–4960, 2020.

[140] Hadoop. Project Website: `http://hadoop.apache.org/core`.

[141] Zhao Haitao, Ding Yi, Zhang Mengkang, Wang Qin, Shi Xinyue, and Zhu Hongbo. Multipath transmission workload balancing optimization scheme based on mobile edge computing in vehicular heterogeneous network. *IEEE Access*, 7:116047–116055, 2019.

[142] Bin Han, Stan Wong, Christian Mannweiler, Marcos Rates Crippa, and Hans D. Schotten. Context-awareness enhances 5g multi-access edge computing reliability. *IEEE Access*, 7:21290–21299, 2019.

[143] Guangjie Han, Juntao Tu, Li Liu, Miguel Martínez-García, and Chang Choi. An intelligent signal processing data denoising method for control systems protection in the industrial internet of things. *IEEE Transactions on Industrial Informatics*, 18(4):2684–2692, 2022.

[144] Jiaxuan Han, Qiteng Hong, Mazheruddin H. Syed, Md Asif Uddin Khan, Guangya Yang, Graeme Burt, and Campbell Booth. Cloud-edge hosted digital twins for coordinated control of distributed energy resources. *IEEE Transactions on Cloud Computing*, 2022.

[145] Rui Han, Shilin Wen, Chi Harold Liu, Ye Yuan, Guoren Wang, and Lydia Y. Chen. Edgetuner: Fast scheduling algorithm tuning for dynamic edge-cloud workloads and resources. In *IEEE INFOCOM 2022 - IEEE Conference on Computer Communications*, pages 880–889, 2022.

[146] Tianshu Hao, Jianfeng Zhan, Kai Hwang, Wanling Gao, and Xu Wen. Ai-oriented workload allocation for cloud-edge computing. In *2021 IEEE/ACM 21st International Symposium on Cluster, Cloud and Internet Computing (CCGrid)*, pages 555–564, 2021.

[147] Yixue Hao, Yingying Jiang, Tao Chen, Donggang Cao, and Min Chen. itaskoffloading: intelligent task offloading for a cloud-edge collaborative system. *IEEE Network*, 33(5):82–88, 2019.

[148] Jude Haris, Perry Gibson, José Cano, Nicolas Bohm Agostini, and David Kaeli. Secdatflite: A toolkit for efficient development of fpga-based dnn accelerators for edge inference. *Journal of Parallel and Distributed Computing*, 173:140–151, 2023.

[149] Jianhua He, Jian Wei, Kai Chen, Zuoyin Tang, Yi Zhou, and Yan Zhang. Multitier fog computing with large-scale IoT data analytics for smart cities. *IEEE Internet of Things Journal*, 5(2):677–686, 2018.

[150] Zhichen He, Dingguo Liang, and Ying Yang. Fault diagnosis scheme for the rotary machine group: A deep mutual learning-based approach with cloud-edge-end collaboration. *IEEE Transactions on Circuits and Systems II: Express Briefs*, 2023.

[151] Rajeshwari Hegde, Geetishree Mishra, and Gurumurthy Kargal. An insight into the hardware and software complexity of ECUs in vehicles. *Communications in Computer and Information Science*, 198:99–106, 2011.

[152] Arvin Hekmati, Peyvand Teymoori, Terence D. Todd, Dongmei Zhao, and George Karakostas. Optimal mobile computation offloading with hard deadline constraints. *IEEE Transactions on Mobile Computing*, 19(9):2160–2173, 2020.

[153] Chien-Ju Ho and Jennifer Vaughan. Online task assignment in crowdsourcing markets. In *Proceedings of the AAAI Conference on Artificial Intelligence*, pages 45–51, 2012.

[154] W. Hoara and S. Tixeuil. A language-driven tool for fault injection in distributed systems. In *The 6th IEEE/ACM International Workshop on Grid Computing, 2005*, pages 194–201, 2005.

[155] Mohammad Mobarak Hossain, Mohammod Abdul Kashem, Nasim Mahmud Nayan, and Mohammad Asaduzzaman Chowdhury. A medical cyber-physical system for predicting maternal health in developing countries using machine learning. *Healthcare Analytics*, 5:100285, 2024.

[156] Seyedmehdi Hosseinimotlagh and Hyoseung Kim. Thermal-aware servers for real-time tasks on multi-core gpu-integrated embedded systems. In *2019 IEEE Real-Time and Embedded Technology and Applications Symposium (RTAS)*, pages 254–266, 2019.

[157] Xiangwang Hou, Zhiyuan Ren, Jingjing Wang, Wenchi Cheng, Yong Ren, Kwang-Cheng Chen, and Hailin Zhang. Reliable computation offloading for edge-computing-enabled software-defined IoV. *IEEE Internet of Things Journal*, 7(8):7097–7111, 2020.

[158] Xiangwang Hou, Zhiyuan Ren, Kun Yang, Chen Chen, Hailin Zhang, and Yao Xiao. Iiot-mec: A novel mobile edge computing framework for 5g-enabled iiot. In *2019 IEEE Wireless Communications and Networking Conference (WCNC)*, pages 1–7, 2019.

[159] Xiaolan Hou, Wu Muqing, Lv Bo, and Liu Yifeng. Multi-controller deployment algorithm in hierarchical architecture for sdwan. *IEEE Access*, 7:65839–65851, 2019.

[160] Zakaria Abou El Houda, Bouziane Brik, Adlen Ksentini, Lyes Khoukhi, and Mohsen Guizani. When federated learning meets game theory: A cooperative framework to secure iiot applications on edge computing. *IEEE Transactions on Industrial Informatics*, 18(11):7988–7997, 2022.

[161] Bin Hu and Hamid Gharavi. Artificial intelligence-assisted edge computing for wide area monitoring. *IEEE Open Journal of the Communications Society*, 4:1659–1667, 2023.

[162] Long Hu, Wei Li, Jun Yang, Giancarlo Fortino, and Min Chen. A sustainable multi-modal multi-layer emotion-aware service at the edge. *IEEE Transactions on Sustainable Computing*, 7(2):324–333, 2019.

[163] Shihong Hu, Guanghui Li, and Weisong Shi. Lars: A latency-aware and real-time scheduling framework for edge-enabled internet of vehicles. *IEEE Transactions on Services Computing*, 16(1):398–411, 2023.

[164] Xiping Hu, Terry H. S. Chu, Henry C. B. Chan, and Victor C. M. Leung. Vita: A crowdsensing-oriented mobile cyber-physical system. *IEEE Transactions on Emerging Topics in Computing*, 1(1):148–165, 2013.

[165] Xiping Hu, Jidi Zhao, Boon-Chong Seet, Victor C. M. Leung, Terry H. S. Chu, and Henry Chan. S-aframe: Agent-based multilayer framework with context-aware semantic service for vehicular social networks. *IEEE Transactions on Emerging Topics in Computing*, 3(1):44–63, 2015.

[166] Zhaoming Hu, Chao Fang, Zhuwei Wang, Shu-Ming Tseng, and Mianxiong Dong. Many-objective optimization based-content popularity prediction for cache-assisted cloud-edge-end collaborative iot networks. *IEEE Internet of Things Journal*, 2023.

[167] Can Huang, Yuang Yan, Yi Zhang, and Jin Sun. A metaheuristic algorithm for mobility-aware task offloading for edge computing using device-to-device cooperation. In *IEEE Smartworld, Ubiquitous Intelligence & Computing, Scalable Computing & Communications, Digital Twin, Privacy Computing, Metaverse, Autonomous & Trusted Vehicles (SmartWorld/UIC/ScalCom/DigitalTwin/PriComp/Meta)*, pages 656–663, 2022.

[168] Chun-Hsian Huang, Po-Jung Chen, Yi-Jie Lin, Bo-Wei Chen, and Jia-Xuan Zheng. A robot-based intelligent management design for agricultural cyber-physical systems. *Computers and Electronics in Agriculture*, 181:105967, 2021.

[169] Ding-Hsiang Huang, Cheng-Fu Huang, and Yi-Kuei Lin. A reliability prediction model for a multistate cloud/edge-based network based on a deep neural network. *Annals of Operations Research*, pages 1–17, 2022.

[170] Qingqing Huang, Rui Wen, Yan Han, Chao Li, and Yan Zhang. Intelligent fault identification for industrial internet of things via prototype-guided partial domain adaptation with momentum weight. *IEEE Internet of Things Journal*, 10(18):16381–16391, 2023.

[171] Qinlong Huang, Lixuan Chen, and Chao Wang. A parallel secure flow control framework for private data sharing in mobile edge cloud. *IEEE Transactions on Parallel and Distributed Systems*, 33(12):4638–4653, 2022.

[172] Tiansheng Huang, Weiwei Lin, Xiaobin Hong, Xiumin Wang, Qingbo Wu, Rui Li, Ching-Hsien Hsu, and Albert Y. Zomaya. Adaptive processor frequency adjustment for mobile-edge computing with intermittent energy supply. *IEEE Internet of Things Journal*, 9(10):7446–7462, 2022.

[173] Yaodong Huang, Jiarui Zhang, Jun Duan, Bin Xiao, Fan Ye, and Yuanyuan Yang. Resource allocation and consensus on edge blockchain in pervasive edge computing environments. In *2019 IEEE 39th International Conference on Distributed Computing Systems (ICDCS)*, pages 1476–1486, 2019.

[174] Zhaowu Huang, Fang Dong, Dian Shen, Huitian Wang, Xiaolin Guo, and Shucun Fu. Enabling latency-sensitive dnn inference via joint optimization of model surgery and resource allocation in heterogeneous edge. In *Proceedings of the 51st International Conference on Parallel Processing (ICPP'22)*, 2023.

[175] C. N. Höfer and G. Karagiannis. Cloud computing services: taxonomy and comparison. *Journal of Internet Services and Applications*, page 81–94, 2011.

[176] IBM. *IBM ILOG CPLEX Optimizer*, 2022.

[177] Muhammad Ibrar, Aiman Erbad, Mohammed Abegaz, Aamir Akbar, Mahdi Houchati, and Juan M. Corchado. REED: Enhanced resource allocation and energy management in SDN-enabled edge computing-based smart buildings. In *International Wireless Communications and Mobile Computing*, pages 860–865, 2023.

[178] Muhammed Tawfiqul Islam, Huaming Wu, Shanika Karunasekera, and Rajkumar Buyya. Sla-based scheduling of spark jobs in hybrid cloud computing environments. *IEEE Transactions on Computers*, 71(5):1117–1132, 2022.

[179] Samuel Isuwa, Somdip Dey, Amit Kumar Singh, and Klaus McDonald-Maier. Teem: Online thermal- and energy-efficiency management on cpu-gpu mpsocs. In *2019 Design, Automation & Test in Europe Conference & Exhibition (DATE)*, pages 438–443, 2019.

[180] Viacheslav Izosimov, Alexandros Asvestopoulos, Oscar Blomkvist, and Martin Törngren. Security-aware development of cyber-physical systems illustrated with automotive case study. In *Design, Automation & Test in Europe Conference & Exhibition*, pages 818–821, 2016.

[181] Zijie Ji, Phee Lep Yeoh, Gaojie Chen, Junqing Zhang, Yan Zhang, Zunwen He, Hao Yin, and Yonghui Li. Physical-layer-based secure communications for static and low-latency industrial internet of things. *IEEE Internet of Things Journal*, 9(19):18392–18405, 2022.

[182] Lin Jia, Zhi Zhou, Fei Xu, and Hai Jin. Cost-efficient continuous edge learning for artificial intelligence of things. *IEEE Internet of Things Journal*, 9(10):7325–7337, 2022.

[183] Mingzhi Jiang, Tianhao Wu, Zhe Wang, Yi Gong, Lin Zhang, and Ren Ping Liu. A multi-intersection vehicular cooperative control based on end-edge-cloud computing. *IEEE Transactions on Vehicular Technology*, 71(3):2459–2471, 2022.

[184] Yuna Jiang, Yi Zhong, and Xiaohu Ge. Iiot data sharing based on blockchain: A multileader multifollower stackelberg game approach. *IEEE Internet of Things Journal*, 9(6):4396–4410, 2022.

[185] Hai Jin, Lin Jia, and Zhi Zhou. Boosting edge intelligence with collaborative cross-edge analytics. *IEEE Internet of Things Journal*, 8(4):2444–2458, 2021.

[186] Jianzhi Jin, Ruiling Li, Xiaolian Yang, Mengyuan Jin, and Fang Hu. A network slicing algorithm for cloud-edge collaboration hybrid computing in 5g and beyond networks. *Computers and Electrical Engineering*, 109:108750, 2023.

[187] Caihong Kai, Hao Zhou, Yibo Yi, and Wei Huang. Collaborative cloud-edge-end task offloading in mobile-edge computing networks with limited communication capability. *IEEE Transactions on Cognitive Communications and Networking*, 7(2):624–634, 2020.

[188] Miller Katherine. Communication theories, perspectives, processes and contexts, 2002.

[189] Ruimin Ke, Zhiyong Cui, Yanlong Chen, Meixin Zhu, Hao Yang, Yifan Zhuang, and Yinhai Wang. Lightweight edge intelligence empowered near-crash detection towards real-time vehicle event logging. *IEEE Transactions on Intelligent Vehicles*, 8(4):2737–2747, 2023.

[190] Fazlullah Khan, Mian Ahmad Jan, Ateeq ur Rehman, Spyridon Mastorakis, Mamoun Alazab, and Paul Watters. A secured and intelligent communication scheme for iiot-enabled pervasive edge computing. *IEEE Transactions on Industrial Informatics*, 17(7):5128–5137, 2021.

[191] Latif U. Khan, Ibrar Yaqoob, Nguyen H. Tran, S. M. Ahsan Kazmi, Tri Nguyen Dang, and Choong Seon Hong. Edge-computing-enabled smart cities: A comprehensive survey. *IEEE Internet of Things Journal*, 7(10):10200–10232, 2020.

[192] Quy Vu Khanh, Nam Vi Hoai, Anh Dang Van, and Quy Nguyen Minh. An integrating computing framework based on edge-fog-cloud for internet of healthcare things applications. *Internet of Things*, 23:100907, 2023.

[193] Haneul Ko, Joonwoo Kim, Dongkyun Ryoo, Inho Cha, and Sangheon Pack. A belief-based task offloading algorithm in vehicular edge computing. *IEEE Transactions on Intelligent Transportation Systems*, 24(5):5467–5476, 2023.

[194] Ovunc Kocabas, Tolga Soyata, and Mehmet K. Aktas. Emerging security mechanisms for medical cyber physical systems. *IEEE/ACM Transactions on Computational Biology and Bioinformatics*, 13(3):401–416, 2016.

[195] Kyriakos Kritikos, Tom Kirkham, Bartosz Kryza, and Philippe Massonet. Towards a security-enhanced paas platform for multi-cloud applications. *Future Generation Computer Systems*, 67:206–226, Feb 2017.

[196] Zhikai Kuang, Yawei Shi, Songtao Guo, Jingpei Dan, and Bin Xiao. Multi-user offloading game strategy in ofdma mobile cloud computing system. *IEEE Transactions on Cloud Computing*, 68(12):12190–12201, 2019.

[197] Shreyas Kulkarni, Qinchen Gu, Eric Myers, Lalith Polepeddi, Szilárd Lipták, Raheem Beyah, and Deepak Divan. Enabling a decentralized smart grid using autonomous edge control devices. *IEEE Internet of Things Journal*, 6(5):7406–7419, 2019.

[198] Jinshan Lai, Xiaotong Song, Ruijin Wang, and Xiong Li. Edge intelligent collaborative privacy protection solution for smart medical. *Cyber Security and Applications*, 1:100010, 2023.

[199] Qifeng Lai, Jinyu Tian, Wei Wang, and Xiping Hu. Spatial-temporal attention graph convolution network on edge cloud for traffic flow prediction. *IEEE Transactions on Intelligent Transportation Systems*, 24(4):4565–4576, 2023.

[200] Yuanjun Laili, Fuqiang Guo, Lei Ren, Xiang Li, Yulin Li, and Lin Zhang. Parallel scheduling of large-scale tasks for industrial cloud-edge collaboration. *IEEE Internet of Things Journal*, 2021.

[201] Isaac Lera, Carlos Guerrero, and Carlos Juiz. Availability-aware service placement policy in fog computing based on graph partitions. *IEEE Internet of Things Journal*, 6(2):3641–3651, 2019.

[202] Chao Li, Hui Yang, Zhengjie Sun, Qiuyan Yao, Jie Zhang, Ao Yu, Athanasios V. Vasilakos, Sheng Liu, and Yunbo Li. High-precision cluster federated learning for smart home: An edge-cloud collaboration approach. *IEEE Access*, 11:102157–102168, 2023.

[203] Chunlin Li, Yong Zhang, and Youlong Luo. A federated learning-based edge caching approach for mobile edge computing-enabled intelligent connected vehicles. *IEEE Transactions on Intelligent Transportation Systems*, 24(3):3360–3369, 2023.

[204] Houjun Li, Chunxia Dou, Dong Yue, Gerhard P. Hancke, Zeng Zeng, Wei Guo, and Lei Xu. End-edge-cloud collaboration based false data injection attack detection in distribution networks. *IEEE Transactions on Industrial Informatics*, 2023.

[205] Keqin Li. Design and analysis of heuristic algorithms for energy-constrained task scheduling with device-edge-cloud fusion. *IEEE Transactions on Sustainable Computing*, 8(2):208–221, 2023.

[206] Liying Li, Peijin Cong, Kun Cao, Junlong Zhou, Tongquan Wei, Mingsong Chen, Shiyan Hu, and Xiaobo Sharon Hu. Game theoretic feedback control for reliability enhancement of ethercat-based networked systems. *IEEE Transactions on Computer-Aided Design of Integrated Circuits and Systems*, 38(9):1599–1610, 2018.

[207] Liying Li, Junlong Zhou, Tongquan Wei, Mingsong Chen, and Xiaobo Sharon Hu. Learning-based modeling and optimization for real-time system availability. *IEEE Transactions on Computers*, 70(4):581–594, 2020.

[208] Meng Li, Pan Pei, F. Richard Yu, Pengbo Si, Yu Li, Enchang Sun, and Yanhua Zhang. Cloud–edge collaborative resource allocation for blockchain-enabled internet of things: A collective reinforcement learning approach. *IEEE Internet of Things Journal*, 9(22):23115–23129, 2022.

[209] Min Li, Yu Li, Ye Tian, Li Jiang, and Qiang Xu. Appealnet: An efficient and highly-accurate edge/cloud collaborative architecture for dnn inference. In *2021 58th ACM/IEEE Design Automation Conference (DAC)*, pages 409–414, 2021.

[210] Minghao Li and Wei Cai. A blockchain-based profiling system for exploring human factors in cloud-edge-end orchestration. In *2022 IEEE 42nd International Conference on Distributed Computing Systems Workshops (ICDCSW)*, pages 13–18. IEEE, 2022.

[211] Peilong Li, Chen Xu, Hao Jin, Chunyang Hu, Yan Luo, Yu Cao, Jomol Mathew, and Yunsheng Ma. ChainSDI: A software-defined infrastructure for regulation-compliant home-based healthcare services secured by blockchains. *IEEE Systems Journal*, 14(2):2042–2053, 2020.

[212] Qiang Li, Liang Huang, Zhao Tong, Ting-Ting Du, Jin Zhang, and Sheng-Chun Wang. Dissec: A distributed deep neural network inference scheduling strategy for edge clusters. *Neurocomputing*, 500:449–460, 2022.

[213] Qiang Li, Yuanmei Zhang, Ashish Pandharipande, Yong Xiao, and Xiaohu Ge. Edge caching in wireless infostation networks: Deployment and cache content placement. In *IEEE INFOCOM 2019 - IEEE Conference on Computer Communications Workshops (INFOCOM WKSHPS)*, pages 1–6, 2019.

[214] Qianmu Li, Xudong Wang, Pengchuan Wang, Weibin Zhang, and Jie Yin. FARDA: A fog-based anonymous reward data aggregation security scheme in smart buildings. *Building and Environment*, 225:109578, 2022.

[215] Qing Li, Shangguang Wang, Ao Zhou, Xiao Ma, Fangchun Yang, and Alex X. Liu. QoS driven task offloading with statistical guarantee in mobile edge computing. *IEEE Transactions on Mobile Computing*, 21(1):278–290, 2022.

[216] Ruixuan Li, Chenglin Shen, Heng He, Xiwu Gu, Zhiyong Xu, and Cheng Zhong Xu. A lightweight secure data sharing scheme for mobile cloud computing. *IEEE Transactions on Cloud Computing*, 6(2):344–357, 2018.

[217] Xiuhua Li, Zhenghui Xu, Fang Fang, Qilin Fan, Xiaofei Wang, and Victor C. M. Leung. Task offloading for deep learning empowered automatic speech analysis in mobile edge-cloud computing networks. *IEEE Transactions on Cloud Computing*, 11(2):1985–1998, 2023.

[218] Yong Li and Wei Gao. Muvr: Supporting multi-user mobile virtual reality with resource constrained edge cloud. In *2018 IEEE/ACM Symposium on Edge Computing (SEC)*, pages 1–16, 2018.

[219] Yuepeng Li, Deze Zeng, Lin Gu, Quan Chen, Song Guo, Albert Zomaya, and Minyi Guo. Efficient and secure deep learning inference in trusted processor enabled edge clouds. *IEEE Transactions on Parallel and Distributed Systems*, 33(12):4311–4325, 2022.

[220] Yun Li, Shichao Xia, Mengyan Zheng, Bin Cao, and Qilie Liu. Lyapunov optimization-based trade-off policy for mobile cloud offloading in heterogeneous wireless networks. *IEEE Transactions on Cloud Computing*, 10(1):491–505, 2022.

[221] Yuqing Li, Wenkuan Dai, Xiaoying Gan, Haiming Jin, Luoyi Fu, Huadong Ma, and Xinbing Wang. Cooperative service placement and scheduling in edge clouds: A deadline-driven approach. *IEEE Transactions on Mobile Computing*, 21(10):3519–3535, 2022.

[222] Zhiyue Li and Guangyan Zhang. A globally shared resource paradigm for encoded storage systems in the public cloud. *Fundamental Research*, 2023.

[223] Zhongjin Li, Victor Chang, Jidong Ge, Linxuan Pan, Haiyang Hu, and Binbin Huang. Energy-aware task offloading with deadline constraint in mobile edge computing. *EURASIP Journal on Wireless Communications and Networking*, 2021:1–24, 2021.

[224] Zezu Liang, Yuan Liu, Tat-Ming Lok, and Kaibin Huang. Multiuser computation offloading and downloading for edge computing with virtualization. *IEEE Transactions on Wireless Communications*, 18(9):4298–4311, 2019.

[225] Haijun Liao, Zehan Jia, Zhenyu Zhou, Yang Wang, Hui Zhang, and Shahid Mumtaz. Cloud-edge-end collaboration in air–ground integrated power iot: A semidistributed learning approach. *IEEE Transactions on Industrial Informatics*, 18(11):8047–8057, 2022.

[226] Weiwei Lie, Bin Jiang, and Wenjing Zhao. Obstetric imaging diagnostic platform based on cloud computing technology under the background of smart medical big data and deep learning. *IEEE Access*, 8:78265–78278, 2020.

[227] Li Lin and Xiaoying Zhang. Ppverifier: A privacy-preserving and verifiable federated learning method in cloud-edge collaborative computing environment. *IEEE Internet of Things Journal*, 10(10):8878–8892, 2022.

[228] Chih Wei Ling, Anwitaman Datta, and Jun Xu. A case for distributed multilevel storage infrastructure for visual surveillance in intelligent transportation networks. *IEEE Internet Computing*, 22(1):42–51, 2018.

[229] Boxi Liu, Yang Cao, Yue Zhang, and Tao Jiang. A distributed framework for task offloading in edge computing networks of arbitrary topology. *IEEE Transactions on Wireless Communications*, 19(4):2855–2867, 2020.

[230] Guanxiong Liu, Hang Shi, Abbas Kiani, Abdallah Khreishah, Joyoung Lee, Nirwan Ansari, Chengjun Liu, and Mustafa Mohammad Yousef. Smart traffic monitoring system using computer vision and edge computing. *IEEE Transactions on Intelligent Transportation Systems*, 23(8):12027–12038, 2022.

[231] Haolin Liu, Le Cao, Tingrui Pei, Qingyong Deng, and Jiang Zhu. A fast algorithm for energy-saving offloading with reliability and latency requirements in multi-access edge computing. *IEEE Access*, 8:151–161, 2019.

[232] Jia-Nan Liu, Jian Weng, Anjia Yang, Yizhao Chen, and Xiaodong Lin. Enabling efficient and privacy-preserving aggregation communication and function query for fog computing-based smart grid. *IEEE Transactions on Smart Grid*, 11(1):247–257, 2020.

[233] Jiagang Liu, Ju Ren, Wei Dai, Deyu Zhang, Pude Zhou, Yaoxue Zhang, Geyong Min, and Noushin Najjari. Online multi-workflow scheduling under uncertain task execution time in iaas clouds. *IEEE Transactions on Cloud Computing*, 9(3):1180–1194, Jul 2021.

[234] Jiagang Liu, Ju Ren, Yongmin Zhang, Xuhong Peng, Yaoxue Zhang, and Yuanyuan Yang. Efficient dependent task offloading for multiple applications in mec-cloud system. *IEEE Transactions on Mobile Computing*, 22(4):2147–2162, 2023.

[235] Jingwei Liu, Yating Li, Rong Sun, Qingqi Pei, Ning Zhang, Mianxiong Dong, and Victor CM Leung. Emk-abse: Efficient multikeyword attribute-based searchable encryption scheme through cloud-edge coordination. *IEEE Internet of Things Journal*, 9(19):18650–18662, 2022.

[236] Jingxian Liu, Ke Xiong, Derrick Wing Kwan Ng, Pingyi Fan, Zhangdui Zhong, and Khaled Ben Letaief. Max-min energy balance in wireless-powered hierarchical fog-cloud computing networks. *IEEE Transactions on Wireless Communications*, 19(11):7064–7080, 2020.

[237] Linfeng Liu, Yaoze Zhou, and Jia Xu. A cloud-edge-end collaboration framework for cruising route recommendation of vacant taxis. pages 1–16, 2023. *IEEE Transactions on Mobile Computing*, 23(5):4678-4693.

[238] Tong Liu, Lu Fang, Yanmin Zhu, Weiqin Tong, and Yuanyuan Yang. A near-optimal approach for online task offloading and resource allocation in edge-cloud orchestrated computing. *IEEE Transactions on Mobile Computing*, 21(8):2687–2700, 2020.

[239] Weihong Liu, Jiawei Geng, Zongwei Zhu, Jing Cao, and Zirui Lian. Sniper: Cloud-edge collaborative inference scheduling with neural network similarity modeling. In *Proceedings of the 59th ACM/IEEE Design Automation Conference*, pages 505–510, 2022.

[240] Yajuan Liu, Zhen Wang, Xiaolun Wu, Fang Fang, and Ali Syed Saqlain. Cloud-edge-end cooperative detection of wind turbine blade surface damage based on lightweight deep learning network. *IEEE Internet Computing*, 27(1):43–51, 2022.

[241] Ye Liu, Peishan Huang, Fan Yang, Kai Huang, and Lei Shu. Quasyncfl: Asynchronous federated learning with quantization for cloud-edge-terminal collaboration enabled aiot. *IEEE Internet of Things Journal*, 2023.

[242] Yizhong Liu, Xinxin Xing, Ziheng Tong, Xun Lin, Jing Chen, Zhenyu Guan, Qianhong Wu, and Willy Susilo. Secure and scalable cross-domain data sharing in zero-trust cloud-edge-end environment based on sharding blockchain. *IEEE Transactions on Dependable and Secure Computing*, 2023.

[243] Yuan Liu, Peng Liu, Weipeng Jing, and Houbing Herbert Song. Pd2s: A privacy-preserving differentiated data sharing scheme based on blockchain and federated learning. *IEEE Internet of Things Journal*, 2023.

[244] Saiqin Long, Ying Zhang, Qingyong Deng, Tingrui Pei, Jinzhi Ouyang, and Zhihua Xia. An efficient task offloading approach based on multi-objective evolutionary algorithm in cloud-edge collaborative environment. *IEEE Transactions on Network Science and Engineering*, 10(2):645–657, 2022.

[245] Yinghan Long, Indranil Chakraborty, Gopalakrishnan Srinivasan, and Kaushik Roy. Complexity-aware adaptive training and inference for edge-cloud distributed ai systems. In *2021 IEEE 41st International Conference on Distributed Computing Systems (ICDCS)*, pages 573–583, 2021.

[246] Jiaheng Lu, Zhenzhe Qu, Anfeng Liu, Shaobo Zhang, and Neal N Xiong. Mlm-wr: A swarm intelligence-based cloud-edge-terminal collaboration data collection scheme in the era of aiot. *IEEE Internet of Things Journal*, 2023.

[247] Junwen Lu, Hao Yongsheng, Kesou Wua, Yuming Chen, and Qin Wang. Dynamic offloading for energy-aware scheduling in a mobile cloud. *Journal of King Saud University - Computer and Information Sciences*, 34(6, B):3167–3177, 2022.

[248] Youshui Lu, Jingning Zhang, Yong Qi, Saiyu Qi, Yuanqing Zheng, Yuhao Liu, Hongyu Song, and Wei Wei. Accelerating at the edge: A storage-elastic blockchain for latency-sensitive vehicular edge computing. *IEEE Transactions on Intelligent Transportation Systems*, 23(8):11862–11876, 2022.

[249] Zhihui Lu, Xueying Wang, Jie Wu, and Patrick C. K. Hung. Instechah: Cost-effectively autoscaling smart computing hadoop cluster in private cloud. *Journal of Systems Architecture*, 80:1–16, 2017.

[250] Quyuan Luo, Changle Li, Tom H. Luan, and Weisong Shi. Collaborative data scheduling for vehicular edge computing via deep reinforcement learning. *IEEE Internet of Things Journal*, 7(10):9637–9650, 2020.

[251] Quyuan Luo, Changle Li, Tom H. Luan, and Weisong Shi. Minimizing the delay and cost of computation offloading for vehicular edge computing. *IEEE Transactions on Services Computing*, 15(5):2897–2909, 2022.

[252] Zhihan Lv, Jingyi Wu, Yuxi Li, and Houbing Song. Cross-layer optimization for industrial internet of things in real scene digital twins. *IEEE Internet of Things Journal*, 9(17):15618–15629, 2022.

[253] Feng Lyu, Hongzi Zhu, Nan Cheng, Haibo Zhou, Wenchao Xu, Minglu Li, and Xuemin Shen. Characterizing urban vehicle-to-vehicle communications for reliable safety applications. *IEEE Transactions on Intelligent Transportation Systems*, 21(6):2586–2602, 2020.

[254] Guifu Ma, Xiaowei Wang, Manjiang Hu, Wenjie Ouyang, Xiaolong Chen, and Yang Li. DRL-based computation offloading with queue stability for vehicular-cloud-assisted mobile edge computing systems. *IEEE Transactions on Intelligent Vehicles*, 8(4):2797–2809, 2023.

[255] Xiao Ma, Ao Zhou, Shan Zhang, and Shangguang Wang. Cooperative service caching and workload scheduling in mobile edge computing. In *IEEE INFOCOM 2020 - IEEE Conference on Computer Communications*, pages 2076–2085, 2020.

[256] Yong Ma, Mengxuan Dai, Shiyun Shao, Yunni Xia, Fan Li, Yulong Shen, Jianqi Li, Yin Li, and Hemeng Peng. A performance and reliability-guaranteed predictive approach to service migration path selection in mobile computing. *IEEE Internet of Things Journal*, 2023.

[257] Yue Ma, Thidapat Chantem, Robert P. Dick, Shige Wang, and X. Sharon Hu. An on-line framework for improving reliability of real-time systems on "big-little" type mpsocs. In *Design, Automation & Test in Europe Conference & Exhibition (DATE), 2017*, pages 446–451, 2017.

[258] Yue Ma, Junlong Zhou, Thidapat Chantem, Robert P. Dick, Shige Wang, and Xiaobo Sharon Hu. Improving reliability of soft real-time embedded systems on integrated cpu and gpu platforms. *IEEE Transactions on Computer-Aided Design of Integrated Circuits and Systems*, page 2218–2229, 2020.

[259] Redowan Mahmud, Satish Narayana Srirama, Kotagiri Ramamohanarao, and Rajkumar Buyya. Quality of experience (qoe)-aware placement of applications in fog computing environments. *Journal of Parallel and Distributed Computing*, 132:190–203, 2019.

[260] Yuyi Mao, Jun Zhang, and Khaled B. Letaief. Joint task offloading scheduling and transmit power allocation for mobile-edge computing systems. In *2017 IEEE Wireless Communications and Networking Conference (WCNC)*, pages 1–6. IEEE, 2017.

[261] Angelos-Christos Maroudis, Theodoros Theodoropoulos, John Violos, Aris Leivadeas, and Konstantinos Tserpes. Leveraging graph neural networks for sla violation prediction in cloud computing. *IEEE Transactions on Network and Service Management*, 21(1):605–620, 2024.

[262] Wiem Matoussi and Tarek Hamrouni. A new temporal locality-based workload prediction approach for saas services in a cloud environment. *Journal of King Saud University - Computer and Information Sciences*, 34(7):3973–3987, Jul 2022.

[263] Satoshi Matsushita, Teruo Tanimoto, Satoshi Kawakami, Takatsugu Ono, and Koji Inoue. An edge autonomous lamp control with camera feedback. In *IEEE World Forum on Internet of Things*, pages 1–7, 2022.

[264] Sa Meng, Liang Luo, Xiwei Qiu, and Yuanshun Dai. Service-oriented reliability modeling and autonomous optimization of reliability for public cloud computing systems. *IEEE Transactions on Reliability*, 71(2):527–538, 2022.

[265] Matteo Metta, Stefano Ciliberti, Chinedu Obi, Fabio Bartolini, Laurens Klerkx, and Gianluca Brunori. An integrated socio-cyber-physical system framework to assess responsible digitalisation in agriculture: A first application with living labs in europe. *Agricultural Systems*, 203:103533, 2022.

[266] Yuchang Mo and Liudong Xing. Efficient analysis of resource availability for cloud computing systems to reduce sla violations. *IEEE Transactions on Dependable and Secure Computing*, 19(6):3699–3710, Nov 2022.

[267] Mehdi Mohammadi, Ala Al-Fuqaha, Mohsen Guizani, and Jun-Seok Oh. Semisupervised deep reinforcement learning in support of IoT and smart city services. *IEEE Internet of Things Journal*, 5(2):624–635, 2018.

[268] ChnarMustafa Mohammed and SubhiR.M Zeebaree. Sufficient comparison among cloud computing services: Iaas, paas, and saas: A review. 5(2):17–30, 2021.

[269] David Michael Mothershed, Robert Lugner, Shahabaz Afraj, Gerald Joy Sequeira, Kilian Schneider, Thomas Brandmeier, and Valentin Soloiu. Comparison and evaluation of algorithms for LiDAR-based contour estimation in integrated vehicle safety. *IEEE Transactions on Intelligent Transportation Systems*, 23(5):3925–3942, 2022.

[270] Carla Mouradian, Fereshteh Ebrahimnezhad, Yassine Jebbar, Jasmeen Kaur Ahluwalia, Seyedeh Negar Afrasiabi, Roch H. Glitho, and Ashok Moghe. An iot platform-as-a-service for nfv-based hybrid cloud/fog systems. *IEEE Internet of Things Journal*, 7(7):6102–6115, 2020.

[271] Khan Muhammad, Salman Khan, Mohamed Elhoseny, Syed Hassan Ahmed, and Sung Wook Baik. Efficient fire detection for uncertain surveillance environment. *IEEE Transactions on Industrial Informatics*, 15(5):3113–3122, 2019.

[272] Meghna Manoj Nair, Amit Kumar Tyagi, and Richa Goyal. Medical cyber physical systems and its issues. *Procedia Computer Science*, 165:647–655, 2019.

[273] Ahmed Nasrallah, Akhilesh S. Thyagaturu, Ziyad Alharbi, Cuixiang Wang, Xing Shao, Martin Reisslein, and Hesham ElBakoury. Ultra-low latency (ULL) networks: The IEEE TSN and IETF DetNet standards and related 5G ULL research. *IEEE Communications Surveys & Tutorials*, 21(1):88–145, 2019.

[274] Samrat Nath, Yaze Li, Jingxian Wu, and Pingzhi Fan. Multi-user multi-channel computation offloading and resource allocation for mobile edge computing. In *IEEE International Conference on Communications*, pages 1–6, 2020.

[275] Azin Neishaboori, Ahmed Saeed, Khaled Harras, and Amr Mohamed. Low complexity target coverage heuristics using mobile cameras. In *2014 IEEE 11th International Conference on Mobile Ad Hoc and Sensor Systems*, pages 217–221, 2014.

[276] Balázs Németh, Nuria Molner, Jorge Martín-Pérez, Carlos J Bernardos, Antonio De la Oliva, and Balázs Sonkoly. Delay and reliability-constrained vnf placement on mobile and volatile 5g infrastructure. *IEEE Transactions on Mobile Computing*, 21(9):3150–3162, 2021.

[277] Vinod Nigade, Pablo Bauszat, Henri Bal, and Lin Wang. Jellyfish: Timely inference serving for dynamic edge networks. In *2022 IEEE Real-Time Systems Symposium (RTSS)*, pages 277–290, 2022.

[278] Seyed Yahya Nikouei, Yu Chen, Sejun Song, and Timothy R. Faughnan. Kerman: A hybrid lightweight tracking algorithm to enable smart surveillance as an edge service. In *IEEE Annual Consumer Communications & Networking Conference*, pages 1–6, 2019.

[279] Xudong Niu, Sujie Shao, Chen Xin, Jun Zhou, Shaoyong Guo, Xingyu Chen, and Feng Qi. Workload allocation mechanism for minimum service delay in edge computing-based power Internet of Things. *IEEE Access*, 7:83771–83784, 2019.

[280] Mohanad Odema, Nafiul Rashid, Berken Utku Demirel, and Mohammad Abdullah Al Faruque. Lens: Layer distribution enabled neural architecture search in edge-cloud hierarchies. In *2021 58th ACM/IEEE Design Automation Conference (DAC)*, pages 403–408. IEEE, 2021.

[281] Adrián Orive, Aitor Agirre, Hong-Linh Truong, Isabel Sarachaga, and Marga Marcos. Quality of service aware orchestration for cloud–edge continuum applications. *Sensors*, 22(5):1755, 2022.

[282] Guadalupe Ortiz, José Antonio Caravaca, Alfonso García-de Prado, Frància Chavez de la O, and Juan Boubeta-Puig. Real-time context-aware microservice architecture for predictive analytics and smart decision-making. *IEEE Access*, 7:183177–183194, 2019.

[283] Daniele Jahier Pagliari, Roberta Chiaro, Yukai Chen, Sara Vinco, Enrico Macii, and Massimo Poncino. Input-dependent edge-cloud mapping of recurrent neural networks inference. In *2020 57th ACM/IEEE Design Automation Conference (DAC)*, pages 1–6, 2020.

[284] Fawaz Paraiso, Nicolas Haderer, Philippe Merle, Romain Rouvoy, and Lionel Seinturier. A federated multi-cloud paas infrastructure. In 2012 *IEEE Fifth International Conference on Cloud Computing*, pages 392-399, 2012.

[285] Kemao Peng, Guowei Cai, Ben Chen, Miaobo Dong, Kai-Yew Lum, and Tong Lee. Design and implementation of an autonomous flight control law for a uav helicopter. *Automatica*, 45:2333–2338, 2009.

[286] Qinglan Peng, Yunni Xia, Zeng Feng, Jia Lee, Chunrong Wu, Xin Luo, Wanbo Zheng, Shanchen Pang, Hui Liu, Yidan Qin, et al. Mobility-aware and migration-enabled online edge user allocation in mobile edge computing. In *2019 IEEE International Conference on Web Services (ICWS)*, pages 91–98. IEEE, 2019.

[287] Drew Penney, Bin Li, Lizhong Chen, Jaroslaw J. Sydir, Anna Drewek-Ossowicka, Ramesh Illikkal, Charlie Tai, Ravi Iyer, and Andrew Herdrich. Rapid: Enabling fast online policy learning in dynamic public cloud environments. *Neurocomputing*, 558:126737, 2023.

[288] Andreas P. Plageras, Kostas E. Psannis, Christos Stergiou, Haoxiang Wang, and Brij B. Gupta. Efficient IoT-based sensor BIG data collection–processing and analysis in smart buildings. *Future Generation Computer Systems*, 82:349–357, 2018.

[289] Dan Popa, Florin Pop, Cristina Serbanescu, and Aniello Castiglione. Deep learning model for home automation and energy reduction in a smart home environment platform. *Neural Computing and Applications*, 31:1317–1337, 2019.

[290] Heng Qi, Member, Junxiao Wang, Wenxin Li, Yuxin Wang, and Tie Qiu. A blockchain-driven iiot traffic classification service for edge computing. *Internet of Things Journal*, 8(4):2124–2134, 2021.

[291] Junjie Qi, Heli Zhang, Xi Li, Hong Ji, and Xun Shao. Edge-edge collaboration based micro-service deployment in edge computing networks. In *2023 IEEE Wireless Communications and Networking Conference (WCNC)*, pages 1–6, 2023.

[292] Fuli Qiao, Jun Wu, Jianhua Li, Ali Kashif Bashir, Shahid Mumtaz, and Usman Tariq. Trustworthy edge storage orchestration in intelligent transportation systems using reinforcement learning. *IEEE Transactions on Intelligent Transportation Systems*, 22(7):4443–4456, 2021.

[293] Tian Qin, Guang Cheng, Yichen Wei, and Zifan Yao. Hier-sfl: Client-edge-cloud collaborative traffic classification framework based on hierarchical federated split learning. *Future Generation Computer Systems*, 149:12–24, 2023.

[294] Chao Qiu, Xiaofei Wang, Haipeng Yao, Jianbo Du, F. Richard Yu, and Song Guo. Networking integrated cloud–edge–end in iot: A blockchain-assisted collective q-learning approach. *IEEE Internet of Things Journal*, 8(16):12694–12704, 2020.

[295] Xiaofeng Qu and Huiqiang Wang. Emergency task offloading strategy based on cloud-edge-end collaboration for smart factories. *Computer Networks*, 234:109915, 2023.

[296] Xidi Qu, Qin Hu, and Shengling Wang. Privacy-preserving model training architecture for intelligent edge computing. *Computer Communications*, 162:94–101, 2020.

[297] QuestMobile Research. 2023 smart home insights, 2023. `https://www.questmobile.com.cn/research/report/1643139516402339842`.

[298] M. Reza Rahimi, Nalini Venkatasubramanian, Sharad Mehrotra, and Athanasios V. Vasilakos. On optimal and fair service allocation in mobile cloud computing. *IEEE Transactions on Cloud Computing*, 6(3):815–828, 2018.

[299] Md Abdur Rahman, M. Shamim Hossain, Ahmad J. Showail, Nabil A. Alrajeh, and Ahmed Ghoneim. AI-enabled IIoT for live smart city event monitoring. *IEEE Internet of Things Journal*, 10(4):2872–2880, 2023.

[300] Md. Abdur Rahman, Md. Mamunur Rashid, M. Shamim Hossain, Elham Hassanain, Mohammed F. Alhamid, and Mohsen Guizani. Blockchain and iot-based cognitive edge framework for sharing economy services in a smart city. *IEEE Access*, 7:18611–18621, 2019.

[301] S. Rajasoundaran, A. V. Prabu, Sidheswar Routray, S. V. N. Santhosh Kumar, Prince Priya Malla, Suman Maloji, Amrit Mukherjee, and Uttam Ghosh. Machine learning based deep job exploration and secure transactions in virtual private cloud systems. *Computers & Security*, 109:102379, 2021.

[302] Aldmour Rakan, Yousef Sufian, Baker Thar, and Benkhelifa Elhadj. An approach for offloading in mobile cloud computing to optimize power consumption and processing time. *Sustainable Computing: Informatics and Systems*, 31:100562, 2021.

[303] Lakshmana Kumar Ramasamy, Firoz Khan, Mohammad Shah, Balusupati Veera Venkata Siva Prasad, Celestine Iwendi, and Cresantus Biamba. Secure smart wearable computing through artificial intelligence-enabled Internet of Things and cyber-physical systems for health monitoring. *Sensors*, 22(3):1076, 2022.

[304] Geetanjali Rathee, Sahil Garg, Georges Kaddoum, Bong Jun Choi, Moham-mad Mehedi Hassan, and Salman A. AlQahtani. Trustsys: Trusted decision mak-ing scheme for collaborative artificial intelligence of things. *IEEE Transactions on Industrial Informatics*, 19(1):1059–1068, 2023.

[305] Chao Ren, Chao Gong, and Luchuan Liu. Task-oriented multi-modal communication based on cloud-edge-uav collaboration. *IEEE Internet of Things Journal*, 2023.

[306] Ju Ren, Deyu Zhang, Shiwen He, Yaoxue Zhang, and Tao Li. A survey on end-edge-cloud orchestrated network computing paradigms: Transparent computing, mobile edge computing, fog computing, and cloudlet. *ACM Computing Surveys*, 52(6), 2019.

[307] Carlos Oberdan Rolim, Fernando Luiz Koch, Carlos Becker Westphall, Jorge Werner, Armando Fracalossi, and Giovanni Schmitt Salvador. A cloud computing solution for patient's data collection in health care institutions. In *International Conference on eHealth, Telemedicine, and Social Medicine*, pages 95–99, 2010.

[308] Sampa Sahoo, Kshira Sagar Sahoo, Bibhudatta Sahoo, and Amir H. Gandomi. An auction based edge resource allocation mechanism for IoT-enabled smart cities. In *IEEE Symposium Series on Computational Intelligence*, pages 1280–1286, 2020.

[309] Borja Bordel Sánchez, Ramón Alcarria, Álvaro Sánchez-Picot, and Diego Sánchez-de Rivera. A methodology for the design of application-specific cyber-physical social sensing co-simulators. *Sensors*, 17(10):2177, 2017.

[310] Linwei Sang, Qinran Hu, Yinliang Xu, and Zaijun Wu. Privacy-preserving hybrid cloud framework for real-time tcl-based demand response. *IEEE Transactions on Cloud Computing*, 11(2):1182–1193, 2023.

[311] Juan Ramón Santana, Luis Sánchez, Pablo Sotres, Jorge Lanza, Tomás Llorente, and Luis Muñoz. A privacy-aware crowd management system for smart cities and smart buildings. *IEEE Access*, 8:135394–135405, 2020.

[312] Soumik Sarkar, Baskar Ganapathysubramanian, Arti Singh, Fateme Fotouhi, Soumyashree Kar, Koushik Nagasubramanian, Girish Chowdhary, Sajal K. Das, George Kantor, Adarsh Krishnamurthy, Nirav Merchant, and Asheesh K. Singh. Cyber-agricultural systems for crop breeding and sustainable production. *Trends in Plant Science*, 2023.

[313] Andreas Seitz, Jan Ole Johanssen, Bernd Bruegge, Vivian Loftness, Volker Hartkopf, and Monika Sturm. A fog architecture for decentralized decision making in smart buildings. In *International Workshop on Science of Smart City Operations and Platforms Engineering*, page 34–39, 2017.

[314] Mohamed Sellami, Sami Yangui, Mohamed Mohamed, and Samir Tata. Paas-independent provisioning and management of applications in the cloud. In *2013 IEEE Sixth International Conference on Cloud Computing*, 2013.

[315] Concetta Semeraro, Mariateresa Caggiano, and Michele Dassisti. Sustainability aspects and impacts in cyber-physical social systems. In *International Conference on Cyber-Physical Social Intelligence*, pages 1–6, 2021.

[316] Tanmoy Sen and Haiying Shen. Distributed training for deep learning models on an edge computing network using shielded reinforcement learning. In *2022 IEEE 42nd International Conference on Distributed Computing Systems (ICDCS)*, pages 581–591, 2022.

[317] Mirza Mohd Shahriar Maswood, MD. Rahinur Rahman, Abdullah G. Alharbi, and Deep Medhi. A novel strategy to achieve bandwidth cost reduction and load balancing in a cooperative three-layer fog-cloud computing environment. *IEEE Access*, 8:113737–113750, 2020.

[318] Ashutosh Sharma, Geetanjali Rathee, Rajiv Kumar, Hemraj Saini, Vijayakumar Varadarajan, Yunyoung Nam, and Naveen Chilamkurti. A secure, energy- and sla-efficient (SESE) E-Healthcare framework for quickest data transmission using cyber-physical system. *Sensors*, 19(9):2119, 2019.

[319] Kanika Sharma, Bernard Butler, and Brendan Jennings. Scaling and placing distributed services on vehicle clusters in urban environments. *IEEE Transactions on Services Computing*, 16(2):1402–1416, 2023.

[320] Fanfan Shen, Chao Xu, and Jun Zhang. Statistical behavior guided block allocation in hybrid cache-based edge computing for cyber-physical-social systems. *IEEE Access*, 8:29055–29063, 2020.

[321] Weisong Shi, Jie Cao, Quan Zhang, Youhuizi Li, and Lanyu Xu. Edge Computing: Vision and Challenges. *IEEE Internet of Things Journal*, 3(5):637–646, 2016.

[322] Younghwan Shin, Wonsik Yang, Sangdo Kim, and Jong-Moon Chung. Multiple adaptive-resource-allocation real-time supervisor (mars) for elastic iiot hybrid cloud services. *IEEE Transactions on Network Science and Engineering*, 9(3):1462–1476, 2022.

[323] Isma Farah Siddiqui, Scott Uk-Jin Lee, Asad Abbas, and Ali Kashif Bashir. Optimizing lifespan and energy consumption by smart meters in green-cloud-based smart grids. *IEEE Access*, 5:20934–20945, 2017.

[324] Alessandro Simeone, Rebecca Grant, Weilin Ye, and Alessandra Caggiano. Operator 4.0 intelligent health monitoring: a cyber-physical approach. *Procedia CIRP*, 118:1033–1038, 2023.

[325] Sukhpal Singh, Inderveer Chana, and Rajkumar Buyya. Star: Sla-aware autonomic management of cloud resources. *IEEE Transactions on Cloud Computing*, 8(4):1040–1053, 2020.

[326] Kevin Skadron, Tarek Abdelzaher, and Mircea R. Stan. Control-theoretic techniques and thermal-rc modeling for accurate and localized dynamic thermal management. In *Proceedings Eighth International Symposium on High Performance Computer Architecture*, pages 17–28, 2002.

[327] Kevin Skadron, Mircea R. Stan, Karthik Sankaranarayanan, Wei Huang, Sivakumar Velusamy, and David Tarjan. Temperature-aware microarchitecture: Modeling and implementation. *ACM Transactions on Architecture and Code Optimization*, 1(1):94–125, 2004.

[328] Georgios Skourletopoulos, Constandinos X. Mavromoustakis, George Mastorakis, Jordi Mongay Batalla, Houbing Song, John N. Sahalos, and Evangelos Pallis. Elasticity debt analytics exploitation for green mobile cloud computing: An equilibrium model. *IEEE Transactions on Green Communications and Networking*, 3(1):122–131, 2019.

[329] Oueida Soraia, Aloqaily Moayad, and Ionescu Sorin. A smart healthcare reward model for resource allocation in smart city. *Multimedia Tools and Applications*, 78:24573–24594, 2019.

[330] Alireza Souri, Yanlei Zhao, Mingliang Gao, Asghar Mohammadian, Jin Shen, and Eyhab Al-Masri. A trust-aware and authentication-based collaborative method for resource management of cloud-edge computing in social internet of things. *IEEE Transactions on Computational Social Systems*, 2023.

[331] J. Srinivasan, S. V. Adve, P. Bose, and J. A. Rivers. The impact of technology scaling on lifetime reliability. In *International Conference on Dependable Systems and Networks, 2004*, Jan 2004.

[332] Jayanth Srinivasan, Sarita V. Adve, Pradip Bose, and Jude A. Rivers. Exploiting structural duplication for lifetime reliability enhancement. *Computer Architecture News, Computer architecture news*, May 2005.

[333] Alexandru Stanciu. Blockchain based distributed control system for edge computing. In *2017 21st International Conference on Control Systems and Computer Science (CSCS)*, pages 667–671, 2017.

[334] AlessandroDi Stefano, AntonellaDi Stefano, and Giovanni Morana. Improving qos through network isolation in paas. *Future Generation Computer Systems*, 131:91–105, Jun 2022.

[335] Qian Su, Qinghui Zhang, Weidong Li, and Xuejie Zhang. Primal-dual-based computation offloading method for energy-aware cloud-edge collaboration. *IEEE Transactions on Mobile Computing*, 2023.

[336] Xin Su, Li An, Zhen Cheng, and Yajuan Weng. Cloud–edge collaboration-based bi-level optimal scheduling for intelligent healthcare systems. *Future Generation Computer Systems*, 141:28–39, 2023.

[337] Zhou Su, Qifan Qi, Qichao Xu, Song Guo, and Xiaowei Wang. Incentive scheme for cyber physical social systems based on user behaviors. *IEEE Transactions on Emerging Topics in Computing*, 8(1):92–103, 2020.

[338] Zhou Su, Yuntao Wang, Tom H. Luan, Ning Zhang, Feng Li, Tao Chen, and Hui Cao. Secure and efficient federated learning for smart grid with edge-cloud collaboration. *IEEE Transactions on Industrial Informatics*, 18(2):1333–1344, 2022.

[339] Suhardi, Anastasia Rose Adellina, Marini Wulandari, Jaka Sembiring, and Leonardi Paris Hasugian. Service innovation for a sustainable fuel supply using cyber physical social system technology. In *2017 6th International Conference on Electrical Engineering and Informatics (ICEEI)*, pages 1–7, 2017.

[340] Mohammed Eisa Suliman. A brief analysis of cloud computing infrastructure as a service (iaas). *International Journal of Innovative Science and Research Technology*, 6(1):1409–1412, 2021.

[341] Danfeng Sun, Shan Xue, Huifeng Wu, and Jia Wu. A data stream cleaning system using edge intelligence for smart city industrial environments. *IEEE Transactions on Industrial Informatics*, 18(2):1165–1174, 2022.

[342] Gang Sun, Yayu Li, Yao Li, Dan Liao, and Victor Chang. Low-latency orchestration for workflow-oriented service function chain in edge computing. *Future Generation Computer Systems*, 85:116–128, 2018.

[343] Gang Sun, Gungyang Zhu, Dan Liao, Hongfang Yu, Xiaojiang Du, and Mohsen Guizani. Cost-efficient service function chain orchestration for low-latency applications in nfv networks. *IEEE Systems Journal*, 13(4):3877–3888, 2018.

[344] Yifei Sun, Jigang Wu, Long Chen, Tonglai Liu, Mianyang Yao, and Weijun Sun. Latency optimization for mobile edge computing with dynamic energy harvesting. In *2019 IEEE Intl Conf on Parallel & Distributed Processing with Applications, Big Data & Cloud Computing, Sustainable Computing & Communications, Social Computing & Networking (ISPA/BDCloud/SocialCom/SustainCom)*, pages 79–83, 2019.

[345] Yuxuan Sun, Sheng Zhou, and Jie Xu. Emm: Energy-aware mobility management for mobile edge computing in ultra dense networks. *IEEE Journal on Selected Areas in Communications*, 35(11):2637–2646, 2017.

[346] Yuyan Sun, Zexiang Cai, Caishan Guo, Guolong Ma, Ziyi Zhang, Haizhu Wang, Jianing Liu, Yiqun Kang, and Jianwen Yang. Collaborative dynamic task allocation with demand response in cloud-assisted multiedge system for smart grids. *IEEE Internet of Things Journal*, 9(4):3112–3124, 2022.

[347] Ee-Leng Tan, Furi Andi Karnapi, Linus Junjia Ng, Kenneth Ooi, and Woon-Seng Gan. Extracting urban sound information for residential areas in smart cities using an end-to-end iot system. *IEEE Internet of Things Journal*, 8(18):14308–14321, 2021.

[348] Haisheng Tan, Zhenhua Han, Xiang-Yang Li, and Francis C. M. Lau. Online job dispatching and scheduling in edge-clouds. In *IEEE INFOCOM 2017-IEEE Conference on Computer Communications*, pages 1–9. IEEE, 2017.

[349] Weiqian Tan and Binwei Wu. Long-distance deterministic transmission among TSN networks: Converging CQF and DIP. In *IEEE International Conference on Network Protocols*, pages 1–6, 2021.

[350] Chaogang Tang, Chunsheng Zhu, Xianglin Wei, Huaming Wu, Qing Li, and Joel J. P. C. Rodrigues. Intelligent resource allocation for utility optimization in rsu-empowered vehicular network. *IEEE Access*, 8:94453–94462, 2020.

[351] Huijun Tang, Huaming Wu, Guanjin Qu, and Ruidong Li. Double deep q-network based dynamic framing offloading in vehicular edge computing. *IEEE Transactions on Network Science and Engineering*, 10(3):1297–1310, 2023.

[352] Tiantian Tang, Chao Li, and Fagui Liu. Collaborative cloud-edge-end task offloading with task dependency based on deep reinforcement learning. *Computer Communications*, 2023.

[353] Xiaoyong Tang, Yi Liu, Zeng Zeng, and Bharadwaj Veeravalli. Service cost effective and reliability aware job scheduling algorithm on cloud computing systems. *IEEE Transactions on Cloud Computing*, 11(2):1461–1473, 2023.

[354] Yuqing Tang, Yujie Yuan, and Yan Liu. Cost-aware reliability task scheduling of automotive cyber-physical systems. *Microprocessors and Microsystems*, 87:103507, 2021.

[355] Zhiqing Tang, Fuming Zhang, Xiaojie Zhou, Weijia Jia, and Wei Zhao. Pricing model for dynamic resource overbooking in edge computing. *IEEE Transactions on Cloud Computing*, 11(2):1970–1984, 2023.

[356] Zhixuan Tang, Haibo Zhou, Ting Ma, Kai Yu, and Xuemin Sherman Shen. Leveraging leo assisted cloud-edge collaboration for energy efficient computation offloading. In *2021 IEEE Global Communications Conference (GLOBECOM)*, pages 1–6. IEEE, 2021.

[357] Ehsan Tanghatari, Mehdi Kamal, Ali Afzali-Kusha, and Massoud Pedram. Distributing dnn training over iot edge devices based on transfer learning. *Neurocomputing*, 467:56–65, 2022.

[358] Xiaoyi Tao, Kaoru Ota, Mianxiong Dong, Heng Qi, and Keqiu Li. Performance guaranteed computation offloading for mobile-edge cloud computing. *IEEE Wireless Communications Letters*, 6(6):774–777, 2017.

[359] Minh-Tuan Thai, Ying-Dar Lin, Yuan-Cheng Lai, and Hsu-Tung Chien. Workload and capacity optimization for cloud-edge computing systems with vertical and horizontal offloading. *IEEE Transactions on Network and Service Management*, 17(1):227–238, 2020.

[360] Aleteng Tian, Bohao Feng, Huachun Zhou, Yunxue Huang, Keshav Sood, Shui Yu, and Hongke Zhang. Efficient federated drl-based cooperative caching for mobile edge networks. *IEEE Transactions on Network and Service Management*, 20(1):246–260, 2023.

[361] Hui Tian, Wenwen Sheng, Hong Shen, and Can Wang. Truth finding by reliability estimation on inconsistent entities for heterogeneous data sets. *Knowledge-Based Systems*, 187:104828, 2020.

[362] Xianzhong Tian, Huixiao Meng, Yanjun Li, Pingting Miao, and Pengcheng Xu. Dynamic computation offloading for green things-edge-cloud computing with local caching. In *2022 IEEE International Parallel and Distributed Processing Symposium (IPDPS)*, pages 1018–1028, 2022.

[363] Xing Tong, Zhao Zhang, Cheqing Jin, and Aoying Zhou. Blockchain for end-edge-cloud architecture:a survey. *Chinese Journal of Computers*, 44(12):2345–2366, 2021.

[364] Zhao Tong, Xiaomei Deng, Jing Mei, Bilan Liu, and Keqin Li. Response time and energy consumption co-offloading with slrta algorithm in cloud–edge collaborative computing. *Future Generation Computer Systems*, 129:64–76, 2022.

[365] Simon Tsch'oke, Frederic Lynker, Hauke Buhr, Florian Schreiner, Alexander Willner, Axel Vick, and Moritz Chemnitz. Time-sensitive networking over metropolitan area networks for remote industrial control. In *IEEE/ACM International Symposium on Distributed Simulation and Real Time Applications*, pages 1–4, 2021.

[366] Shreshth Tuli, Redowan Mahmud, Shikhar Tuli, and Rajkumar Buyya. Fogbus: A blockchain-based lightweight framework for edge and fog computing. *The Journal of Systems and Software*, 154:22–36, 2019.

[367] Nikolaos Tzanis, Eleftherios Mylonas, Panagiotis Papaioannou, Michael Birbas, Alexios Birbas, Christos Tranoris, Spyros Denazis, and Alex Papalexopoulos. Cloud-edge architecture with virtualized hardware functionality for real-time diagnosis of transients in smart grids. *IEEE Transactions on Cloud Computing*, 11(2):1230–1241, 2023.

[368] Fath U. Min Ullah, Khan Muhammad, Ijaz Ul Haq, Noman Khan, Ali Asghar Heidari, Sung Wook Baik, and Victor Hugo C. de Albuquerque. AI-assisted edge vision for violence detection in IoT-based industrial surveillance networks. *IEEE Transactions on Industrial Informatics*, 18(8):5359–5370, 2022.

[369] Steven E. Underwood. Integrated assessment for automated driving systems in the united states. In *Road Vehicle Automation 2*, page 119—138, 2015.

[370] Narseo Vallina-Rodriguez and Jon Crowcroft. Energy management techniques in modern mobile handsets. *IEEE Communications Surveys & Tutorials*, 15(1):179–198, 2013.

[371] Luis M. Vaquero, Luis Rodero-Merino, Juan Caceres, and Maik Lindner. A break in the clouds: towards a cloud definition. *ACM Sigcomm Computer Communication Review*, 39(1):50–55, 2008.

[372] Rafael Oliveira Vasconcelos, Igor Vasconcelos, and Markus Endler. Management of mobile dynamic adaptation in cyber-physical systems. In *International Conference on Network and Service Management and Workshop*, pages 272–275, 2014.

[373] Johannes Vater, Lars Harscheidt, and Alois Knoll. Smart manufacturing with prescriptive analytics. In *International Conference on Industrial Technology and Management*, pages 224–228, 2019.

[374] Karima Velasquez, David Perez Abreu, Marilia Curado, and Edmundo Monteiro. Service placement for latency reduction in the fog using application profiles. *IEEE Access*, 9:80821–80834, 2021.

[375] Yiannis Verginadis, Ioannis Patiniotakis, Panagiotis Gouvas, Spyros Mantzouratos, Simeon Veloudis, Sebastian Thomas Schork, Ludwig Seitzluwig, Iraklis Paraskakis, and Gregoris Mentzas. Context-aware policy enforcement for paas-enabled access control. *IEEE Transactions on Cloud Computing*, 10(1):276–291, 2022.

[376] Prabal Verma, Aditya Gupta, Mohit Kumar, and Sukhpal Singh Gill. Fcmcps-covid: Ai propelled fog–cloud inspired scalable medical cyber-physical system, specific to coronavirus disease. *Internet of Things*, 23:100828, 2023.

[377] Prabal Verma and Sandeep K. Sood. Fog assisted-IoT enabled patient health monitoring in smart homes. *IEEE Internet of Things Journal*, 5(3):1789–1796, 2018.

[378] Manas Wakchaure, B. K. Patle, and A. K. Mahindrakar. Application of AI techniques and robotics in agriculture: A review. *Artificial Intelligence in the Life Sciences*, 3:100057, 2023.

[379] Jiang Wan, Arquimedes Canedo, and Mohammad Abdullah Al Faruque. Cyber–physical codesign at the functional level for multidomain automotive systems. *IEEE Systems Journal*, 11(4):2949–2959, 2017.

[380] Fei Wang, Lei Jiao, Konglin Zhu, Xiaojun Lin, and Lei Li. Toward sustainable ai: Federated learning demand response in cloud-edge systems via auctions. In *IEEE INFOCOM 2023-IEEE Conference on Computer Communications*, pages 1–10. IEEE, 2023.

[381] Hao Wang, Hong Qin, Minghao Zhao, Xiaochao Wei, Hua Shen, and Willy Susilo. Blockchain-based fair payment smart contract for public cloud storage auditing. *Information Sciences*, 519:348–362, 2020.

[382] Huaqun Wang, Debiao He, Yanfei Sun, Neeraj Kumar, and Kim-Kwang Raymond Choo. Pat: A precise reward scheme achieving anonymity and traceability for crowd-computing in public clouds. *Future Generation Computer Systems*, 79:262–270, 2018.

[383] Huaqun Wang, Debiao He, and Shaohua Tang. Identity-based proxy-oriented data uploading and remote data integrity checking in public cloud. *IEEE Transactions on Information Forensics and Security*, 11(6):1165–1176, 2016.

[384] Huaqun Wang, Zhiwei Wang, and Josep Domingo-Ferrer. Anonymous and secure aggregation scheme in fog-based public cloud computing. *Future Generation Computer Systems*, 78:712–719, 2018.

[385] Jing Wang, Libing Wu, Kim-Kwang Raymond Choo, and Debiao He. Blockchain-based anonymous authentication with key management for smart grid edge computing infrastructure. *IEEE Transactions on Industrial Informatics*, 16(3):1984–1992, 2020.

[386] Jingjing Wang, Chunxiao Jiang, Kai Zhang, Xiangwang Hou, Yong Ren, and Yi Qian. Distributed q-learning aided heterogeneous network association for energy-efficient iiot. *IEEE Transactions on Industrial Informatics*, 16(4):2756–2764, 2020.

[387] Kuan Wang, Jun Wu, Xi Zheng, Jianhua Li, Wu Yang, and Athanasios V. Vasilakos. Cloud-edge orchestrated power dispatching for smart grid with distributed energy resources. *IEEE Transactions on Cloud Computing*, 11(2):1194–1203, 2023.

[388] Liang Wang, Kezhi Wang, Cunhua Pan, Wei Xu, Nauman Aslam, and Lajos Hanzo. Multi-agent deep reinforcement learning-based trajectory planning for multi-uav assisted mobile edge computing. *IEEE Transactions on Cognitive Communications and Networking*, 7(1):73–84, 2021.

[389] Pu Wang, Tao Ouyang, Guocheng Liao, Jie Gong, Shuai Yu, and Xu Chen. Edge intelligence in motion: Mobility-aware dynamic dnn inference service migration with downtime in mobile edge computing. *Journal of Systems Architecture*, 130:102664, 2022.

[390] Puming Wang, Laurence T. Yang, and Jintao Li. An edge cloud-assisted CPSS framework for smart city. *IEEE Cloud Computing*, 5(5):37–46, 2018.

[391] Rongkai Wang, Chaojie Gu, Shibo He, Zhiguo Shi, and Wenchao Meng. An interoperable and flat industrial internet of things architecture for low latency data collection in manufacturing systems. *Journal of Systems Architecture*, 129:102631, 2022.

[392] Shiqiang Wang, Rahul Urgaonkar, Murtaza Zafer, Ting He, Kevin Chan, and Kin K. Leung. Dynamic service migration in mobile edge-clouds. In *2015 IFIP Networking Conference (IFIP Networking)*, pages 1–9, 2015.

[393] Shiqiang Wang, Murtaza Zafer, and Kin K. Leung. Online placement of multi-component applications in edge computing environments. *IEEE Access*, 5:2514–2533, 2017.

[394] Tian Wang, Yucheng Lu, Jianhuang Wang, Hong-Ning Dai, Xi Zheng, and Weijia Jia. Eihdp: Edge-intelligent hierarchical dynamic pricing based on cloud-edge-client collaboration for iot systems. *IEEE Transactions on Computers*, 70(8):1285–1298, 2021.

[395] Tian Wang, Hao Luo, Weijia Jia, Anfeng Liu, and Mande Xie. MTES: An intelligent trust evaluation scheme in sensor-cloud-enabled industrial Internet of Things. *IEEE Transactions on Industrial Informatics*, 16(3):2054–2062, 2020.

[396] Tian Wang, Dan Zhao, Shaobin Cai, Weijia Jia, and Anfeng Liu. Bidirectional prediction-based underwater data collection protocol for end-edge-cloud orchestrated system. *IEEE Transactions on Industrial Informatics*, 16(7):4791–4799, 2020.

[397] Tian Wang, Junlong Zhou, Gongxuan Zhang, Tongquan Wei, and Shiyan Hu. Customer perceived value- and risk-aware multiserver configuration for profit maximization. *IEEE Transactions on Parallel and Distributed Systems*, 31(5):1074–1088, 2020.

[398] Tingting Wang, Jianqing Li, Wei Wei, Wei Wang, and Kai Fang. Deep-learning-based weak electromagnetic intrusion detection method for zero touch networks on industrial iot. *IEEE Network*, 36(6):236–242, 2022.

[399] Y. Wang, X. Gao, X. Wen, H. Zhou, H. Li, S. Wang, T. Yu, and Z. Lv. Design and implementation of user oriented power ubiquitous Internet of Things access. In *CIRED 2022 Shanghai Workshop*, volume 2022, pages 131–135, 2022.

[400] Yue Wang, Xiaofeng Tao, Xuefei Zhang, Ping Zhang, and Y Thomas Hou. Cooperative task offloading in three-tier mobile computing networks: An admm framework. *IEEE Transactions on Vehicular Technology*, 68(3):2763–2776, 2019.

[401] Zhao Wang, Zhenyu Zhou, Hui Zhang, Geng Zhang, Huixia Ding, and Ahmed Farouk. Ai-based cloud-edge-device collaboration in 6g space-air-ground integrated power iot. *IEEE Wireless Communications*, 29(1):16–23, 2022.

[402] Zhaoyang Wang, Song Wang, Zhiyao Zhao, and Muyi Sun. Trustworthy localization with em-based federated control scheme for iiots. *IEEE Transactions on Industrial Informatics*, 19(1):1069–1079, 2022.

[403] Zhiyuan Wang, Hongli Xu, Jianchun Liu, He Huang, Chunming Qiao, and Yangming Zhao. Resource-efficient federated learning with hierarchical aggregation in edge computing. In *IEEE INFOCOM 2021 - IEEE Conference on Computer Communications*, pages 1–10, 2021.

[404] Wenting Wei, Ruying Yang, Huaxi Gu, Weike Zhao, Chen Chen, and Shaohua Wan. Multi-objective optimization for resource allocation in vehicular cloud computing networks. *IEEE Transactions on Intelligent Transportation Systems*, 23(12):25536–25545, 2022.

[405] Xinliang Wei and Yu Wang. Popularity-based data placement with load balancing in edge computing. *IEEE Transactions on Cloud Computing*, 11(1):397–411, 2023.

[406] Yunkai Wei, Sipei Zhou, Supeng Leng, Sabita Maharjan, and Yan Zhang. Federated learning empowered end-edge-cloud cooperation for 5g hetnet security. *IEEE Network*, 35(2):88–94, 2021.

[407] Simon S. Woo and Jelena Mirkovic. Optimal application allocation on multiple public clouds. *Computer Networks*, 68:138–148, 2014.

[408] Dapeng Wu, Puning Zhang, and Ruyan Wang. Smart internet of things aided by "terminal-edge-cloud" cooperation. *Chinese Journal on Internet of Things*, 2(3):21, 2018.

[409] Huaming Wu, Yi Sun, and Katinka Wolter. Energy-efficient decision making for mobile cloud offloading. *IEEE Transactions on Cloud Computing*, 8(2):570–584, 2020.

[410] Huaming Wu, Katinka Wolter, Pengfei Jiao, Yingjun Deng, Yubin Zhao, and Minxian Xu. Eedto: An energy-efficient dynamic task offloading algorithm for blockchain-enabled iot-edge-cloud orchestrated computing. *IEEE Internet of Things Journal*, 8(4):2163–2176, 2020.

[411] Ya-Jie Wu, Ricardo Brito, Wai-Hei Choi, Chi-Seng Lam, Man-Chung Wong, Sai-Weng Sin, and Rui Paulo Martins. IoT cloud-edge reconfigurable mixed-signal smart meter platform for arc fault detection. *IEEE Internet of Things Journal*, 10(2):1682–1695, 2023.

[412] Yuxin Wu, Changjun Cai, Xuanming Bi, Junjuan Xia, Chongzhi Gao, Yajuan Tang, and Shiwei Lai. Intelligent resource allocation scheme for cloud-edge-end framework aided multi-source data stream. *EURASIP Journal on Advances in Signal Processing*, 2023(1):56, 2023.

[413] Qiufen Xia, Weifa Liang, Zichuan Xu, and Bingbing Zhou. Online algorithms for location-aware task offloading in two-tiered mobile cloud environments. In *IEEE/ACM International Conference on Utility and Cloud Computing*, pages 109–116, 2014.

[414] Qiufen Xia, Wenhao Ren, Zichuan Xu, Xin Wang, and Weifa Liang. When edge caching meets a budget: Near optimal service delivery in multi-tiered edge clouds. *IEEE Transactions on Services Computing*, 15(6):3634–3648, 2022.

[415] Shichao Xia, Zhixiu Yao, Guangfu Wu, and Yun Li. Distributed offloading for cooperative intelligent transportation under heterogeneous networks. *IEEE Transactions on Intelligent Transportation Systems*, 23(9):16701–16714, 2022.

[416] Xiaoyu Xia, Feifei Chen, Qiang He, Guangming Cui, John Grundy, Mohamed Abdelrazek, Athman Bouguettaya, and Hai Jin. Ol-medc: An online approach for cost-effective data caching in mobile edge computing systems. *IEEE Transactions on Mobile Computing*, 22(3):1646–1658, 2023.

[417] Hui Xiao, Jiawei Huang, Zhigang Hu, Meiguang Zheng, and Keqin Li. Collaborative cloud-edge-end task offloading in mec-based small cell networks with distributed wireless backhaul. *IEEE Transactions on Network and Service Management*, 2023.

[418] Huizi Xiao, Jun Zhao, Qingqi Pei, Jie Feng, Lei Liu, and Weisong Shi. Vehicle selection and resource optimization for federated learning in vehicular edge computing. *IEEE Transactions on Intelligent Transportation Systems*, 23(8):11073–11087, 2022.

[419] Tingting Xiao, Chen Chen, and Shaohua Wan. Mobile-edge-platooning cloud: a lightweight cloud in vehicular networks. *IEEE Wireless Communications*, 29(3):87–94, 2022.

[420] Wenjing Xiao, Yiming Miao, Giancarlo Fortino, Di Wu, Min Chen, and Kai Hwang. Collaborative cloud-edge service cognition framework for dnn configuration toward smart iiot. *IEEE Transactions on Industrial Informatics*, 18(10):7038–7047, 2021.

[421] Guoqi Xie, Kehua Yang, Cheng Xu, Renfa Li, and Shiyan Hu. Digital twinning based adaptive development environment for automotive cyber-physical systems. *IEEE Transactions on Industrial Informatics*, 18(2):1387–1396, 2022.

[422] Yong Xie, Gang Zeng, Kurachi Ryo, Guoqi Xie, Yong Dou, and Zhili Zhou. An optimized design of CAN FD for automotive cyber-physical systems. *Journal of Systems Architecture*, 81:101–111, 2017.

[423] Baogui Xin and Yanying Wang. Stability and hopf bifurcation of a stochastic cournot duopoly game in a blockchain cloud services market driven by brownian motion. *IEEE Access*, 8:41432–41438, 2020.

[424] Ying Xiong, Yulin Sun, Li Xing, and Ying Huang. Extend cloud to edge with kubeedge. *2018 IEEE/ACM Symposium on Edge Computing (SEC)*, pages 373–377, 2018.

[425] Chenglin Xu, Cheng Xu, Bo Li, Siqi Li, and Tao Li. Load-aware dynamic controller placement based on deep reinforcement learning in sdn-enabled mobile cloud-edge computing networks. *Computer Networks*, page 109900, 2023.

[426] Chunmei Xu, Shengheng Liu, Cheng Zhang, Yongming Huang, Zhaohua Lu, and Luxi Yang. Multi-agent reinforcement learning based distributed transmission in collaborative cloud-edge systems. *IEEE Transactions on Vehicular Technology*, 70(2):1658–1672, 2021.

[427] Dianlei Xu, Tong Li, Yong Li, Xiang Su, Sasu Tarkoma, Tao Jiang, Jon Crowcroft, and Pan Hui. Edge intelligence: Empowering intelligence to the edge of network. *Proceedings of the IEEE*, 109(11):1778–1837, 2021.

[428] Jian Xu, Laiwen Wei, Wei Wu, Andi Wang, Yu Zhang, and Fucai Zhou. Privacy-preserving data integrity verification by using lightweight streaming authenticated data structures for healthcare cyber–physical system. *Future Generation Computer Systems*, 108:1287–1296, 2020.

[429] Lingwei Xu, Xinpeng Zhou, Xingwang Li, Rutvij H. Jhaveri, Thippa Reddy Gadekallu, and Yuan Ding. Mobile collaborative secrecy performance prediction for artificial iot networks. *IEEE Transactions on Industrial Informatics*, 18(8):5403–5411, 2022.

[430] Linli Xu, Jing Han, Tian Wang, and Lianfa Bai. An efficient CNN to realize speckle correlation imaging based on cloud-edge for cyber-physical-social-system. *IEEE Access*, 8:54154–54163, 2020.

[431] Minghui Xu, Zongrui Zou, Ye Cheng, Qin Hu, Dongxiao Yu, and Xiuzhen Cheng. Spdl: A blockchain-enabled secure and privacy-preserving decentralized learning system. *IEEE Transactions on Computers*, 72(2):548–558, 2023.

[432] Siya Xu, Yimin Li, Shaoyong Guo, Chenghao Lei, Di Liu, and Xuesong Qiu. Cloud–edge collaborative sfc mapping for industrial iot using deep reinforcement learning. *IEEE Transactions on Industrial Informatics*, 18(6):4158–4168, 2022.

[433] Xiaobin Xu, Qi Wang, Yanzhao Hou, and Shangguang Wang. Ai-space: A cloud-edge aggregated artificial intelligent architecture for tiansuan constellation-assisted space-terrestrial integrated networks. *IEEE Network*, 37(2):22–28, 2023.

[434] Xiaolong Xu, Shucun Fu, Qing Cai, Wei Tian, Wenjie Liu, Wanchun Dou, Xingming Sun, Alex X. Liu, et al. Dynamic resource allocation for load balancing in fog environment. *Wireless Communications and Mobile Computing*, 2018, 2018.

[435] Xiaolong Xu, Qihe Huang, Xiaochun Yin, Mahdi Abbasi, Mohammad Reza Khosravi, and Lianyong Qi. Intelligent offloading for collaborative smart city services in edge computing. *IEEE Internet of Things Journal*, 7(9):7919–7927, 2020.

[436] Xiaolong Xu, Qihe Huang, Yiwen Zhang, Shancang Li, Lianyong Qi, and Wanchun Dou. An lsh-based offloading method for iomt services in integrated cloud-edge environment. *ACM Transactions on Multimedia Computing, Communications, and Applications (TOMM)*, 16(3s):1–19, 2021.

[437] Xiaolong Xu, Xihua Liu, Xiaochun Yin, Shoujin Wang, Quan Qi, and Lianyong Qi. Privacy-aware offloading for training tasks of generative adversarial network in edge computing. *Information Sciences*, 532:1–15, 2020.

[438] Zichuan Xu, Lizhen Zhou, Sid Chi-Kin Chau, Weifa Liang, Qiufen Xia, and Pan Zhou. Collaborate or separate? distributed service caching in mobile edge clouds. In *IEEE INFOCOM 2020 - IEEE Conference on Computer Communications*, pages 2066–2075, 2020.

[439] Shi Xuewei, Shi Xuefang, Dong Wenqi, Zang Peng, Jia Hongyan, Wu Jinfang, and Wang Yang. Research on energy storage configuration method based on wind and solar volatility. In *2020 10th International Conference on Power and Energy Systems (ICPES)*, pages 464–468, 2020.

[440] Yijie Xun, Junman Qin, and Jiajia Liu. Deep learning enhanced driving behavior evaluation based on vehicle-edge-cloud architecture. *IEEE Transactions on Vehicular Technology*, 70(6):6172–6177, 2021.

[441] Nuzhat Yamin and Ganapati Bhat. Near-optimal energy management for energy harvesting iot devices using imitation learning. *IEEE Transactions on Computer-Aided Design of Integrated Circuits and Systems*, 41(11):4551–4562, 2022.

[442] Linjie Yan, Haiming Chen, Youpeng Tu, and Xinyan Zhou. A task offloading algorithm with cloud edge jointly load balance optimization based on deep reinforcement learning for unmanned surface vehicles. *IEEE Access*, 10:16566–16576, 2022.

[443] Chao Yang, Baichuan Liu, Haoyu Li, Bo Li, Kan Xie, and Shengli Xie. Learning based channel allocation and task offloading in temporary UAV-assisted vehicular edge computing networks. *IEEE Transactions on Vehicular Technology*, 71(9):9884–9895, 2022.

[444] Chen Yang, Qunjian Chen, Zexuan Zhu, Zhi-An Huang, Shulin Lan, and Liehuang Zhu. Evolutionary multitasking for costly task offloading in mobile edge computing networks. *IEEE Transactions on Evolutionary Computation*, 2023.

[445] Chen Yang, Yingchao Wang, Shulin Lan, Lihui Wang, Weiming Shen, and George Q Huang. Cloud-edge-device collaboration mechanisms of deep learning models for smart robots in mass personalization. *Robotics and Computer-Integrated Manufacturing*, 77:102351, 2022.

[446] Chun Yang, Hongliu Xu, Shixiao Fan, Xuan Cheng, Minghui Liu, and Xiaomin Wang. Efficient resource allocation policy for cloud edge end framework by reinforcement learning. In *2022 IEEE 8th International Conference on Computer and Communications (ICCC)*, pages 1363–1367. IEEE, 2022.

[447] Dongxu Yang, Hua Wei, Yun Zhu, Peijie Li, and Jian-Cheng Tan. Virtual private cloud based power-dispatching automation system—architecture and application. *IEEE Transactions on Industrial Informatics*, 15(3):1756–1766, 2019.

[448] Huan Yang, Sheng Sun, Min Liu, Qiuping Zhang, and Yuwei Wang. Mjoa-mu: End-to-edge collaborative computation for dnn inference based on model uploading. *Computer Networks*, 231:109801, 2023.

[449] Junchao Yang, Zhiwei Guo, Jiangtao Luo, Yu Shen, and Keping Yu. Cloud-edge-end collaborative caching based on graph learning for cyber-physical virtual reality. *IEEE Systems Journal*, 2023.

[450] Shusen Yang, Zhanhua Zhang, Cong Zhao, Xin Song, Siyan Guo, and Hailiang Li. Cnnpc: End-edge-cloud collaborative cnn inference with joint model partition and compression. *IEEE Transactions on Parallel and Distributed Systems*, 33(12):4039–4056, 2022.

[451] Xiang Yang, Qi Qi, Jingyu Wang, Song Guo, and Jianxin Liao. Towards efficient inference: Adaptively cooperate in heterogeneous iot edge cluster. In *2021 IEEE 41st International Conference on Distributed Computing Systems (ICDCS)*, pages 12–23, 2021.

[452] Xiaolong Yang, Zesong Fei, Jianchao Zheng, Ning Zhang, and Alagan Anpalagan. Joint multi-user computation offloading and data caching for hybrid mobile cloud/edge computing. *IEEE Transactions on Vehicular Technology*, 68(11):11018–11030, 2019.

[453] Xueqing Yang, Xin Guan, Ning Wang, Yongnan Liu, Huayang Wu, and Yan Zhang. Cloud-edge-end intelligence for fault-tolerant renewable energy accommodation in smart grid. *IEEE Transactions on Cloud Computing*, 11(2):1144–1156, 2023.

[454] Zheming Yang, Dieli Hu, Qi Guo, Lulu Zuo, and Wen Ji. Visual E2C: AI-driven visual end-edge-cloud architecture for 6G in low-carbon smart cities. *IEEE Wireless Communications*, 30(3):204–210, 2023.

[455] Zhigang Yang, Ruyan Wang, Dapeng Wu, Honggang Wang, Haina Song, and Xinqiang Ma. Local trajectory privacy protection in 5g enabled industrial intelligent logistics. *IEEE Transactions on Industrial Informatics*, 18(4):2868–2876, 2021.

[456] Su Yao, Mu Wang, Qiang Qu, Ziyi Zhang, Yi-Feng Zhang, Ke Xu, and Mingwei Xu. Blockchain-empowered collaborative task offloading for cloud-edge-device computing. *IEEE Journal on Selected Areas in Communications*, 40(12):3485–3500, 2022.

[457] Xingjia Yao, Shu Liu, Xiaodong Wang, Hongliang Jiang, Faming Sui, and Zuoxia Xing. Research on maximum wind energy capture control strategy. In *2010 International Conference on Optics, Photonics and Energy Engineering (OPEE)*, volume 1, pages 34–37, 2010.

[458] Anila Yasmeen, Nadeem Javaid, Obaid Ur Rehman, Hina Iftikhar, Muhammad Faizan Malik, and Fatima J Muhammad. Efficient resource provisioning for smart buildings utilizing fog and cloud based environment. In *2018 14th International Wireless Communications & Mobile Computing Conference (IWCMC)*, pages 811–816. IEEE, 2018.

[459] Shengyuan Ye, Liekang Zeng, Qiong Wu, Ke Luo, Qingze Fang, and Xu Chen. Eco-fl: Adaptive federated learning with efficient edge collaborative pipeline training. In *Proceedings of the 51st International Conference on Parallel Processing (ICPP'22)*, 2023.

[460] Shanhe Yi, Zijiang Hao, Zhengrui Qin, and Qun Li. Fog computing: Platform and applications. In *2015 Third IEEE Workshop on Hot Topics in Web Systems and Technologies (HotWeb)*, pages 73–78, 2015.

[461] Bereket Abera Yilma, Hervé Panetto, and Yannick Naudet. Systemic formalisation of cyber-physical-social system (CPSS): A systematic literature review. *Computers in Industry*, 129:103458, 2021.

[462] Jianwei Yin, Yan Tang, Shuiguang Deng, Bangpeng Zheng, and Albert Y. Zomaya. Muse: A multi-tierd and sla-driven deduplication framework for cloud storage systems. *IEEE Transactions on Computers*, 70(5):759–774, 2021.

[463] Lu Yin, Jin Sun, Junlong Zhou, Zonghua Gu, and Keqin Li. Ecfa: An efficient convergent firefly algorithm for solving task scheduling problems in cloud-edge computing. *IEEE Transactions on Services Computing*, 2023.

[464] Daniel Yokoyama, Bruno Schulze, Henrique Kloh, Matheus Bandini, and Vinod Rebello. Affinity aware scheduling model of cluster nodes in private clouds. *Journal of Network and Computer Applications*, 95:94–104, 2017.

[465] Ayman Younis, Brian Qiu, and Dario Pompili. Latency-aware hybrid edge cloud framework for mobile augmented reality applications. In *2020 17th Annual IEEE International Conference on Sensing, Communication, and Networking (SECON)*, pages 1–9, 2020.

[466] Wenjin Yu, Yuehua Liu, Tharam Dillon, and Wenny Rahayu. Edge computing-assisted IoT framework with an autoencoder for fault detection in manufacturing predictive maintenance. *IEEE Transactions on Industrial Informatics*, 19(4):5701–5710, 2023.

[467] Haitao Yuan, Jing Bi, and MengChu Zhou. Multiqueue scheduling of heterogeneous tasks with bounded response time in hybrid green iaas clouds. *IEEE Transactions on Industrial Informatics*, 15(10):5404–5412, 2019.

[468] Haitao Yuan, Qinglong Hu, Meijia Wang, Jing Bi, and MengChu Zhou. Cost-minimized user association and partial offloading for dependent tasks in hybrid cloud–edge systems. In *2022 IEEE 18th International Conference on Automation Science and Engineering (CASE)*, pages 1059–1064. IEEE, 2022.

[469] Jingling Yuan, Yao Xiang, Yuhui Deng, Yi Zhou, and Geyong Min. Upoa: A user preference based latency and energy aware intelligent offloading approach for cloud-edge systems. *IEEE Transactions on Cloud Computing*, 2022.

[470] Xiaoqun Yuan, Mengting Sun, and Wenjing Lou. A dynamic deep-learning-based virtual edge node placement scheme for edge cloud systems in mobile environment. *IEEE Transactions on Cloud Computing*, 10(2):1317–1328, 2022.

[471] Dongdong Yue, Ruixuan Li Li, Yan Zhang, Wenlong Tian, and Yongfeng Huang. Blockchain-based verification framework for data integrity in edge-cloud storage. *Journal of Parallel and Distributed Computing*, 146:1–14, 2020.

[472] Sheng Yue, Ju Ren, Nan Qiao, Yongmin Zhang, Hongbo Jiang, Yaoxue Zhang, and Yuanyuan Yang. TODG: Distributed task offloading with delay guarantees for edge computing. *IEEE Transactions on Parallel and Distributed Systems*, 33(7):1650–1665, 2022.

[473] Alessio Zappone, Emil Björnson, Luca Sanguinetti, and Eduard Jorswieck. Globally optimal energy-efficient power control and receiver design in wireless networks. *IEEE Transactions on Signal Processing*, 65(11):2844–2859, 2017.

[474] Feng Zeng, Kanwen Zhang, Lin Wu, and Jinsong Wu. Efficient caching in vehicular edge computing based on edge-cloud collaboration. *IEEE Transactions on Vehicular Technology*, 72(2):2468–2481, 2023.

[475] Xuezhi Zeng, Saurabh Garg, Mutaz Barika, Sanat Bista, Deepak Puthal, Albert Y. Zomaya, and Rajiv Ranjan. Detection of sla violation for big data analytics applications in cloud. *IEEE Transactions on Computers*, 70(5):746–758, May 2021.

[476] Wenhan Zhan, Chunbo Luo, Jin Wang, Chao Wang, Geyong Min, Hancong Duan, and Qingxin Zhu. Deep-reinforcement-learning-based offloading scheduling for vehicular edge computing. *IEEE Internet of Things Journal*, 7(6):5449–5465, 2020.

[477] Fan Zhang, Guangjie Han, Li Liu, Miguel Martinez-Garcia, and Yan Peng. Deep reinforcement learning based cooperative partial task offloading and resource allocation for iiot applications. *IEEE Transactions on Network Science and Engineering*, 10(5):2991–3006, 2023.

[478] Feng Zhang, Hao Wang, Lu Zhou, Dequan Xu, and Liang Liu. A blockchain-based security and trust mechanism for ai-enabled iiot systems. *Future Generation Computer Systems*, 146:78–85, 2023.

[479] Guanglin Zhang, Wenqian Zhang, Yu Cao, Demin Li, and Lin Wang. Energy-delay tradeoff for dynamic offloading in mobile-edge computing system with energy harvesting devices. *IEEE Transactions on Industrial Informatics*, 14(10):4642–4655, 2018.

[480] Hang Zhang, Jinsong Wang, Hongwei Zhang, and Chao Bu. Security computing resource allocation based on deep reinforcement learning in serverless multi-cloud edge computing. *Future Generation Computer Systems*, 151:152–161, 2024.

[481] Jiawei Zhang, Peng Wang, and Ning Zhang. Frequency regulation from distributed energy resource using cloud-edge collaborations under wireless environments. *IEEE Transactions on Smart Grid*, 13(1):367–380, 2021.

[482] Jinyi Zhang and Xinglin Zhang. Multi-task allocation in mobile crowd sensing with mobility prediction. *IEEE Transactions on Mobile Computing*, 22(2):1081–1094, 2023.

[483] Liang Zhang and Jacob Chakareski. Uav-assisted edge computing and streaming for wireless virtual reality: Analysis, algorithm design, and performance guarantees. *IEEE Transactions on Vehicular Technology*, 71(3):3267–3275, 2022.

[484] Lichen Zhang and Jingyong Liu. Specify and model automotive cyber physical systems using hybrid relation calculus. In *International Conference on Automation and Computing*, pages 1–6, 2021.

[485] Lin Zhang, Jianhao Peng, Jiabao Zheng, and Ming Xiao. Intelligent cloud-edge collaborations assisted energy-efficient power control in heterogeneous networks. *IEEE Transactions on Wireless Communications*, 2023.

[486] Minjia Zhang, Hai Jin, Xuanhua Shi, and Song Wu. Virtcft: A transparent vm-level fault-tolerant system for virtual clusters. In *2010 IEEE 16th International Conference on Parallel and Distributed Systems*, pages 147–154, 2010.

[487] Peiying Zhang, Chao Wang, Chunxiao Jiang, and Zhu Han. Deep reinforcement learning assisted federated learning algorithm for data management of iiot. *IEEE Transactions on Industrial Informatics*, 17(12):8475–8484, 2021.

[488] Peiying Zhang, Pan Yang, Neeraj Kumar, and Mohsen Guizani. Space-air-ground integrated network resource allocation based on service function chain. *IEEE Transactions on Vehicular Technology*, 71(7):7730–7738, 2022.

[489] Peiying Zhang, Yi Zhang, Neeraj Kumar, Mohsen Guizani, Ahmed Barnawi, and Wei Zhang. Energy-aware positioning service provisioning for cloud-edge-vehicle collaborative network based on drl and service function chain. *IEEE Transactions on Mobile Computing*, 2023.

[490] Puning Zhang, Meiyu Sun, Yanli Tu, Xuefang Li, Zhigang Yang, and Ruyan Wang. Device–edge collaborative differentiated data caching strategy toward aiot. *IEEE Internet of Things Journal*, 10(13):11316–11325, 2023.

[491] Sai Qian Zhang, Jieyu Lin, and Qi Zhang. Adaptive distributed convolutional neural network inference at the network edge with adcnn. In *Proceedings of the 49th International Conference on Parallel Processing (ICPP'20)*, 2020.

[492] Sunxuan Zhang, Zhao Wang, Zhenyu Zhou, Yang Wang, Hui Zhang, Geng Zhang, Huixia Ding, Shahid Mumtaz, and Mohsen Guizani. Blockchain and federated deep reinforcement learning based secure cloud-edge-end collaboration in power iot. *IEEE Wireless Communications*, 29(2):84–91, 2022.

[493] Wei Zhang, Xiaohui Chen, Yueqi Liu, and Qian Xi. A distributed storage and computation k-nearest neighbor algorithm based cloud-edge computing for cyber-physical-social systems. *IEEE Access*, 8:50118–50130, 2020.

[494] Wenyu Zhang, Zhenjiang Zhang, Sherali Zeadally, Han-Chieh Chao, and Victor CM Leung. Energy-efficient workload allocation and computation resource configuration in distributed cloud/edge computing systems with stochastic workloads. *IEEE Journal on Selected Areas in Communications*, 38(6):1118–1132, 2020.

[495] Xinglin Zhang, Zhenjiang Li, Chang Lai, and Junna Zhang. Joint edge server placement and service placement in mobile-edge computing. *IEEE Internet of Things Journal*, 9(13):11261–11274, 2022.

[496] Yameng Zhang, Tong Liu, Yanmin Zhu, and Yuanyuan Yang. A deep reinforcement learning approach for online computation offloading in mobile edge computing. In *2020 IEEE/ACM 28th International Symposium on Quality of Service (IWQoS)*, pages 1–10. IEEE, 2020.

[497] Yi Zhang, Chunxiao Jiang, and Peiying Zhang. Security-aware resource allocation scheme based on drl in cloud-edge-terminal cooperative vehicular network. *IEEE Internet of Things Journal*, 2023.

[498] Yi Zhang and Hung-Yu Wei. Risk-aware cloud-edge computing framework for delay-sensitive industrial iots. *IEEE Transactions on Network and Service Management*, 18(3):2659–2671, 2021.

[499] Yin Zhang, Meikang Qiu, Chun-Wei Tsai, Mohammad Mehedi Hassan, and Atif Alamri. Health-cps: Healthcare cyber-physical system assisted by cloud and big data. *IEEE Systems Journal*, 11(1):88–95, 2017.

[500] Yongmin Zhang, Xiaolong Lan, Ju Ren, and Lin Cai. Efficient computing resource sharing for mobile edge-cloud computing networks. *IEEE/ACM Transactions on Networking*, 28(3):1227–1240, 2020.

[501] Zheng Zhang and Feng Zeng. Efficient task allocation for computation offloading in vehicular edge computing. *IEEE Internet of Things Journal*, 10(6):5595–5606, 2023.

[502] Zhenkai Zhang, Emeka Eyisi, Xenofon Koutsoukos, Joseph Porter, Gabor Karsai, and Janos Sztipanovits. A co-simulation framework for design of time-triggered automotive cyber physical systems. *Simulation Modelling Practice and Theory*, 43:16–33, 2014.

[503] Fengjun Zhao, Ying Chen, Yongchao Zhang, Zhiyong Liu, and Xin Chen. Dynamic offloading and resource scheduling for mobile-edge computing with energy harvesting devices. *IEEE Transactions on Network and Service Management*, 18(2):2154–2165, 2021.

[504] Gongming Zhao, Hongli Xu, Yangming Zhao, Chunming Qiao, and Liusheng Huang. Offloading tasks with dependency and service caching in mobile edge computing. *IEEE Transactions on Parallel and Distributed Systems*, 32(11):2777–2792, 2021.

[505] Jie Zhao, Yifeng Zheng, Hejiao Huang, Jing Wang, Xiaojun Zhang, and Daojing He. Lightweight certificateless privacy-preserving integrity verification with conditional anonymity for cloud-assisted medical cyber–physical systems. *Journal of Systems Architecture*, 138:102860, 2023.

[506] Mingxiong Zhao, Jun-Jie Yu, Wen-Tao Li, Di Liu, Shaowen Yao, Wei Feng, Changyang She, and Tony Q. S. Quek. Energy-aware task offloading and resource allocation for time-sensitive services in mobile edge computing systems. *IEEE Transactions on Vehicular Technology*, 70(10):10925–10940, 2021.

[507] Yali Zhao, Rodrigo N. Calheiros, Graeme Gange, James Bailey, and Richard O. Sinnott. SLA-based profit optimization resource scheduling for big data analytics-as-a-service platforms in cloud computing environments. *IEEE Transactions on Cloud Computing*, 9(3):1236–1253, 2021.

[508] Yingying Zheng, Junlong Zhou, Yufan Shen, Peijin Cong, and Zebin Wu. Time and energy-sensitive end-edge-cloud resource provisioning optimization method for collaborative vehicle-road systems. *Journal of Computer Research and Development*, 60(5):1037–1052, 2023.

[509] Zhengyi Zhong, Weidong Bao, Ji Wang, Guanlin Wu, and Xiang Zhao. A hierarchically heterogeneous federated learning method for cloud-edge-end system. *Journal of Computer Research and Development*, 59(11):2408–2422, 2022.

[510] Ao Zhou, Shangguang Wang, Bo Cheng, Zibin Zheng, Fangchun Yang, Rong N. Chang, Michael R. Lyu, and Rajkumar Buyya. Cloud service reliability enhancement via virtual machine placement optimization. *IEEE Transactions on Services Computing*, 10(6):902–913, 2017.

[511] Huan Zhou, Zhenning Wang, Nan Cheng, Deze Zeng, and Pingzhi Fan. Stackelberg-game-based computation offloading method in cloud–edge computing networks. *IEEE Internet of Things Journal*, 9(17):16510–16520, 2022.

[512] Huan Zhou, Tong Wu, Xin Chen, Shibo He, Deke Guo, and Jie Wu. Reverse auction-based computation offloading and resource allocation in mobile cloud-edge computing. *IEEE Transactions on Mobile Computing*, 2022.

[513] Huan Zhou, Tong Wu, Xin Chen, Shibo He, Deke Guo, and Jie Wu. Reverse auction-based computation offloading and resource allocation in mobile cloud-edge computing. *IEEE Transactions on Mobile Computing*, 22(10):6144–6159, 2023.

[514] Huan Zhou, Zhenyu Zhang, Dawei Li, and Zhou Su. Joint optimization of computing offloading and service caching in edge computing-based smart grid. *IEEE Transactions on Cloud Computing*, 11(2):1122–1132, 2023.

[515] Jingwen Zhou, Feifei Chen, Qiang He, Xiaoyu Xia, Rui Wang, and Yong Xiang. Data caching optimization with fairness in mobile edge computing. *IEEE Transactions on Services Computing*, 16(3):1750–1762, 2023.

[516] Junlong Zhou, Kun Cao, Xiumin Zhou, Mingsong Chen, Tongquan Wei, and Shiyan Hu. Throughput-conscious energy allocation and reliability-aware task assignment for renewable powered in-situ server systems. *IEEE Transactions on Computer-Aided Design of Integrated Circuits and Systems*, 41(3):516–529, Mar 2022.

[517] Junlong Zhou, X. Sharon Hu, Yue Ma, and Tongquan Wei. Balancing lifetime and soft-error reliability to improve system availability. In *2016 21st Asia and South Pacific Design Automation Conference (ASP-DAC)*, pages 685–690, 2016.

[518] Junlong Zhou, Jin Sun, Mingyue Zhang, and Yue Ma. Dependable scheduling for real-time workflows on cyber–physical cloud systems. *IEEE Transactions on Industrial Informatics*, 17(11):7820–7829, 2020.

[519] Junlong Zhou, Jianming Yan, Kun Cao, Yanchao Tan, Tongquan Wei, Mingsong Chen, Gongxuan Zhang, Xiaodao Chen, and Shiyan Hu. Thermal-aware correlated two-level scheduling of real-time tasks with reduced processor energy on heterogeneous mpsocs. *Journal of Systems Architecture*, 82:1–11, 2018.

[520] Tianqing Zhou, Yali Yue, Dong Qin, Xuefang Nie, Xuan Li, and Chunguo Li. Joint device association, resource allocation, and computation offloading in ultradense multidevice and multitask iot networks. *IEEE Internet of Things Journal*, 9(19):18695–18709, 2022.

[521] Xiaokang Zhou, Guangquan Xu, Jianhua Ma, and Ivan Ruchkin. Scalable platforms and advanced algorithms for IoT and cyber-enabled applications. *Journal of Parallel and Distributed Computing*, 118:1–4, 2018.

[522] Zhi Zhou, Xu Chen, En Li, Liekang Zeng, Ke Luo, and Junshan Zhang. Edge intelligence: Paving the last mile of artificial intelligence with edge computing. *Proceedings of the IEEE*, 107(8):1738–1762, 2019.

[523] Zhi Zhou, Shuai Yu, Wuhui Chen, and Xu Chen. Ce-iot: Cost-effective cloud-edge resource provisioning for heterogeneous iot applications. *IEEE Internet of Things Journal*, 7(9):8600–8614, 2020.

[524] Anqi Zhu, Zhiwen Zeng, Songtao Guo, Huimin Lu, Mingfang Ma, and Zongtan Zhou. Game-theoretic robotic offloading via multi-agent learning for agricultural applications in heterogeneous networks. *Computers and Electronics in Agriculture*, 211:108017, 2023.

[525] Sha Zhu, Kaoru Ota, and Mianxiong Dong. Energy-efficient artificial intelligence of things with intelligent edge. *IEEE Internet of Things Journal*, 9(10):7525–7532, 2022.

[526] Tongxin Zhu, Jianzhong Li, Zhipeng Cai, Yingshu Li, and Hong Gao. Computation scheduling for wireless powered mobile edge computing networks. In *IEEE INFOCOM 2020 - IEEE Conference on Computer Communications*, pages 596–605, 2020.

[527] Tongxin Zhu, Tuo Shi, Jianzhong Li, Zhipeng Cai, and Xun Zhou. Task scheduling in deadline-aware mobile edge computing systems. *IEEE Internet of Things Journal*, 6(3):4854–4866, 2019.

[528] Xiaomin Zhu, Ji Wang, Hui Guo, Dakai Zhu, Laurence T. Yang, and Ling Liu. Fault-tolerant scheduling for real-time scientific workflows with elastic resource provisioning in virtualized clouds. *IEEE Transactions on Parallel and Distributed Systems*, 27(12):3501–3517, 2016.

For Product Safety Concerns and Information please contact our EU
representative GPSR@taylorandfrancis.com
Taylor & Francis Verlag GmbH, Kaufingerstraße 24, 80331 München, Germany

www.ingramcontent.com/pod-product-compliance
Lightning Source LLC
Chambersburg PA
CBHW082006190326
41458CB00010B/3092

9 781032 884578